EUCLIDEAN

AND TRANSF

Clayton W. Dodge
University of Maine

DOVER PUBLICATIONS, INC.
Mineola, New York

Bibliographical Note

This Dover edition, first published in 2004, is an unabridged, corrected republication of the work originally published by Addison-Wesley Publishing Company, Inc., Reading, Massachusetts, in 1972. Some minor corrections have been made within the text and a Supplement to pages 112–113 has been added on page 296.

Library of Congress Cataloging-in-Publication Data

Dodge, Clayton W.
Euclidean geometry and transformations / Clayton W. Dodge.
p. cm.
Originally published : Reading, Mass. : Addison-Wesley Pub. Co., 1972, in series: Addison-Wesley series in mathematics.
Includes bibliographical references and index.
ISBN-13: 978-0-486-43476-6 (pbk.)
ISBN-10: 0-486-43476-1 (pbk.)
1. Geometry. 2. Transformations (Mathematics) I. Title.

QA453.D67 2004
516.2—dc22

2004041357

Manufactured in the United States by LSC Communications
43476106 2018
www.doverpublications.com

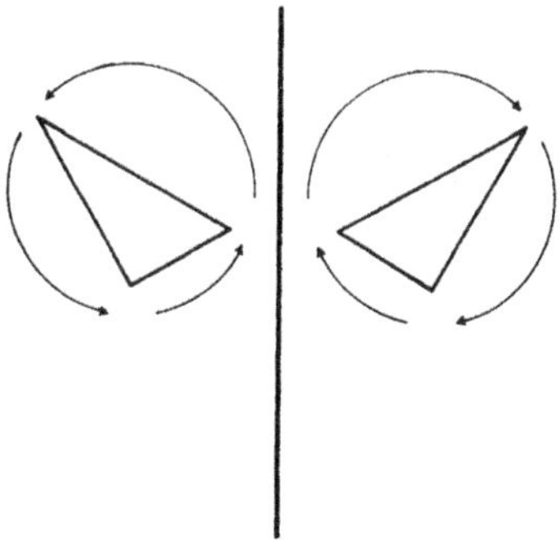

To our daughter, Kathy

Although we sometimes go around in circles,
we still reflect each other.

PREFACE

Just as analytic geometry is recognized today as an important tool in geometry, so also are isometries and similarities important geometric tools. It is well known that Euclidean geometry is the study of those properties of points that are invariant under isometries and similarities, but just how such properties are exhibited using these transformations has not been widely discussed in textbooks. A primary purpose of this book is to provide a source for both the theory and the practical application to geometry of these transformations for college students of mathematics in general, and for teachers and prospective teachers of geometry in particular.

The spirit of modern elementary geometry is also presented with topics such as Menelaus' and Ceva's theorems, Euclidean constructions, and the geometry of special lines and points associated with a triangle, thereby reviewing and refreshing the reader's memory for high school geometry and preparing him to *do* geometry. The high school geometry teacher who has mastered this text can be confident that he is prepared to handle the geometry problems that arise in high school classes.

Prerequisites for this material include high school algebra, geometry, and elementary trigonometry. In addition, some familiarity with the concept of function will prove helpful.

The primary goal of this book is to prepare the reader to do Euclidean geometry. Hence much of it is written in the style of the classic *College Geometry* by N. A. Court. The reader is given many opportunities to work exercises, for such is the key to understanding mathematics. It is suggested that the reader pause a moment after reading the statement of each theorem in the text, draw an appropriate figure, and attempt a proof of the theorem before reading further. Compare the attempted proof with the proof given in the text. Work an abundance of exercises. Look first in the section of "Hints" when unable to obtain a solution, then look at the "Answers" section only as a last resort. Steady progress toward genuine understanding will result.

Geometry, when understood, is indeed a fascinating study.

Each chapter begins with a section of history or commentary which need not be assigned for formal class study. Although exercises are provided for these sections, their purposes are to whet the appetite of the student and to provide some enrichment material.

Following the commentary section in each of Chapters 2, 3, 4, and 6, one or two sections entitled "Introduction . . ." inform the reader of the theory that is developed in the sections that follow. The casual reader may wish to skip over the formal development at that time and look directly at the "Applications" sections, returning later to fill in theoretic gaps.

Hints for the solutions of about half of the exercises are provided in the back of the book, followed by a section of answers to alternate parts of all multi-part exercises and to all other odd-numbered exercises. The bibliography, which precedes the "Hints" section, contains full information on all books referred to in the text and on other selected sources.

Those items preceded by a solid triangle (▶) within a section, or sections preceded by solid triangles, may be omitted without loss of continuity.

Although it is divided into six chapters, the book is numbered according to sections and items or paragraphs within sections. Thus Definition 15.3 refers to the third numbered paragraph in Section 15, and that paragraph is a definition. Similarly, 15.4 refers to the fourth numbered paragraph in Section 15. Exercise 15.3 is the third exercise in Exercise Set 15, which follows Section 15. Such double numbers always refer to text items unless the word "Exercise" is specifically stated. Furthermore, please note that the index lists item numbers instead of page numbers. The reader should find it easier and faster to use this index than a page index.

Except for Section 43 and a small portion of Section 44, which are easily omitted, Chapters 4, 5, and 6 are independent of one another. Thus a reasonable one-semester (45-hour) course for students with little or no background in college geometry might include Chapters 1 to 3, omitting Sections 21, 22, and 30, covered at the rate of about two sections each three hours. Enough time should remain to study one of Chapters 4, 5, and 6. Historical sections may be assigned as outside reading.

The author extends his deep thanks to Professors Henrik Bresinsky, George Cunningham, and Howard Eves for their inspiration and kind words of advice, to 34 students in three classes who aided the author in class-testing this material, and to the staff at Addison-Wesley for their patient understanding of an author's idiosyncrasies.

Orono, Maine C.W.D.
January 1972

CONTENTS

CHAPTER 1 MODERN ELEMENTARY GEOMETRY

1 *The Beginnings of Geometry* 1
2 Directed segments and angles 4
3 Ideal points and ratios 9
4 The theorem of Menelaus 12
5 Ceva's theorem 19
6 Some geometry of the triangle 25
7 More geometry of the triangle 33
8 Geometric constructions 40

CHAPTER 2 ISOMETRIES IN THE PLANE

9 *The Amazing Greeks* 48
10 Introduction to translations, rotations, and reflections 50
11 Introduction to isometries 55
12 Transformation theory 59
13 Isometries as products of reflections 63
14 Translations and rotations 68
15 Halfturns 72
16 Products of reflections 74
17 Properties of isometries; a summary 77
18 Applications of isometries to elementary geometry 79
19 Further elementary applications 83
20 Advanced applications 87
21 Analytic representations of direct isometries 93
22 Analytic representations of opposite isometries 97

CHAPTER 3 SIMILARITIES IN THE PLANE

23 *The rebirth of mathematical thinking* 101
24 Introduction to similarities 104
25 Homothety 106
26 Similarity 110
27 Applications of similarities to elementary geometry 114
28 Further elementary applications 119
29 Advanced applications 124
30 Analytic representations of similarities 128

CHAPTER 4 VECTORS AND COMPLEX NUMBERS IN GEOMETRY

31 *The search for the meaning of complex numbers* 131
32 Introduction to complex numbers 134
33 Vectors 138
34 Vector multiplication 143
35 Vectors and complex numbers 150
36 Triangles in the Gauss plane 155
37 Lines in the Gauss plane 161
38 The circle 165
39 Isometries and similarities in the Gauss plane 168

CHAPTER 5 INVERSION

40 *Matchless modern mathematics* 171
41 Inversion 175
42 Progressions, ratios, and Peaucellier's cell 180
43 Inversion and complex geometry 185
44 Applications of inversion 189

CHAPTER 6 ISOMETRIES IN SPACE

45 *What next?* 196
46 Introduction to three dimensions 201
47 Reflection in a plane 204
48 Basic space isometries 208
49 More space isometries 211
50 Some applications 218
51 Analytic representations 222

Appendixes

A. A Summary of Book I of Euclid's *Elements* 226
B. Basic Ruler and Compass Constructions 228

Bibliography 232

Hints for Selected Exercises 235

Answers 248

Index 288

1 MODERN ELEMENTARY GEOMETRY

SECTION 1 THE BEGINNINGS OF GEOMETRY

1.1 The first section in each chapter of this book is devoted to a discussion of the history of geometry, specifically a history of the type of material covered by this text. These sections, although they contain a few exercises appropriate to the history discussed, are not an integral part of the general text material, so they may be read at any convenient time.

With the exception of Sections 31 and 45 in Chapters 5 and 6, these historical sections progress chronologically, so reading them in their given order is suggested. Section 45, which is less historical and more editorial in form, may be read at any time, but will be more meaningful if the student reads it after he studies the contents of Chapter 2.

1.2 The geometry, indeed all the mathematics, which has come down to us through Europe had its origins in the practical engineering and agriculture of the ancient Babylonians and Egyptians from about 5000 to 2000 B.C. These earliest "practical mathematicians" were concerned only with the solutions to problems: how much grain a certain granary can hold, how much area in a farmer's land for tax purposes, etc. The height of this early mathematical skill is quite visible in the great Egyptian pyramids and other structures. The pyramid of Gizeh, for example, was built about 2900 B.C., using about two million huge stones, as heavy as 54 tons each, hauled 600 miles and cut to an accuracy greater than one part in ten thousand! Great admiration is due these hard-working early peoples for such magnificent structures. Of course, the heavy manual labor was done by as many as 100,000 slaves working for as long as 30 years, but much careful mathematical thought certainly preceded such projects.

1.3 In the Rhind papyrus, deciphered in 1877 and copied about 1700 B.C. by the scribe Ahmes from an earlier work of about 3400 B.C., we find "Directions for Obtaining Knowledge of all Dark Things." Here the area of an isosceles triangle of side 10 and base 4 is taken as 20; that is, half the base times the *side*. The area of a circle is given as the square of eight-ninths of the diameter, a good approximation which assumes that $\pi = 3.1604\ldots$. The area of a quadrilateral is given as $(a+c)(b+d)/4$, which is correct for a rectangle, but too much for any other quadrilateral.

1.4 Many correct formulas were given, such as the areas of a trapezoid and of a triangle, and the volume of a right circular cylinder. Most amazing of all is the correct formula

$$V = \tfrac{1}{3}(B^2 + Bb + b^2)h$$

for the volume of the frustum of a square pyramid of (lower) base edge B, summit (upper base) edge b, and altitude h, given in the Moscow papyrus (*ca.* 1850 B.C.). "The greatest Egyptian pyramid" is how E. T. Bell refers to the Egyptian's knowledge of this formula. It is surely curious that the Egyptians should have known this formula and not a correct formula for the area of a quadrilateral.

1.5 Mathematics in Egypt declined after about 2000 B.C. Poor notation and the complete lack of any evidence of logical reasoning seem the most probable causes for this stagnation. Although they used a knotted rope to form a 3–4–5 triangle to obtain their right angles, there is no evidence whatever that they were aware of even one instance of the Pythagorean theorem.

1.6 The mathematics of ancient China was very similar to that of Egypt, but it did continue to develop over the succeeding centuries to bring forth an occasional theorem—such as Horner's method for reducing each of the roots of a polynomial equation by a constant—a full 500 years before it was discovered in the West.

1.7 The Babylonians were better mathematicians, if the term "mathematician" can really be applied to any of these early peoples. It was the Babylonians who divided the circle into 360 parts. They knew that the altitude from the base of an isosceles triangle bisects the base, that an angle inscribed in a semicircle is a right angle; they knew the Pythagorean theorem, and that the sides of similar triangles are proportional. In various places they have π equal to 3 and to $3\frac{1}{8}$. The Bible (I Kings 7:23 and II Chronicles 4:2) also gives the approximation $\pi = 3$.

1.8 By constructing a table of values for $n^3 + n^2$, they were enabled to solve cubic equations of the form $n^3 + n^2 = c$. Perhaps the most advanced table of all is that known as *Plimpton 322*, dating from about 1800 B.C. This clay tablet lists Pythagorean triples and the values of $\sec^2\theta$ obtained from them for angles from 45° to 31°, with amazingly regular increments in the function values. Such calculations indicate a fairly advanced understanding of trigonometry and of the Pythagorean theorem.

1.9 The Babylonians never discovered the correct volume of the frustum of a pyramid. By analogy they said that it should be half the sum of the areas of the bases times the altitude, since that is the right idea for the area of a trapezoid. Many mathematicians living a thousand or more years later have fallen into the same trap: Because a formula holds for a certain two-dimensional figure, the same formula is assumed for the corresponding three-dimensional figure.

1.10 All mathematics recorded prior to about 600 B.C. was very practical in nature, lacking in generalizations, and lacking in logical structure. Each special case was treated separately. Several numerical examples would be given, followed by a statement to the effect that "such is the procedure." The reader was to deduce the formula from the many examples. There are times when one is tempted to question whether our teaching today has, in many cases, really improved over the last 2000 to 4000 years, since the method mentioned above is used so often both in the classroom and in the literature. Again, the recent flood of mathematics textbooks includes many works of truly superb quality, clear, concise, accurate, and readable. But alas! there is also a glut of mediocre and even venomous writings that use all the "right" words, but are misleading, and even contain downright lies. So the teacher must be most careful in selecting the texts for his courses. Let us hope that future historians of mathematics will be kind enough to judge us by our best and not by our worst.

Exercise Set 1

1. Find the correct area of an isosceles triangle with base 4 and side 10.
2. Show that when one takes the area of a circle as the square of eight-ninths of the diameter, then one is taking $\pi = 3.1604\ldots$.
3. Show that $(a + c)(b + d)/4$ is greater than the area of a nonrectangular quadrilateral whose successive sides have lengths a, b, c, d. Find a correct formula for this area.
4. Derive the formula $V = \frac{1}{3}(B^2 + Bb + b^2)h$ for the volume of the frustum of a square pyramid of base edge B, summit edge b, and height h.
5. Show how the accompanying figure may be used to prove that a 3–4–5 triangle is a right triangle.

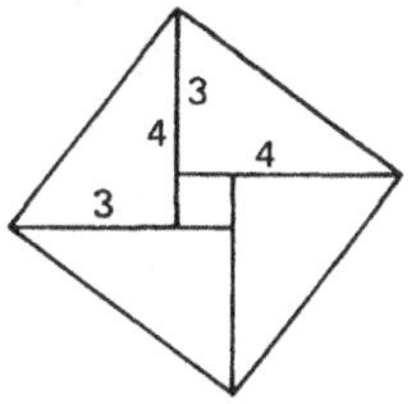

Exercise 1.5

6. Look up I Kings 7:23 and II Chronicles 4:2 in the Bible. Draw and label a figure to show what value of π is assumed there.
7. Construct a table of values for $n^3 + n^2$ for $n = 1, 2, \ldots, 12$. Then use this table to find a root for each of these equations:
   a) $x^3 + x^2 - 1452 = 0$,
   b) $x^6 + 2x^5 + x^4 = 22{,}500$,
   c) $2x^3 + x^2 = 468$,
   d) $x^3 + 3x^2 = 2160$.

SECTION 2 | DIRECTED SEGMENTS AND ANGLES

2.1 We begin with more theorems in high school geometry, one purpose of which is to help ease the reader back into geometric thinking. Thus high school Euclidean geometry is assumed, and no axioms or postulates are stated here. The reader may find it helpful to read the contents of Book I of Euclid's *Elements*, as summarized in Appendix A. The theorems listed therein will provide a sufficient basis for the geometry of this text. A basic knowledge of algebraic manipulation and of the sine and cosine functions is also assumed. The purpose of the entire first chapter is to refresh the reader's memory about high school geometry and to lead him back onto the path of geometrical thinking. This section introduces the concept of a directed segment or angle, an idea most useful in modern geometry, as will be seen especially in the next three sections. Theorems 2.17 and 2.19 will be of particular value to us in the later development.

2.2 Throughout this book, unless otherwise stated, we shall always write the corresponding members of congruent or similar figures in the same order relative to one another. Thus, when we write $\triangle ABC \cong \triangle DEF$ (triangle ABC is congruent to triangle DEF), we require that $\measuredangle A \cong \measuredangle D$, $\measuredangle B \cong \measuredangle E$, $AB \cong DE$, etc. The careful student of geometry will be sure to observe this convention in his own writing.

The understanding of this convention makes clear, for example, the intent in the following proof. The reader is urged to draw a figure to illustrate this theorem.

▶**2.3 Theorem** The base angles of an isosceles triangle are congruent.

Let $AB \cong AC$ in triangle ABC. We have $\measuredangle BAC \cong \measuredangle CAB$ by identity. Since also $AB \cong AC$ and $AC \cong AB$, we have $\triangle BAC \cong \triangle CAB$ by SAS (two sides and the included angle of one triangle each congruent to the corresponding parts of the other triangle). Now $\measuredangle B \cong \measuredangle C$ since they are corresponding parts of congruent figures. □*

2.4 Definition A *line* is properly an undefined term, but we take the word *line* to mean a straight line without beginning or endpoints, infinite in length. If point C is between points A and B, then these three points are distinct and they all lie on a line. Conversely, if A, B, C are three distinct points on a line, then exactly one of these points is between the other two. A *segment* AB is the set of points consisting of points A and B and all points between A and B.

2.5 Since a line or a segment is a set of points, we use the notations $P \in m$ and $Q \notin m$ to denote that point P lies on line m and point Q does not lie on line m. Of course, a line is also a generalization of the physical concept of the edge of a table or of a sheet of paper. Similarly, a point is the idealization of a dot or a spot or a location. In fact, Euclidean geometry is basically the idealized study of certain properties ascribed to the real physical world.

2.6 Definition Points lying on a line are called *collinear*, and they form a *range* of points with the line as *base*. Lines that all pass through one point are called *con-*

---

* The *Halmos symbol* □ indicates the end of a proof.

current, and this point is called their *vertex*. Lines that all concur or are all parallel are said to form a *pencil* of lines (see Fig. 2.6*).

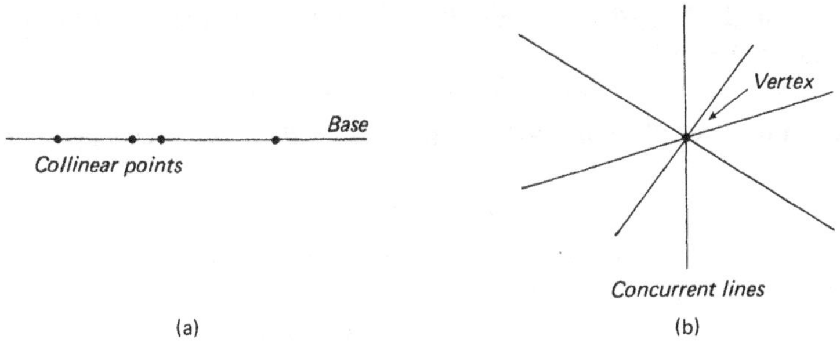

Figure 2.6

2.7 Definition We denote by AB either the line on the points A and B or the segment terminated by A and B. The context will make clear which use is intended. The *measure* (length) of segment AB will be denoted by $m(AB)$. If A and B coincide, we write $A = B$ or $m(AB) = 0$. Similarly $\measuredangle BAC$ denotes the angle formed by the rays AB and AC. The notation $\measuredangle A$ also will be used for $\measuredangle BAC$ when no confusion arises. The measure (generally in degrees) of $\measuredangle BAC$ or $\measuredangle A$ will be denoted by $m(\measuredangle BAC)$ or $m(\measuredangle A)$.

2.8 From Definition 2.7, it follows that we write $\measuredangle A = \measuredangle B$ only when these angles coincide. If these angles have the same measure, we write $\measuredangle A \cong \measuredangle B$ (*$\measuredangle A$ is congruent to $\measuredangle B$*). Similarly, $m(AB) = m(CD)$ or $AB \cong CD$ means that segments AB and CD have the same length, whereas $AB = CD$ indicates that these segments (or lines) coincide.

2.9 Definition Choose a direction along line m as *positive*. We define the *directed length* from A to B, denoted by $d(AB)$, by

$$d(AB) = m(AB)$$

if the direction from A to B is positive, and

$$d(AB) = -m(AB)$$

if the direction from B to A is positive. (See Fig. 2.9.)

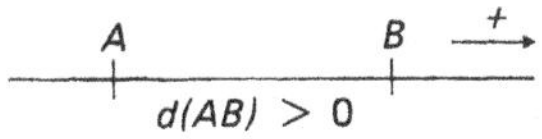

Figure 2.9

2.10 Theorem For any two points A and B,

$$d(AB) + d(BA) = 0.$$

---

* Figures are not numbered consecutively. Instead, they are identified by the number of the item they accompany.

2.11 Theorem If A, B, C are any three collinear points, then

$$d(AB) = d(CB) - d(CA).$$

2.12 Theorem If O, A, B are three collinear points, then the midpoint M of segment AB satisfies the relation

$$d(OM) = \tfrac{1}{2}(d(OA) + d(OB)).$$

2.13 Definition Let the *directed measure* of $\measuredangle BAC$, denoted by $d(\measuredangle BAC)$, be defined by

$$d(\measuredangle BAC) = m(\measuredangle BAC)$$

when $\measuredangle BAC$ is measured counterclockwise (a counterclockwise rotation carries ray AB into ray AC), and

$$d(\measuredangle BAC) = -m(\measuredangle BAC)$$

when $\measuredangle BAC$ is measured clockwise. (See Fig. 2.13.)

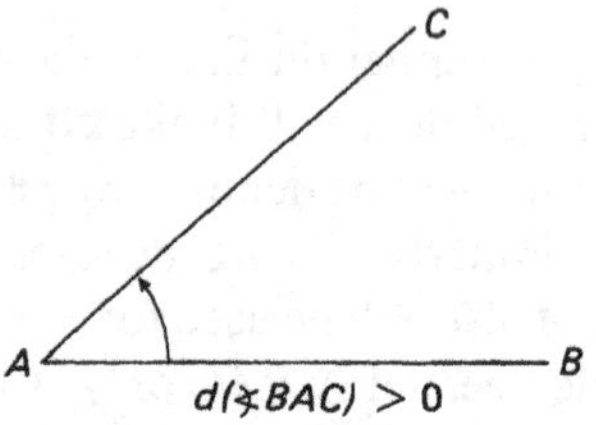

Figure 2.13

2.14 Theorem For any angle BAC, $d(\measuredangle BAC) + d(\measuredangle CAB) = 0$.

2.15 Agreement Since directed distances and directed angles are used quite extensively in this book, we shall denote these sensed magnitudes by boldface type, and undirected magnitudes by lightface italic type, in formulas in which it is clear that distances are implied. In all other cases the m and d notations will be used.* Thus, in formulas, we write

$$m(AB) = AB \quad \text{and} \quad d(AB) = \mathbf{AB},$$
$$m(\measuredangle BAC) = \measuredangle BAC \quad \text{and} \quad d(\measuredangle BAC) = \measuredangle \mathbf{BAC}.$$

▶**2.16 Theorem** *Euler's Theorem.* If A, B, C, D are any four collinear points, then

$$\mathbf{AB} \cdot \mathbf{CD} + \mathbf{AC} \cdot \mathbf{DB} + \mathbf{AD} \cdot \mathbf{BC} = 0.$$

By Theorem 2.11, write

$$\mathbf{AB} = \mathbf{DB} - \mathbf{DA}, \qquad \mathbf{AC} = \mathbf{DC} - \mathbf{DA}, \qquad \text{and} \qquad \mathbf{BC} = \mathbf{DC} - \mathbf{DB}.$$

Then the given expression becomes

$$\begin{aligned} &\mathbf{AB} \cdot \mathbf{CD} + \mathbf{AC} \cdot \mathbf{DB} + \mathbf{AD} \cdot \mathbf{BC} \\ &= (\mathbf{DB} - \mathbf{DA}) \cdot \mathbf{CD} + (\mathbf{DC} - \mathbf{DA}) \cdot \mathbf{DB} + \mathbf{AD} \cdot (\mathbf{DC} - \mathbf{DB}) \\ &= -(\mathbf{DB} - \mathbf{DA}) \cdot \mathbf{DC} + (\mathbf{DC} - \mathbf{DA}) \cdot \mathbf{DB} - \mathbf{DA} \cdot (\mathbf{DC} - \mathbf{DB}) \\ &= 0. \ \square \end{aligned}$$

---

* Since it is quite difficult to *write* in boldface, it is suggested that an overbar be used for directed lengths in written formulas. Then the formula of Theorem 2.11 would be handwritten as $\overline{AB} = \overline{CB} - \overline{CA}$.

2.17 Theorem The area K of triangle ABC is given by

$$K = \tfrac{1}{2} AB \cdot BC \cdot \sin B;$$

that is, the area of a triangle is half the product of any two sides and the sine of the angle included between them.

2.18 For convenience in the formulas that follow, we agree that $a/b = c/d$ shall be termed *true* whenever $ad = bc$ is true, whether or not $b = 0$ or $d = 0$. This convention will prove useful when we are using the algebraic expressions in Menelaus' and Ceva's theorems in Sections 4 and 5. It eliminates many awkward special cases, treating all possibilities at once.

2.19 Theorem Let ABC be any triangle and let L be any point on line BC. Then

$$\frac{\mathbf{BL}}{\mathbf{LC}} = \frac{AB \cdot \sin \measuredangle \mathbf{BAL}}{CA \cdot \sin \measuredangle \mathbf{LAC}}.$$

First note that lengths AB and CA are not directed, but all other measures in this formula are directed.

Let h denote the length of the altitude from vertex A in triangle ABC. (See Fig. 2.19.) The areas K_1 and K_2 of triangles ABL and ALC are given by

$$K_1 = \tfrac{1}{2} BL \cdot h = \tfrac{1}{2} AB \cdot AL \cdot \sin \measuredangle BAL$$
$$K_2 = \tfrac{1}{2} LC \cdot h = \tfrac{1}{2} AL \cdot CA \cdot \sin \measuredangle LAC,$$

from which we obtain, provided $L \neq C$,

$$\frac{K_1}{K_2} = \frac{BL}{LC} = \frac{AB \cdot \sin \measuredangle BAL}{CA \cdot \sin \measuredangle LAC}.$$

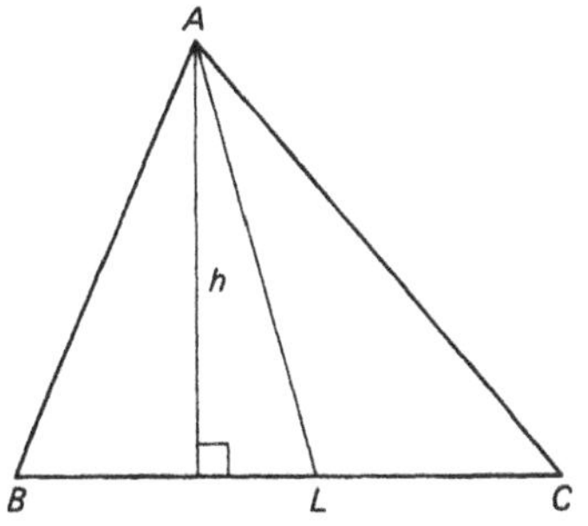

Figure 2.19

Now $\mathbf{BL}/\mathbf{LC}$ and $(\sin \measuredangle \mathbf{BAL})/(\sin \measuredangle \mathbf{LAC})$ are both positive or both negative according as L lies between B and C or outside segment BC. If $L = B$, then both numerators are zero. Thus in all cases it follows that these two fractions have the same sign, so the theorem follows when $L \neq C$.

If $L = C$, the theorem follows, by 2.18, since both denominators are zero. ▯

Exercise Set 2

1. Prove that if A, B, C are any three collinear points, then $d(AB) + d(BC) + d(CA) = 0$.
2. Prove Theorem 2.10.
3. Prove Theorem 2.11.
4. Prove Theorem 2.17.
5. Prove that an internal angle bisector in a triangle divides the opposite side into segments proportional to the adjacent sides.
6. Prove Theorem 2.12.
7. Let A, B, C, D be collinear points. If M and N are the midpoints of AB and CD, show that $2\mathbf{MN} = \mathbf{AC} + \mathbf{BD} = \mathbf{AD} + \mathbf{BC}$.
8. If A, B, C, D are collinear points and if the midpoints of AB and CD coincide, show that $d(AC) = d(DB)$.
9. If A, B, C, D are any four collinear points, then prove that
$$DA^2 \cdot \mathbf{BC} + DB^2 \cdot \mathbf{CA} + DC^2 \cdot \mathbf{AB} + \mathbf{AB} \cdot \mathbf{BC} \cdot \mathbf{CA} = 0.$$
10. *Stewart's theorem.* Prove that the formula of Exercise 2.9 holds even when point D does not lie on line ABC.
11. Prove that if A, B, C, D are collinear points such that $d(AC) = d(DB)$, then the midpoints of AB and CD coincide.
12. Use Exercise 2.10 to find the lengths of the medians of a triangle.
13. Use Exercise 2.10 to find the lengths of the internal angle bisectors of a triangle.
14. Let the lengths of the sides of triangle DBC be a, b, c for sides BC, CD, DB. Let altitude DA have length h, and let $d(BA) = d$ and $d(AC) = e$ so that $a = d + e$. Then $c^2 = d^2 + h^2$. See the accompanying figure. Use these relations along with Exercise 2.10 to show that
$$h^2 = \frac{2(c^2a^2 + a^2b^2 + b^2c^2) - (a^4 + b^4 + c^4)}{4a^2}.$$

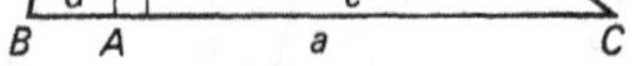

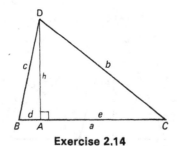

Exercise 2.14

15. Use Exercise 2.14 to prove Heron's formula for the area K of a triangle with sides a, b, c and semiperimeter $s = (a + b + c)/2$:
$$K = \sqrt{s(s-a)(s-b)(s-c)}.$$
This also shows that the altitude h to side a is given by
$$h = \frac{2}{a}\sqrt{s(s-a)(s-b)(s-c)}.$$

16. Use the accompanying figure to prove Euler's theorem, Theorem 2.16.

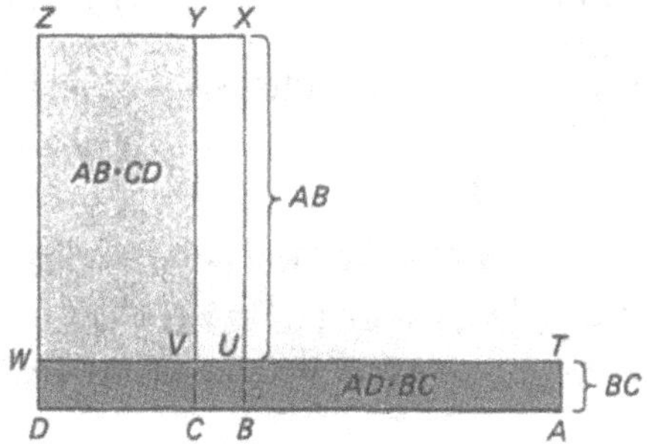

Exercise 2.15

SECTION 3 | IDEAL POINTS AND RATIOS

3.1 Although we shall work primarily in the Euclidean plane, we shall occasionally use Euclidean space of three dimensions. The next definition permits both considerations. No picture of these ideal elements will be presented, since they simply do not appear in ordinary Euclidean figures. The reader is urged to answer carefully and completely Exercises 3.1 and 3.2 to reinforce the concept of ideal elements.

Many properties of Euclidean geometry have rather unfortunate special cases. For example, two distinct points always determine exactly one line (passing through the two points), but two distinct *coplanar* lines (lines lying in the same plane) determine exactly one point (of intersection) only when they are not parallel. This deficiency can be remedied by imagining a *point at infinity* at which the two parallel lines meet. Definition 3.2 provides the details of such infinite elements. Ratios of division of a segment then serve to tie together both the ideas of infinite elements and directed measures. Sections 4 and 5 will use all these ideas.

3.2 Definition To each Euclidean line, hereafter called an *ordinary line*, we add one *ideal point* (or *point at infinity*) having the following properties.

1. Parallel ordinary lines share the same ideal point.
2. Skew or intersecting ordinary lines have distinct ideal points.
3. All the ideal points belonging to the ordinary lines in a given Euclidean plane, hereafter called an *ordinary plane*, form the *ideal line* of that plane.
4. Parallel ordinary planes share the same ideal line.
5. Intersecting ordinary planes have distinct ideal lines.
6. All the ideal points (and ideal lines) in space form the *ideal plane*.
7. Every ideal point is considered to be infinitely far removed from every other (ordinary or ideal) point.

The Euclidean plane thus augmented is called the *extended plane* and Euclidean space thus augmented is called *extended space*. Points, as well as lines and planes, that are not ideal are called *ordinary*.

3.3 Definition Let A and B be ordinary points and let P be any point collinear with A and B. We define the *ratio r in which P divides segment AB* by

$$r = \frac{\mathbf{AP}}{\mathbf{PB}} \quad \text{if } P \text{ is an ordinary point and } P \neq B,$$

$$r = \infty \quad \text{if } P = B,$$

$$r = -1 \quad \text{if } P \text{ is the ideal point on line } AB.$$

If P is between A and B, then P is said to *divide AB internally*; if $P = A$ or $P = B$, then *P divides AB improperly*; otherwise P divides AB *externally*. In all cases we write $r = \mathbf{AP}/\mathbf{PB}$.

3.4 It follows that the ratio r in which P divides segment AB can be any real number; $r = 1$ if P is the midpoint of AB, for example. If P lies between A and B, then $r > 0$; $0 < r < 1$ if P is closer to A; and $r > 1$ if P is closer to B. If B is between A and P, then $r < -1$, and $r \to -\infty$ as $P \to B$. If A is between B and P, then $-1 < r < 0$. If P is ideal, then $r = -1$. If $P = A$, $r = 0$, and if $P = B$, $r = \infty$. The ratios of division of segment AB are indicated for several points in Fig. 3.4.

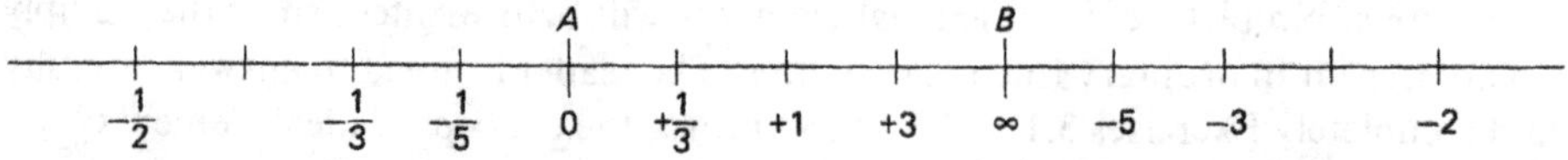

Figure 3.4

3.5 Theorem If $\mathbf{AP}/\mathbf{PB} = \mathbf{AQ}/\mathbf{QB}$, where A and B are distinct ordinary points and P and Q lie on line AB, then $P = Q$.

From the given equation we have

$$\frac{\mathbf{AB}}{\mathbf{PB}} = \frac{\mathbf{AP} + \mathbf{PB}}{\mathbf{PB}} = \frac{\mathbf{AP}}{\mathbf{PB}} + 1 = \frac{\mathbf{AQ}}{\mathbf{QB}} + 1 = \frac{\mathbf{AQ} + \mathbf{QB}}{\mathbf{QB}} = \frac{\mathbf{AB}}{\mathbf{QB}}.$$

Since the first and last fractions have equal nonzero numerators, their denominators are equal too. Thus $\mathbf{PB} = \mathbf{QB}$, so $P = Q$. □

3.6 Definition The *cross ratio* of four collinear points A, B, C, D, denoted by (AB, CD), is defined by

$$(AB, CD) = \left(\frac{\mathbf{AC}}{\mathbf{CB}}\right) \div \left(\frac{\mathbf{AD}}{\mathbf{DB}}\right).$$

3.7 Definition The *cross ratio* of four concurrent lines VA, VB, VC, VD, denoted by $V(AB, CD)$, is defined by

$$V(AB, CD) = \left(\frac{\sin \sphericalangle \mathbf{AVC}}{\sin \sphericalangle \mathbf{CVB}}\right) \div \left(\frac{\sin \sphericalangle \mathbf{AVD}}{\sin \sphericalangle \mathbf{DVB}}\right).$$

3.8 Theorem If an ordinary transversal m cuts four concurrent lines VA, VB, VC,

VD in the four points A, B, C, D, then the cross ratio of the four lines is equal to the cross ratio of the four points. That is, $V(AB, CD) = (AB, CD)$.

3.9 Theorem If A, B, C, D are collinear ordinary points, then

$$(AB, CD) = (CD, AB) = (BA, DC) = (DC, BA).$$

3.10 Definition The equations of Theorem 3.9 are used to define (AB, CD) in case A or B is an ideal point and C and D are distinct ordinary points. That is, $(AB, CD) = (CD, AB)$.

3.11 Definition If $(AB, CD) = -1$, then points C and D are said to *divide AB harmonically*, and D is called the *harmonic conjugate* of C with respect to segment AB.

3.12 Theorem If C and D divide AB harmonically, then A and B divide CD harmonically.

3.13 Theorem The harmonic conjugate of the midpoint of a segment is the ideal point on that line.

Exercise Set 3

1. Decide whether each statement is true or false in extended space.
   a) Each line has exactly one ideal point.
   b) Each plane has exactly one ideal line.
   c) Each two distinct planes meet in just one line.
   d) There are infinitely many ideal lines.
   e) There are infinitely many ideal planes.
   f) There are infinitely many ideal points.

2. Decide whether each statement is true or false in the extended plane.
   a) Each line has exactly one ideal point.
   b) Each line has at least one ideal point.
   c) Each two distinct lines meet (have a point in common).
   d) Through a given point P not on a given line m there passes exactly one line meeting line m in an ideal point (a *parallel* line).
   e) If lines m and n meet at an ideal point, and lines n and p meet at an ideal point, then lines m and p meet at an ideal point.
   f) There are infinitely many ideal points.
   g) There are infinitely many ideal lines.

3. Draw a triangle having exactly:
   a) zero ideal vertices,
   b) one ideal vertex,
   c) two ideal vertices,
   d) three ideal vertices.

4. Locate the point P that divides segment AB in the indicated ratio, where A and B are points with Cartesian coordinates (0, 0) and (1, 0).
   a) 1 b) 2
   c) 0 d) ∞
   e) -1 f) -2
   g) $\frac{1}{2}$ h) $-\frac{1}{2}$
   i) $\frac{1}{10}$ j) $-\frac{1}{10}$

5. Verify the statements in 3.4.
6. Find two equivalent fractions with equal numerators and unequal denominators. Does this invalidate the proof of Theorem 3.5?
7. a) Prove that if two points divide sides AB and AC of triangle ABC in the same ratio, then the line joining these points is parallel to side BC.
   b) Is the theorem still true if "AC" is replaced by "CA"?
8. Prove Theorem 3.8.
9. Prove Theorem 3.9.
10. Prove that if D is the harmonic conjugate of C with respect to AB, then C is the harmonic conjugate of D with respect to AB.
11. a) Prove that if $(AB, CD) = (AB, CE)$, then $D = E$.
    b) Prove that the harmonic conjugate of a given point with respect to a given segment collinear with the given point is unique.
12. Prove Theorem 3.12.
13. Prove Theorem 3.13.
14. There are 24 different ways of taking the cross ratio of four distinct points A, B, C, D, four of which are displayed in Theorem 3.9. Show that these 24 arrangements yield only 6 different cross ratios, and that, if $(AB, CD) = r$, the other five values are $1 - r$, $1/r$, $r/(r - 1)$, $1/(1 - r)$, and $(r - 1)/r$.
15. Given $(AB, CD) = r$, show that:
    a) interchanging the first and second pairs of points, or interchanging the points within the first pair and also those within the second pair, does not change the value of the cross ratio,
    b) interchanging the points within the first pair only, or within the second pair only, changes the value of the cross ratio to $1/r$,
    c) interchanging the first and fourth points only, or the second and third points only, changes the value of the cross ratio to $1 - r$, and
    d) interchanging the first and third points only, or the second and fourth points only, changes the value of the cross ratio to $r/(r - 1)$.
16. In the Cartesian plane, let $A(0, 0)$ and $B(1, 0)$. Show that the point P that divides segment AB in the ratio r has coordinates $(r/(r + 1), 0)$.

SECTION 4 | THE THEOREM OF MENELAUS

4.1 Definition Let ABC be a triangle. A point P lying on the line determined by a side of the triangle is called a *Menelaus point* for that side. If the point is not a vertex of the triangle, it is a *proper Menelaus point.*

4.2 Theorem *Menelaus' theorem.* Let L, M, N be Menelaus points for sides BC, CA, AB of ordinary triangle ABC. Then points L, M, N are collinear iff*

$$\frac{\mathbf{BL}}{\mathbf{LC}}\cdot\frac{\mathbf{CM}}{\mathbf{MA}}\cdot\frac{\mathbf{AN}}{\mathbf{NB}} = -1.$$

* The word *iff* means "if and only if," and is so read.

First, suppose that L, M, N are proper Menelaus points collinear on a base line m. Drop perpendiculars AR, BS, CT of lengths r, s, t, to line m from points A, B, C, as shown in Fig. 4.2. The following pairs of right triangles are similar because they either share an acute angle or have a pair of vertical angles for corresponding acute angles:

$$\triangle BLS \sim \triangle CLT, \qquad \text{so } \frac{BL}{LC} = \frac{s}{t};$$

$$\triangle CMT \sim \triangle AMR, \qquad \text{so } \frac{CM}{MA} = \frac{t}{r};$$

$$\triangle ANR \sim \triangle BNS, \qquad \text{so } \frac{AN}{NB} = \frac{r}{s}.$$

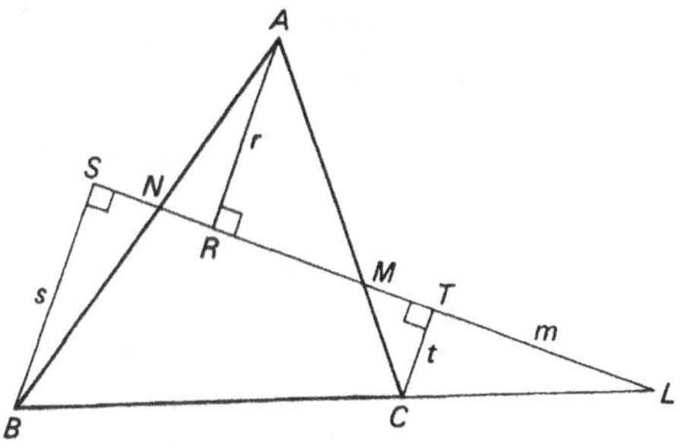

Figure 4.2

Hence

$$\frac{\mathbf{BL}}{\mathbf{LC}} \cdot \frac{\mathbf{CM}}{\mathbf{MA}} \cdot \frac{\mathbf{AN}}{\mathbf{NB}} = \pm \frac{r}{s} \cdot \frac{s}{t} \cdot \frac{t}{r} = \pm 1.$$

Since either one or three of the divisions must be external, this product is negative.

To establish the converse, let L, M, N be proper Menelaus points such that

$$\frac{\mathbf{BL}}{\mathbf{LC}} \cdot \frac{\mathbf{CM}}{\mathbf{MA}} \cdot \frac{\mathbf{AN}}{\mathbf{NB}} = -1.$$

Let the line joining points L and M cut line AB at point N'. Then L, M, N' are three collinear Menelaus points for triangle ABC, so by the first part of this proof,

$$\frac{\mathbf{BL}}{\mathbf{LC}} \cdot \frac{\mathbf{CM}}{\mathbf{MA}} \cdot \frac{\mathbf{AN'}}{\mathbf{N'B}} = -1.$$

Now $\mathbf{AN}/\mathbf{NB} = \mathbf{AN'}/\mathbf{N'B}$, so $N = N'$ by Theorem 3.5. $\square$

4.3 Theorem *Trigonometric form of Menelaus' theorem.* Let L, M, N be Menelaus points for sides BC, CA, AB of ordinary triangle ABC. Then points L, M, N are collinear iff

$$\frac{\sin \measuredangle \mathbf{BAL}}{\sin \measuredangle \mathbf{LAC}} \cdot \frac{\sin \measuredangle \mathbf{CBM}}{\sin \measuredangle \mathbf{MBA}} \cdot \frac{\sin \measuredangle \mathbf{ACN}}{\sin \measuredangle \mathbf{NCB}} = -1.$$

This theorem follows quite readily from Theorems 4.2 and 2.19. ▯

As examples of the application of Menelaus' theorem, the following theorems are given.

4.4 Theorem The line joining the midpoints of two sides of a triangle is parallel to the third side.

Let M and N be the midpoints of sides AB and CA of triangle ABC. (See Fig. 4.4.) Let line MN meet side BC at point L. By Menelaus' theorem,

$$\frac{\mathbf{BL}}{\mathbf{LC}} \cdot \frac{\mathbf{CM}}{\mathbf{MA}} \cdot \frac{\mathbf{AN}}{\mathbf{NB}} = -1.$$

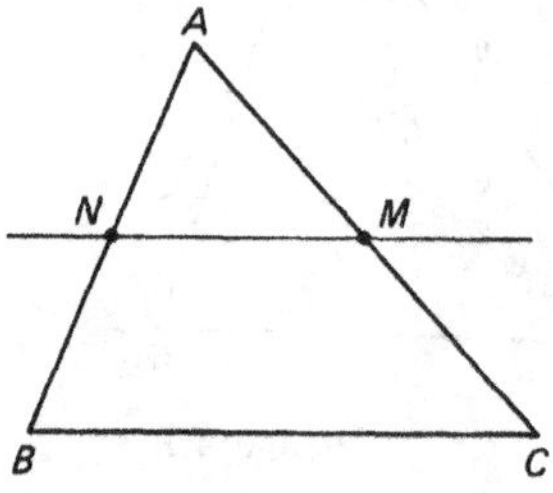

Figure 4.4

But $\mathbf{CM}/\mathbf{MA} = \mathbf{AN}/\mathbf{NB} = 1$, since M and N are midpoints. Thus $\mathbf{BL}/\mathbf{LC} = -1$, so L is an ideal point; that is, lines BC and MN are parallel. ▯

4.5 Theorem The lines tangent to the circumcircle of a triangle at its vertices cut the opposite sides in three collinear points.

Let the tangent to the circumcircle at A meet line BC at L as in Fig. 4.5. Then $\measuredangle BAL \cong \measuredangle C$, since each angle is measured by half of arc AB. Also we have that $\measuredangle LAC = 180° - \measuredangle ABC$, since these angles are measured by halves of the two opposite arcs AC. Then

$$\frac{\mathbf{BL}}{\mathbf{LC}} = \frac{AB \cdot \sin \measuredangle \mathbf{BAL}}{CA \cdot \sin \measuredangle \mathbf{LAC}} = \pm \frac{AB \cdot \sin \measuredangle C}{CA \cdot \sin \measuredangle ABC} = \pm \frac{(AB)^2}{(CA)^2},$$

by Theorem 2.19 and the law of sines. Since the division is clearly external, the minus sign holds. Similarly,

$$\frac{\mathbf{CM}}{\mathbf{MA}} = -\frac{(BC)^2}{(AB)^2} \quad \text{and} \quad \frac{\mathbf{AN}}{\mathbf{NB}} = -\frac{(CA)^2}{(BC)^2},$$

where M and N are the corresponding intersections of the tangents to the circumcircle at B and C with the lines CA and AB. Now

$$\frac{\mathbf{BL}}{\mathbf{LC}}\cdot\frac{\mathbf{CM}}{\mathbf{MA}}\cdot\frac{\mathbf{AN}}{\mathbf{NB}}=-1,$$

so the theorem follows from Menelaus' theorem. □

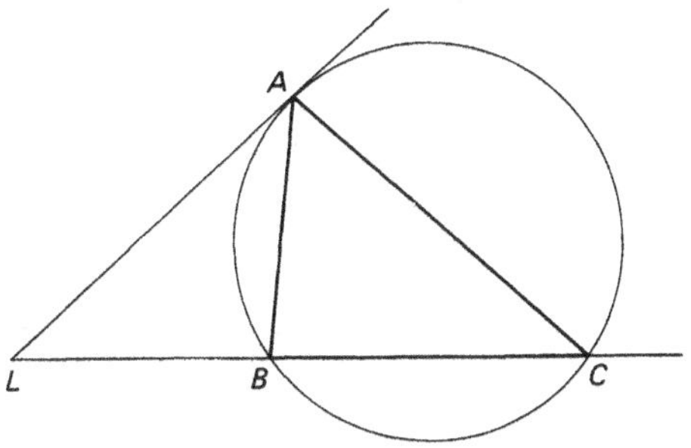

Figure 4.5

4.6 Definition Two triangles ABC and $A'B'C'$ are said to be *copolar* iff the three lines AA', BB', CC' joining corresponding vertices are concurrent (in an ordinary or ideal point); they are *coaxial* iff the three points L, M, N of intersection of corresponding sides BC and $B'C'$, CA and $C'A'$, AB and $A'B'$ are collinear (on an ordinary or ideal line). (See Fig. 4.6.)

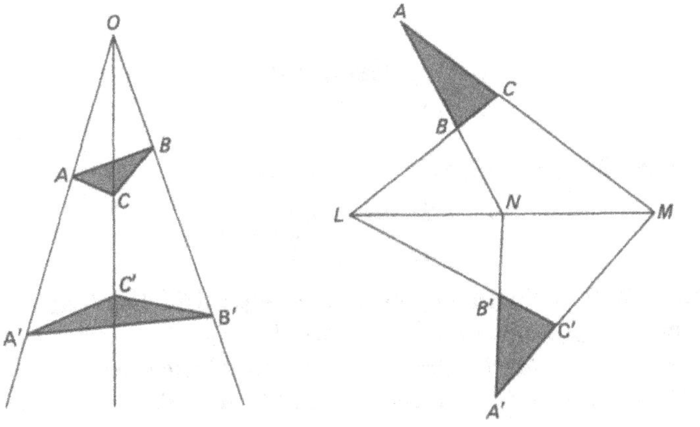

Copolar triangles

(a)

Coaxial triangles

(b)

Figure 4.6

4.7 Theorem *Desargues' two-triangle theorem.* If two triangles are copolar, then they are coaxial; conversely, if two triangles are coaxial, then they are copolar.

Let triangles ABC and $A'B'C'$ be copolar at O. (See Fig. 4.7.) Applying Menelaus' theorem to triangle OBC with collinear Menelaus points L, C', B', to triangle OCA with points M, A', C', and to triangle OAB with points N, B', A', obtain

$$\frac{BL}{LC}\cdot\frac{CC'}{C'O}\cdot\frac{OB'}{B'B}=-1,\qquad \frac{CM}{MA}\cdot\frac{AA'}{A'O}\cdot\frac{OC'}{C'C}=-1,\qquad \frac{AN}{NB}\cdot\frac{BB'}{B'O}\cdot\frac{OA'}{A'A}=-1.$$

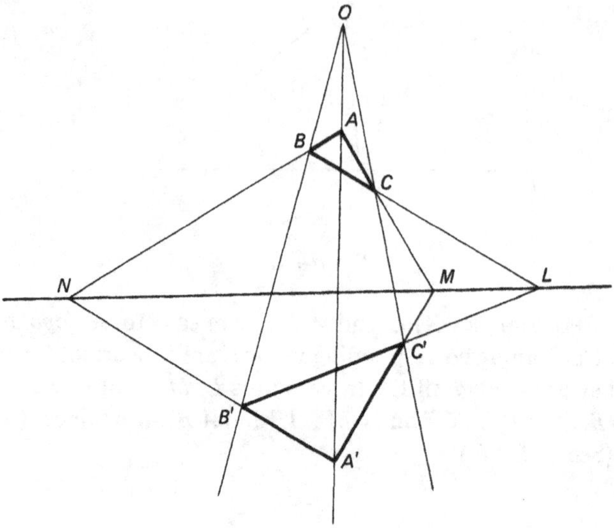

Figure 4.7

Multiply these three equations side for side to get

$$\frac{BL}{LC}\cdot\frac{CM}{MA}\cdot\frac{AN}{NB}=-1.$$

Thus, since L, M, N are Menelaus points for triangle ABC, then L, M, N are collinear; that is, triangles ABC and $A'B'C'$ are coaxial.

Conversely, suppose triangles ABC and $A'B'C'$ are coaxial in points L, M, N. Let BB' and CC' meet at O. Now triangles MCC' and NBB' are copolar at L. Hence, by the first part of this proof, these triangles are coaxial; that is, points A, A', O are collinear. Thus triangles ABC and $A'B'C'$ are copolar at O. ▯

4.8 Theorem *Pappus' theorem.* If hexagon $ABCDEF$ has its vertices A, C, E lying on one line and its vertices B, D, F lying on another line, then the three points L, M, N of intersection of pairs of opposite sides AB and DE, BC and EF, CD and FA are collinear.

Let AB meet CD at P and EF at R, and let CD and EF meet at Q, as shown in Fig. 4.8. Applying Menelaus' theorem to triangle PQR cut in turn by lines FAN, MBC, and ELD, obtain

$$\frac{QF}{FR}\cdot\frac{RA}{AP}\cdot\frac{PN}{NQ}=-1,\qquad \frac{QM}{MR}\cdot\frac{RB}{BP}\cdot\frac{PC}{CQ}=-1,\qquad \frac{QE}{ER}\cdot\frac{RL}{LP}\cdot\frac{PD}{DQ}=-1.$$

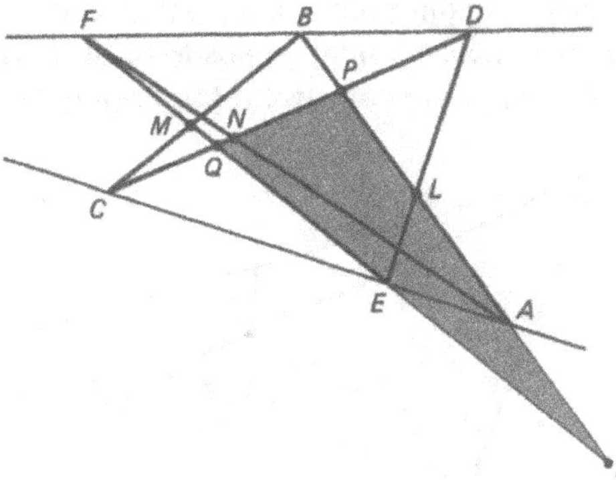

Figure 4.8

Now multiply these three equations side for side, and regroup the factors to obtain

$$\left(\frac{QM}{MR}\cdot\frac{RL}{LP}\cdot\frac{PN}{NQ}\right)\left(\frac{QF}{FR}\cdot\frac{RB}{BP}\cdot\frac{PD}{DQ}\right)\left(\frac{QE}{ER}\cdot\frac{RA}{AP}\cdot\frac{PC}{CQ}\right)=-1.$$

Since F, B, D and E, A, C form two sets of collinear Menelaus points for triangle PQR, then each of the last two triple products in parentheses is equal to -1. Thus this equation reduces to

$$\frac{QM}{MR}\cdot\frac{RL}{LP}\cdot\frac{PN}{NQ}=-1,$$

establishing, by the converse part of Menelaus' theorem, that L, M, N are collinear. ▯

4.9 We have seen in this section a most convenient test for the collinearity of three points. In the next section we shall find a similar test for the concurrence of three lines. These ideas are called *duals* of one another. That is, two statements are called *duals* of one another if each is transformed into the other by the interchange of the words "point" and "line," and, of course, also such associated terms as "collinear" and "concurrent." A striking illustration of duality is the Desargues theorem, for copolar triangles and coaxial triangles are dual concepts. The theorem states that these two dual concepts are equivalent, provided, of course, that we permit ideal elements.

4.10 In projective geometry, in which the Desargues and Pappus theorems are basic, extended space is always used, and the *theorem of duality*, a theorem about theorems, states that the dual of any projective theorem is another projective theorem. Observe that the Desargues theorem is its own dual. We close this section by stating without proof the dual of Pappus' theorem. Compare its statement word for word with that of Pappus' theorem, noting how each concept is dualized.

4.11 Theorem *The dual of Pappus' theorem.* If a hexagon (with sides) a, b, c, d, e, f has its sides a, c, e concurrent in one point P and its sides b, d, f concurrent in another point Q, then the three lines l, m, n joining opposite vertices (at the intersections of) ab and de, bc and ef, cd and fa are concurrent (at a point R). (See Fig. 4.11.)

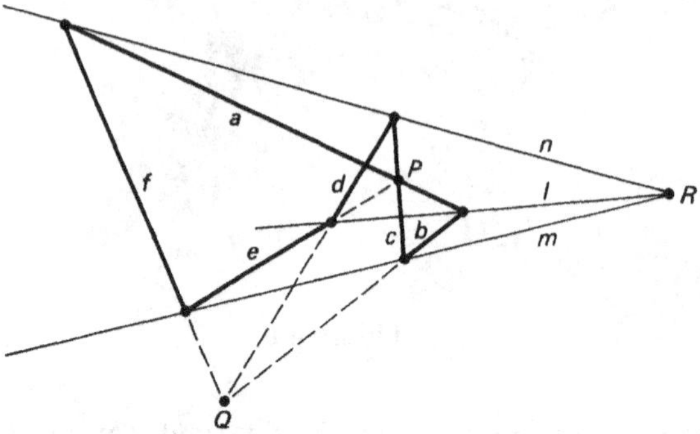

Figure 4.11

4.12 Now that Theorem 4.11 has been stated and illustrated, we find that its wording may be improved. This fact does not at all detract from the usefulness of duality, for we have obtained at least one statement of the theorem, and there are very few long sentences whose wording cannot be clarified. We restate the theorem as follows: If the sides of a hexagon pass alternately through two points, then the three diagonals that join opposite vertices are concurrent.

Exercise Set 4

1. The proof of Menelaus' theorem assumed that the Menelaus points were proper. Complete the proof of this theorem by establishing the theorem for the case in which one or more Menelaus points are not proper.
2. Restate and prove Menelaus' theorem for a triangle having one ideal vertex.
3. Can Menelaus' theorem be applied to a triangle having two or three ideal vertices?
4. Prove Theorem 4.3.
5. Given that P and Q are points on lines AB and AC of triangle ABC, prove that PQ is parallel to BC iff $\mathbf{AP}/\mathbf{PB} = \mathbf{AQ}/\mathbf{QC}$.
6. Choose point P on side AB of triangle ABC. Let the parallel to BC through P meet side AC in Q, the parallel to AB through Q meet BC in R, the parallel to AC through R meet AB in S, the parallel to BC through S meet AC in T, and the parallel to AB through T meet BC in U. Prove that PU is parallel to AC.

7. Prove Desargues' two-triangle theorem (Theorem 4.7) for the case in which O is an ideal point.
8. Prove that the three external angle bisectors of a triangle meet the opposite sides in three collinear points.
9. Prove that two internal angle bisectors of a triangle and the external bisector of the third vertex meet the opposite sides in three collinear points.
10. Points P and Q are *isotomic conjugates* with respect to segment AB iff A, B, P, Q are collinear and $d(AP) = -d(BQ)$; that is, iff P and Q lie on line AB and are symmetric with respect to the midpoint of segment AB. Prove that when L, M, N are collinear Menelaus points for triangle ABC, then their isotomic conjugates L', M', N' are collinear, too.
11. Lines AP and AQ are *isogonal conjugates* with respect to angle BAC iff AP and AQ are symmetric with respect to the bisector of $\measuredangle BAC$; that is, iff $d(\measuredangle BAP) = -d(\measuredangle CAQ)$. Let AL', BM', CN' be the isogonal conjugates of AL, BM, CN, with L and L', M and M', N and N' lying on lines BC, CA, AB. Prove that if L, M, N are collinear, then L', M', N' are collinear.
12. Generalize the concept of a Menelaus point to apply to a quadrilateral. Then prove that Menelaus points L, M, N, O for sides AB, BC, CD, DA of quadrilateral $ABCD$ are collinear only when

$$\frac{\mathbf{AL}}{\mathbf{LB}} \cdot \frac{\mathbf{BM}}{\mathbf{MC}} \cdot \frac{\mathbf{CN}}{\mathbf{ND}} \cdot \frac{\mathbf{DO}}{\mathbf{OA}} = +1.$$

Is the converse true?

13. Let the incircle (circle inscribed) for triangle ABC touch sides BC, CA, AB in points X, Y, Z, and extend YZ to cut BC at K. Prove that $\mathbf{BX/XC} = -\mathbf{BK/KC}$.
14. Let P be the midpoint of median AA' in triangle ABC, and let CP cut AB at Q. Prove that $\mathbf{AQ/QB} = \frac{1}{2}$.
15. Surveyors have the problem of drawing a line between two points having an obstruction (such as a building or a mountain) intervening. Show how Desargues' two-triangle theorem can be used to locate other points on the line determined by two such points.
16. *The ϵ-ruler problem.* Show how to draw line AB when points A and B are much farther apart than the length of the available straightedge.
17. Show how to draw a line through a point P and through the inaccessible point of intersection of two given lines m and n.

SECTION 5 | CEVA'S THEOREM

5.1 Definition A line through a vertex of a triangle is called a *Cevian* for that vertex of the triangle. If it does not coincide with a side of the triangle, the Cevian is called a *proper Cevian*. We agree that when stating that AL is a Cevian, we imply that A is the vertex and L is the point of intersection of the Cevian with the opposite side of the triangle.

5.2 Theorem *Ceva's theorem.* Three Cevians AL, BM, CN for triangle ABC concur iff

$$\frac{BL}{LC}\cdot\frac{CM}{MA}\cdot\frac{AN}{NB}=+1.$$

Suppose proper Cevians AL, BM, CN meet at O, as in Fig. 5.2. Then C, O, N and B, O, M are trios of collinear Menelaus points for triangles ABL and ALC, respectively. So, by Menelaus' theorem,

$$\frac{BC}{CL}\cdot\frac{LO}{OA}\cdot\frac{AN}{NB}=-1 \quad\text{and}\quad \frac{LB}{BC}\cdot\frac{CM}{MA}\cdot\frac{AO}{OL}=-1.$$

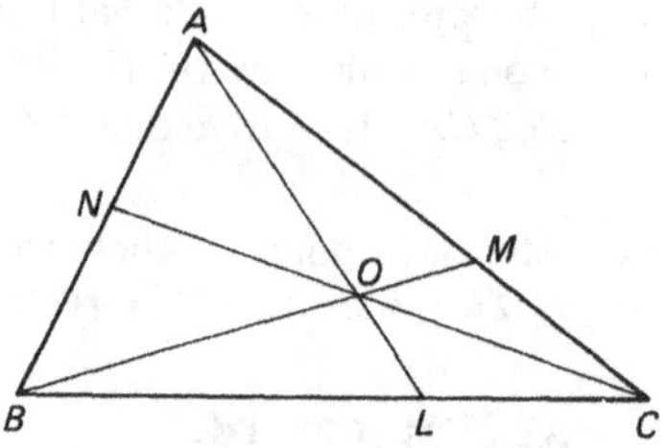

Figure 5.2

Now, multiplying these equations side for side, obtain

$$\frac{BL}{LC}\cdot\frac{CM}{MA}\cdot\frac{AN}{NB}=1.$$

The converse to this theorem is established in the same manner as was the converse to Menelaus' theorem. ▯

5.3 Theorem *Trigonometric form of Ceva's theorem.* Three Cevians AL, BM, CN for triangle ABC concur iff

$$\frac{\sin\measuredangle BAL}{\sin\measuredangle LAC}\cdot\frac{\sin\measuredangle CBM}{\sin\measuredangle MBA}\cdot\frac{\sin\measuredangle ACN}{\sin\measuredangle NCB}=1.$$

▶**5.4 Theorem** Let the incircle for (the circle inscribed in) triangle ABC touch sides BC, CA, AB at points X, Y, Z. Then Cevians AX, BY, CZ concur in a point called the *Gergonne point* for the triangle.

Since BX and BZ are tangents from a point to a circle (see Fig. 5.4), then $ZB = BX$. Similarly $XC = CY$ and $YA = AZ$. Then

$$\frac{BX}{XC}\cdot\frac{CY}{YA}\cdot\frac{AZ}{ZB}=\pm\frac{BX}{CY}\cdot\frac{CY}{AZ}\cdot\frac{AZ}{BX}=\pm 1.$$

The plus sign holds since points X, Y, Z divide the triangle internally. The theorem follows from Ceva's theorem. ▯

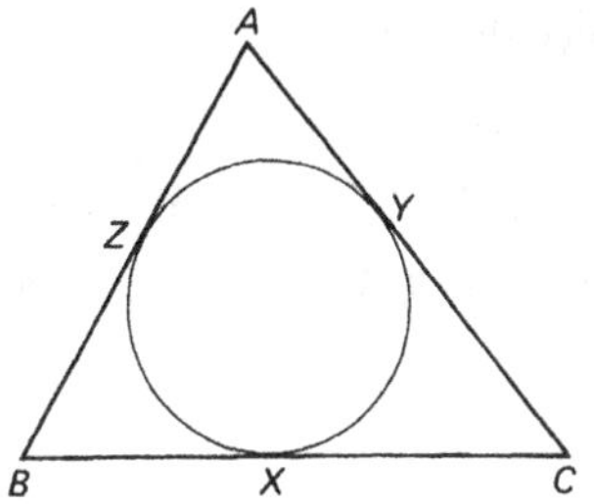

Figure 5.4

5.5 Theorem The three altitudes of a triangle concur.

Referring to Fig. 5.5, let D, E, F be the feet of the altitudes from vertices A, B, C of triangle ABC. From right triangle BAD,

$$90° \pm \sphericalangle BAD = \sphericalangle B,$$

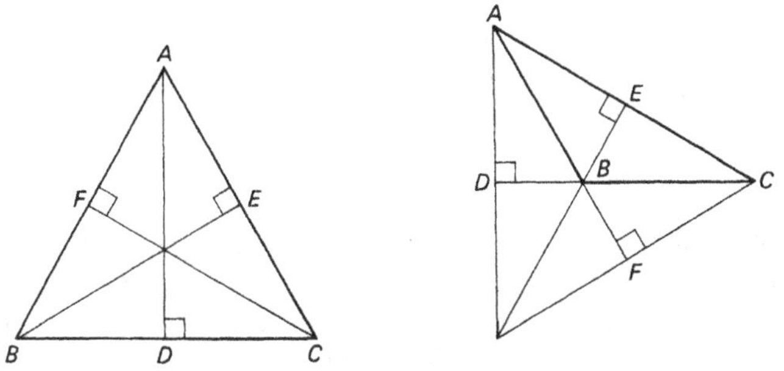

(a) (b)

Figure 5.5

the minus sign is used when $\sphericalangle B$ is acute, the plus sign when $\sphericalangle B$ is obtuse. Thus $\sin \sphericalangle BAD = \pm\cos \sphericalangle B$. Similar relations occurring in the other five right triangles having as a leg an altitude of triangle ABC yield

$$\frac{\sin \sphericalangle \mathbf{BAD}}{\sin \sphericalangle \mathbf{DAC}} \cdot \frac{\sin \sphericalangle \mathbf{CBE}}{\sin \sphericalangle \mathbf{EBA}} \cdot \frac{\sin \sphericalangle \mathbf{ACF}}{\sin \sphericalangle \mathbf{FCB}} = \pm \frac{\cos \sphericalangle B}{\cos \sphericalangle C} \cdot \frac{\cos \sphericalangle C}{\cos \sphericalangle A} \cdot \frac{\cos \sphericalangle A}{\cos \sphericalangle B} = \pm 1.$$

The plus sign holds, since either none or two of these feet D, E, F divide externally the sides on which they lie. □

▶**5.6 Theorem** If AL, BM, CN are three concurrent Cevians for triangle ABC, and if L', M', N' are isotomic conjugates (see Exercise 4.10) of L, M, N with respect to sides BC, CA, AB, then AL', BM', CN' concur. The two points of concurrence are said to be *isotomic conjugate points* for triangle ABC.

Since L and L' are equidistant from the midpoint of side BC, then $\mathbf{BL} = \mathbf{L'C}$ and $\mathbf{BL'} = \mathbf{LC}$. (See Fig. 5.6.) Similar relations hold for M and M' and for N and N'.

From these relations it follows that

$$\frac{\mathrm{BL'}}{\mathrm{L'C}}\cdot\frac{\mathrm{CM'}}{\mathrm{M'A}}\cdot\frac{\mathrm{AN'}}{\mathrm{N'B}}=\frac{\mathrm{LC}}{\mathrm{BL}}\cdot\frac{\mathrm{MA}}{\mathrm{CM}}\cdot\frac{\mathrm{NB}}{\mathrm{AN}}=\frac{1}{+1}=1,$$

by Ceva's theorem, since *AL*, *BM*, *CM* concur. Hence *AL'*, *BM'*, *CN'* concur by Ceva's theorem. ▯

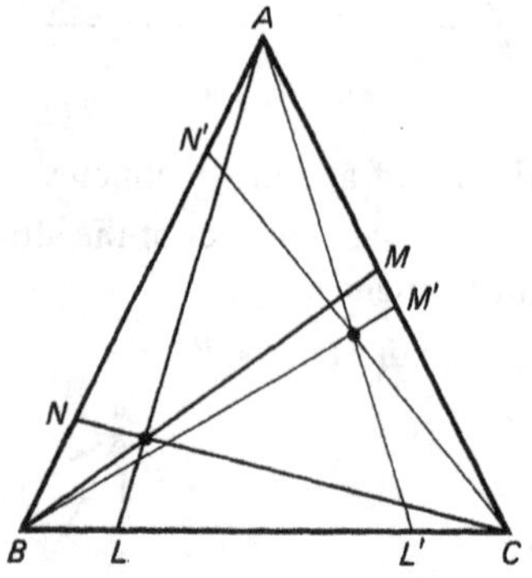

Figure 5.6

5.7 In the last section, Menelaus' theorem was proved by general synthetic means, then in this section Ceva's theorem was shown to be a consequence of Menelaus' theorem. The process may be reversed. In the exercises, you are asked to prove Ceva's theorem synthetically. So, assuming Ceva's theorem has been so established, let us prove Menelaus' theorem. Its converse need not be proved again.

▶**5.8 Theorem** Ceva's theorem implies Menelaus' theorem.

Let *L*, *M*, *N* be three collinear Menelaus points for sides *BC*, *CA*, *AB* of triangle *ABC*, as in Fig. 5.8a. Let *CN* meet *BM* in *P* and *AL* in *Q*, and let *AL* and *BM* meet in

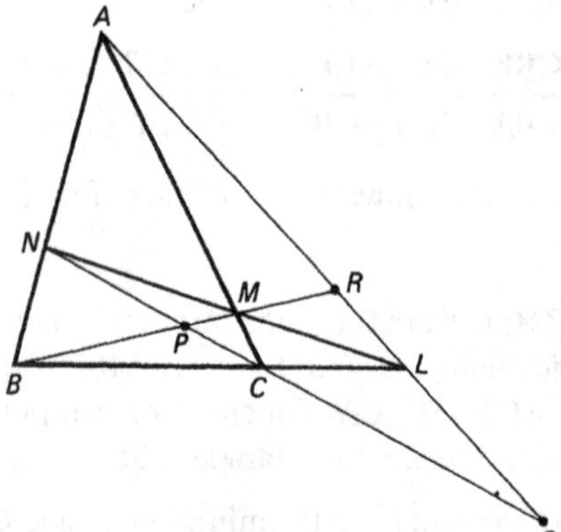

Figure 5.8a

R. We apply Ceva's theorem to each of the six triangles listed in Fig. 5.8b. By multiplying these six equations side for side, we obtain

$$\left(\frac{BL}{LC}\cdot\frac{CM}{MA}\cdot\frac{AN}{NB}\right)^2 = 1,$$

from which the theorem follows, since either one or three of the collinear Menelaus points divide the sides externally. □

Triangle	*Cevians*	*Point of concurrence*	*Resulting equation*
ΔLAB	*LN, AC BR*	*M*	$\frac{AN}{NB}\cdot\frac{BC}{CL}\cdot\frac{LR}{RA} = 1$
ΔMBC	*ML, BA, CP*	*N*	$\frac{BL}{LC}\cdot\frac{CA}{AM}\cdot\frac{MP}{PB} = 1$
ΔNCA	*NM, CB, AQ*	*L*	$\frac{CM}{MA}\cdot\frac{AB}{BN}\cdot\frac{NQ}{QC} = 1$
ΔLMA	*LC, MR, AN*	*B*	$\frac{MC}{CA}\cdot\frac{AR}{RL}\cdot\frac{LN}{NM} = 1$
ΔMNB	*MN, NP, BL*	*C*	$\frac{NA}{AB}\cdot\frac{BP}{PM}\cdot\frac{ML}{LN} = 1$
ΔNLC	*NB, LQ, CM*	*A*	$\frac{LB}{BC}\cdot\frac{CQ}{QN}\cdot\frac{NM}{ML} = 1$

Figure 5.8b

5.9 Theorem If AL, BM, CN are three concurrent Cevians for triangle ABC and if MN meets BC at L', then $(BC, LL') = -1$; that is,

$$\frac{BL}{LC} = -\frac{BL'}{L'C}.$$

Applying Ceva's theorem to triangle ABC of Fig. 5.9 and the concurrent Cevians AL, BM, CN, and applying Menelaus' theorem to triangle ABC and collinear Menelaus points L', M, N, obtain

$$\frac{BL}{LC}\cdot\frac{CM}{MA}\cdot\frac{AN}{NB} = 1 \quad \text{and} \quad \frac{BL'}{L'C}\cdot\frac{CM}{MA}\cdot\frac{AN}{NB} = -1.$$

The theorem follows when we divide these two equations side for side. □

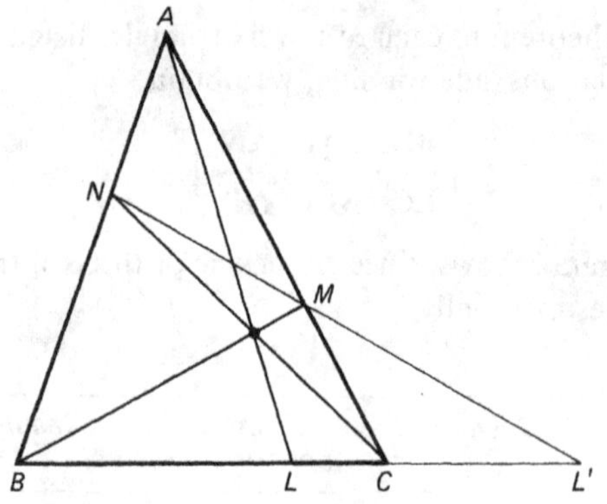

Figure 5.9

5.10 Now that we have stated, proved, and examined Ceva's theorem, it is appropriate to remind you of the remarks in 4.9, that Ceva's and Menelaus' theorems are duals. First, a triangle itself is a self-dual figure: It is the figure formed by three noncollinear points and the three lines joining pairs of the points; it is also the figure formed by three nonconcurrent lines and the three points of intersection of pairs of the lines. Thus a tri*angle* is also a tri*lateral*.

Where Menelaus' theorem discusses the *collinearity* of three *points* on the *lines* (sides) of a triangle, Ceva's theorem considers the *concurrence* of three *lines* on (through) the *points* (vertices) of a triangle. So each is dual to the other. It is odd that, although Menelaus' theorem was known by 100 A.D., Ceva's theorem escaped detection until 1678, more than 1500 years later.

Exercise Set 5

1. Prove the converse to Ceva's theorem.
2. The proof of Ceva's theorem given in the text assumed that the Cevians were proper. Prove the theorem with this restriction removed.
3. Prove Theorem 5.3.
4. Prove Ceva's theorem synthetically. [*Hint:* Draw a line through A parallel to BC to meet BM in S and CN in T. Then use similar triangles.]
5. Prove Theorem 5.5 for the case in which $\measuredangle B = 90°$.
6. Show how to construct the harmonic conjugate of a point C lying on line AB with respect to segment AB.
7. Let AL, BM, CN be Cevians for triangle ABC. Let L', M', N' be the points of intersection of MN and BC, NL and CA, LM and AB. Prove that L', M', N' are collinear iff AL, BM, CN concur.
8. Show that the medians of a triangle concur. Their point of concurrence is called the *centroid* of the triangle.
9. Show that the centroid is the only point whose Cevians divide the sides of the triangle in equal ratios.

10. Show that the centroid is the only point which divides its Cevians in equal ratios.
11. Show that the three internal angle bisectors of a triangle concur.
12. Show that one internal and two external angle bisectors of a triangle concur.
13. Show that if three Cevians concur, then their isogonal conjugates concur (see Exercise 4.11).
14. A circle intersects sides BC, CA, AB of triangle ABC in points L and L', M and M', N and N'. Show that AL, BM, CN concur iff AL', BM', CN' concur.

SECTION 6 | SOME GEOMETRY OF THE TRIANGLE

6.1 It is convenient to introduce some standard terminology and symbolism for certain points, lines, and other objects related to a given triangle ABC. The existence of some of these objects will be established later in this chapter. Following Definition 6.2 and continuing through Section 7, various theorems concerning these defined elements are presented. The purpose of this development is to familiarize the reader with some of the basic properties of triangles and with the methods of proving them. Perhaps the item of greatest interest in Sections 6 and 7 is that which describes the famous ninepoint circle. The relevant information is contained in Definition 6.2 and Theorems 7.13 and 7.14, and summarized in 7.15.

6.2 Definition In triangle ABC the measures of the angles A, B, C will be denoted by A, B, C. The sides opposite these vertices will have lengths a, b, c, respectively, and the semiperimeter $(a + b + c)/2$ is denoted by s. The term *side* of a triangle will refer to either the segment or the entire line on which the segment lies. The context will make the use clear.

The midpoints of the sides a, b, c are denoted by A', B', C'. The lines AA', BB', CC' are called *medians* and have lengths m_a, m_b, m_c. Triangle $A'B'C'$ is called the *medial triangle*. The feet of the altitudes are denoted by D, E, F. The altitudes AD, BE, CF have lengths h_a, h_b, h_c. Triangle DEF is called the *orthic triangle*. The internal angle bisectors meet the opposite sides at U, V, W, and the lengths of the bisectors AU, BV, CW are t_a, t_b, t_c.

The circle inside triangle ABC and tangent to its sides is the *inscribed circle* (or *incircle*) with center I and radius r. The three circles exterior to triangle ABC and tangent to its sides (see Fig. 6.2) are called *excircles* and have centers I_a, I_b, I_c and radii r_a, r_b, r_c. The incircle touches the sides of triangle ABC at X, Y, Z, and the excircle I_j touches the sides at X_j, Y_j, Z_j (for $j = a, b, c$). These four circles are also called *equicircles*.

The circle through the vertices A, B, C is called the *circumcircle* and its center and radius are denoted by O and R.

We use G to denote the *centroid* (the meeting of the medians), H for the *orthocenter* (the meeting of the altitudes), and N for the *ninepoint center* (the center of the circle through the midpoints of the sides). The area of triangle ABC is denoted by K or by K_{ABC}.

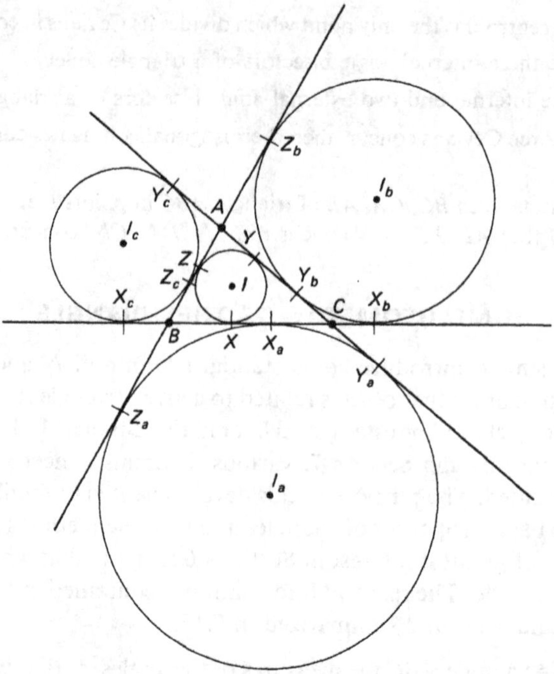

Figure 6.2

6.3 Theorem The medial triangle of a given triangle (assumed to be triangle ABC) has sides parallel to and half the length of the sides of the given triangle.

Since $AC' = AB/2$ and $AB' = AC/2$ and $\measuredangle C'AB' = \measuredangle BAC$ (see Fig. 6.3), then $\triangle BAC \sim$ (is similar to) $\triangle C'AB'$ by SAS (two pairs of corresponding sides proportional and the included angles congruent). Thus $C'B' \cong BC/2 \cong BA'$. Similarly $A'B' \cong BC'$, so $BA'B'C'$ is a parallelogram. The theorem follows. □

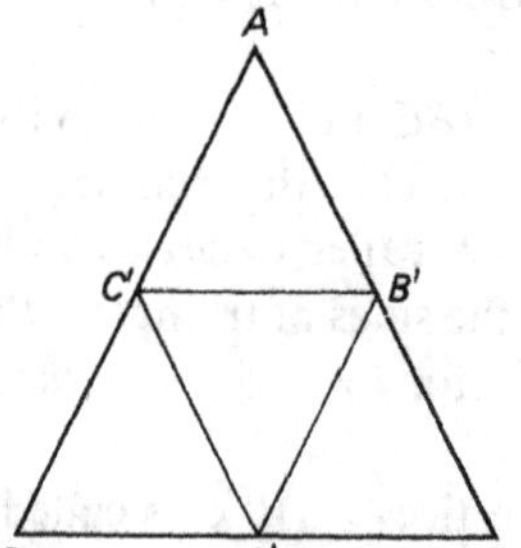

Figure 6.3

6.4 Corollary The medial triangle partitions the given triangle into four congruent triangles.

6.5 Corollary The ninepoint radius is half the circumradius of the triangle.

6.6 Theorem The three medians concur at a point (the centroid) that trisects each median.

Let G denote the intersection of medians BB' and CC', as shown in Fig. 6.6, and let P and Q be the midpoints of segments BG and CG. Since segments $B'C'$ and PQ are each parallel to and half the length of BC by Theorem 6.3, then $PQB'C'$ is a parallelogram, so its diagonals PB' and QC' bisect each other. That is, $\mathbf{PG} = \mathbf{GB'}$ and $\mathbf{QG} = \mathbf{GC'}$. Since also $\mathbf{BP} = \mathbf{PG}$ and $\mathbf{CQ} = \mathbf{QG}$, it follows that BB' and CC' trisect each other. By symmetry, AA' is also trisected by G. (See Exercise 5.8.) ▯

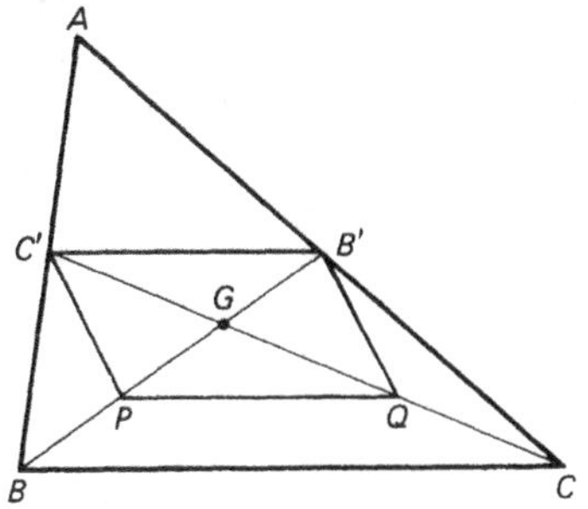

Figure 6.6

▶**6.7 Theorem** The medians of a triangle, as vectors, form a triangle whose area is three-fourths of the area of the given triangle.

Extend median AA' its own length to point P to form $\triangle PCB \cong \triangle ABC$. (See Fig. 6.7.) Let B'' be the midpoint of PC. Now $BB''CC'$ is a parallelogram, so vector

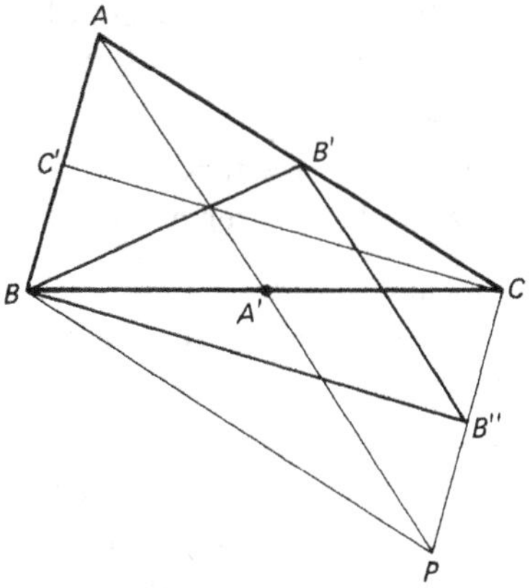

Figure 6.7

CC' is equal to vector $B''B$. Since segment $B'B''$ is parallel to and has length equal to half of AP, then vector $B'B''$ equals vector AA'. Thus triangle $BB'B''$ is formed from vectors equal to the medians of triangle ABC. The proof of the rest of the theorem is left as an exercise. ▯

6.8 Theorem A triangle and its medial triangle have the same centroid.

▶**6.9 Theorem** If two medians of a triangle are congruent, then the triangle is isosceles.

Suppose that $m_b = m_c$. Then use Theorem 6.6 to obtain $\triangle BGC' \cong \triangle CGB'$. ▯

6.10 Theorem The internal angle bisectors of a triangle concur at the incenter of the triangle.

Let I denote the intersection of the internal bisectors of angles A and B in triangle ABC, as in Fig. 6.10. Drop perpendiculars IX, IY, IZ from I to the sides BC, CA, AB. Since BI is the bisector of angle B, then triangles BIX and BIZ are congruent

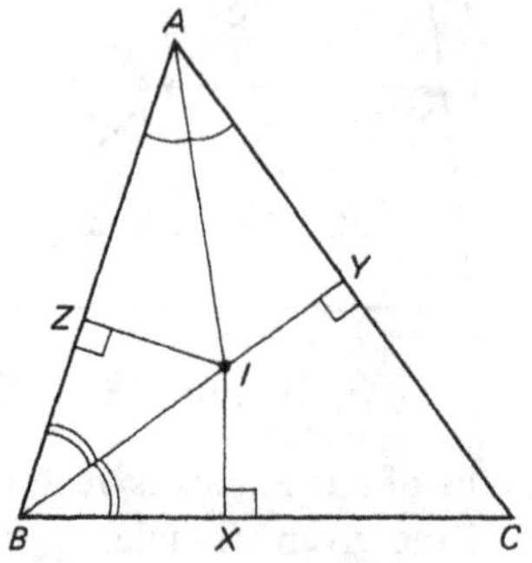

Figure 6.10

(by AAS), so $IX \cong IZ$. Similarly $IZ \cong IY$. Then I is equidistant from the three sides, so I is indeed the incenter of the triangle. Since $\triangle CIY \cong \triangle CIX$ by HL (hypotenuse and leg of one right triangle congruent to the corresponding parts of a second right triangle), then CI bisects angle C. ▯

6.11 Theorem Two external angle bisectors and the internal bisector of the third angle of a triangle concur at each excenter of the triangle.

6.12 Theorem The external and internal bisectors of a given angle of a triangle cut the circumcircle again at the extremities of that circumdiameter perpendicular to the opposite side of the triangle.

If triangle ABC is isosceles with vertex A, then bisector AU is the circumdiameter perpendicular to BC, and the external bisector of angle A is tangent to the circumcircle. So the theorem is true when the point of tangency is considered to be the second point of intersection of the external bisector and the circumcircle.

So assume $AB \ncong AC$. Let the internal and external bisectors of angle A cut the circumcircle again at S and T. (See Fig. 6.12.) Since AS bisects angle BAC, it also bisects arc BC. Thus the circumdiameter through S is the perpendicular bisector of side BC. We need only show then that ST is a diameter. This is true, since $\measuredangle SAT = 90°$. ▯

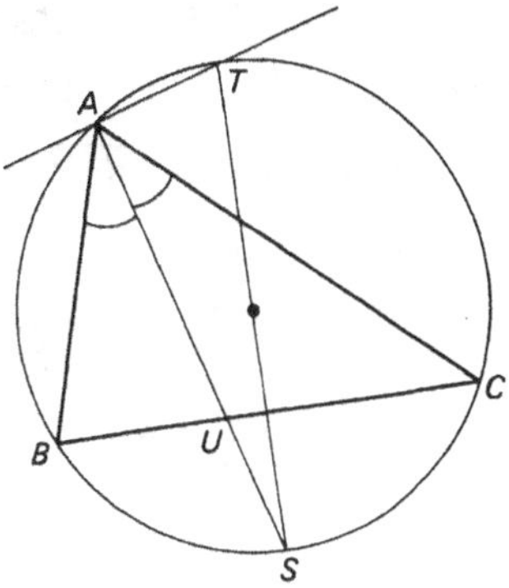

Figure 6.12

6.13 Corollary The incenter of a triangle is the orthocenter of the triangle $I_aI_bI_c$ formed by the three excenters of the given triangle and the vertices of the given triangle are the feet of the altitudes of triangle $I_aI_bI_c$.

6.14 Corollary Each internal angle bisector of a triangle bisects the arc of the circumcircle cut off by the opposite side of the triangle. Hence this bisector and the perpendicular bisector of this opposite side meet on the circumcircle of the triangle.

6.15 Theorem The area of a triangle is equal to the product of its inradius and semiperimeter; that is, $K = rs$.

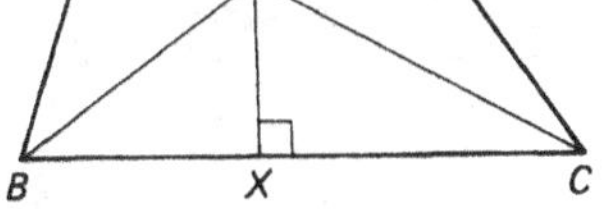

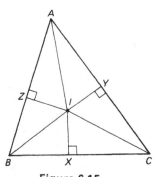

Figure 6.15

In triangle ABC with incenter I (see Fig. 6.15), the areas of triangles BCI, CAI, and ABI are given by

$$\tfrac{1}{2}ar, \qquad \tfrac{1}{2}br, \qquad \text{and} \qquad \tfrac{1}{2}cr.$$

The theorem follows when one adds these three quantities. ▯

▶**6.16 Theorem** The area of a triangle is equal to the product of a given exradius r_j and the semiperimeter diminished by side j. That is,

$$K_{ABC} = r_a(s-a) = r_b(s-b) = r_c(s-c).$$

6.17 Theorem *Heron's formula.* The area of triangle ABC is given by

$$K = \sqrt{s(s-a)(s-b)(s-c)}.$$

In triangle ABC, let $x = m(BD)$, as shown in Fig. 6.17. Then

$$c^2 - x^2 = h_a^2 = b^2 - (a-x)^2,$$

from which one obtains

$$x = \frac{c^2 + a^2 - b^2}{2a}.$$

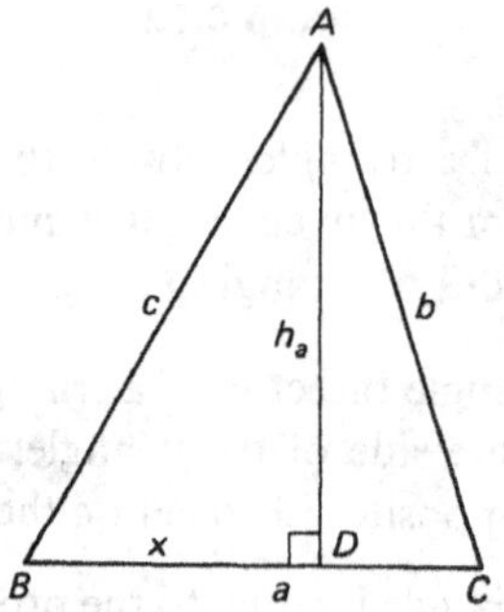

Figure 6.17

Now

$$\begin{aligned} h_a^2 &= c^2 - x^2 \\ &= (c-x)(c+x) \\ &= \left(c - \frac{c^2+a^2-b^2}{2a}\right)\left(c + \frac{c^2+a^2-b^2}{2a}\right) \\ &= \frac{b^2-(a-c)^2}{2a} \cdot \frac{(c+a)^2-b^2}{2a} \\ &= \frac{(b-a+c)(b+a-c)}{2a} \cdot \frac{(c+a+b)(c+a-b)}{2a} \\ &= \frac{(2s-2a)(2s-2c)(2s)(2s-2b)}{4a^2} \\ &= \frac{4}{a^2}\, s(s-a)(s-b)(s-c). \end{aligned}$$

From this equation we obtain

$$\begin{aligned}K_{ABC} &= \tfrac{1}{2}ah_a \\ &= \tfrac{1}{2}a\sqrt{\frac{4}{a^2}\,s(s-a)(s-b)(s-c)} \\ &= \sqrt{s(s-a)(s-b)(s-c)}.\ \square\end{aligned}$$

6.18 For a synthetic geometric proof of Heron's formula, see H. Eves' *An Introduction to the History of Mathematics*, third edition, pages 171–172.

▶**6.19 Corollary** The area of a triangle is given by

$$K=\sqrt{rr_ar_br_c}.$$

6.20 Theorem The product of two sides of a triangle is equal to the product of the circumdiameter and the altitude to the third side.

Let circumdiameter AO of triangle ABC cut the circumcircle again at P. (See Fig. 6.20.) Then $\measuredangle ABC \cong \measuredangle APC$, since each angle is measured by half of arc AC.

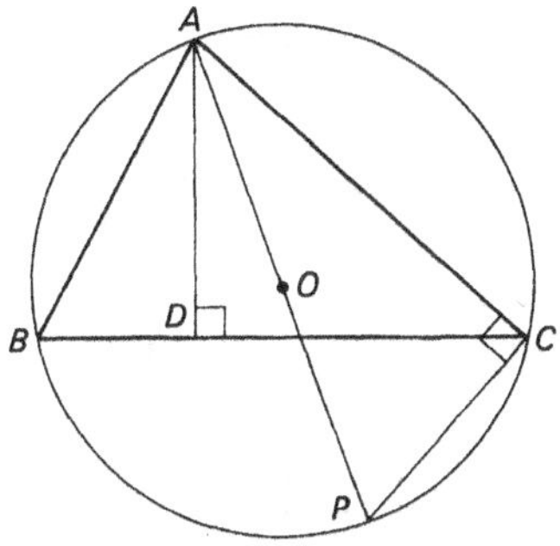

Figure 6.20

Thus right triangles ABD and APC are similar, so

$$\frac{AB}{AD}=\frac{AP}{AC} \qquad \text{and} \qquad AB\cdot AC = AP\cdot AD.$$

That is, $cb = 2Rh_a$. $\square$

▶**6.21 Corollary** The area of triangle ABC is given by

$$K=\frac{abc}{4R}.$$

6.22 Theorem The midpoint of a side of a triangle is the midpoint of the segment on that side whose extremities are the points of contact of the incircle and the excircle named for that side.

In Fig. 6.22, since two tangents from a point to a circle are congruent, we have

$$2s = c + a + b = c + BX_a + X_aC + b$$
$$= c + BZ_a + CY_a + b = AZ_a + AY_a.$$

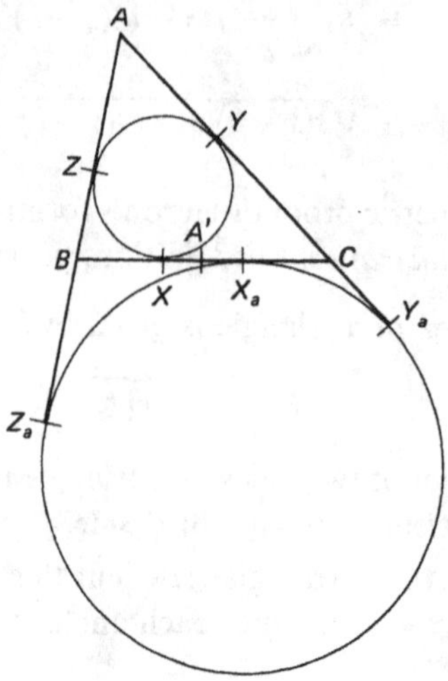

Figure 6.22

But $AZ_a = AY_a$, so

$$s = AZ_a = AY_a.$$

Thus $CX_a = CY_a = AY_a - AC = s - b$. Also $BX = BZ = AB - AZ = c - AY$ and $BX = BC - XC = a - YC$, so

$$2BX = a + c - AY - YC = a + c - b,$$

whence

$$BX = \frac{a + c - b}{2} = s - b = CX_a,$$

from which the theorem follows. □

Exercise Set 6

1. Review the theorems in Book I of Euclid's *Elements*.
2. Prove Corollary 6.4.
3. Prove Corollary 6.5.
4. Prove that the diagonals of a parallelogram bisect each other.
5. Complete the proof of Theorem 6.7.
6. Prove Theorem 6.8.
7. Prove Theorem 6.9.
8. Prove Theorem 6.11.
9. Prove that the internal and external bisectors of a given angle are perpendicular.
10. Prove Corollary 6.13.

11. Prove Corollary 6.14.
12. Prove Theorem 6.16.
13. Prove Corollary 6.19.
14. Prove Corollary 6.21.
15. The diameter of a semicircle is divided into two segments a and b by its point of contact with an inscribed circle as in the accompanying figure. Show that the diameter d of the inscribed circle is the harmonic mean of a and b; that is, $d=2ab/(a+b)$. (*Pi Mu Epsilon Journal*, spring 1970, page 237.)

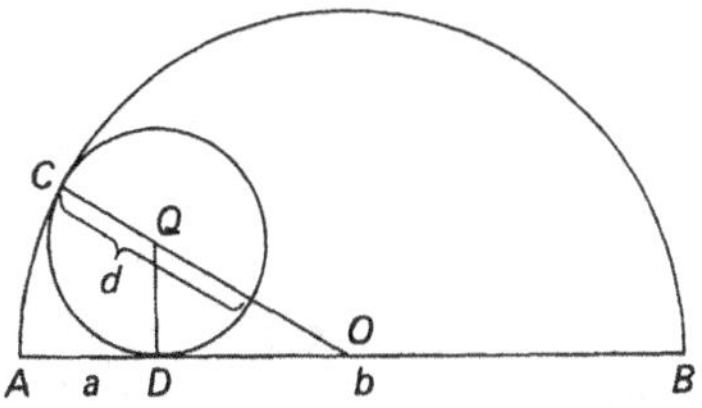

Exercise 6.15

16. Given that a central angle in a circle and the arc it intercepts have the same measure, prove these theorems.
    a) An angle inscribed in a circle or the angle between a chord and a tangent is measured by half its intercepted arc.
    b) The angle between two tangents, two secants, or a tangent and a secant to a circle is measured by half the difference between the intercepted arcs.
    c) The angle between two chords in a circle is measured by half the sum of the arcs intercepted by it and its vertical angle.
17. Prove that the two tangents from a point to a circle are congruent.
18. Prove that the product of the whole secant and its external segment drawn from a point to a circle is equal to the square of the tangent from that point.
19. Prove that if two chords in a circle intersect, then the product of the segments into which one chord is divided is equal to the product of the segments of the other.
20. Prove that $1/r = 1/r_a + 1/r_b + 1/r_c$.
21. Prove that $r_b r_c + r_c r_a + r_a r_b = s^2$.
22. a) Prove that in any triangle $\sin \measuredangle A = a/2R$.
    b) Deduce the law of sines: $a/\sin \measuredangle A = b/\sin \measuredangle B = c/\sin \measuredangle C$.
23. Prove that the bisector of an angle of a triangle also bisects the angle between the altitude and the circumdiameter issuing from that vertex.

SECTION 7 | MORE GEOMETRY OF THE TRIANGLE

7.1 Theorem The perpendicular bisectors of the sides of a triangle concur at the circumcenter of the triangle.

Since the circumcenter O is equidistant from all three vertices, it is equidistant from the endpoints of any side of the triangle. Thus it lies on the perpendicular bisector of each side. □

▶**7.2 Theorem** In any triangle, $R = (r_a + r_b + r_c - r)/4$.

In Fig. 7.2, let IA' meet $I_a X_a$ at P, and let the circumdiameter through A' meet the circumcircle and the internal and external bisectors of angle A at S and T (by Theorem 6.12). Since A' bisects XX_a (by Theorem 6.22) and IX and PX_a are both

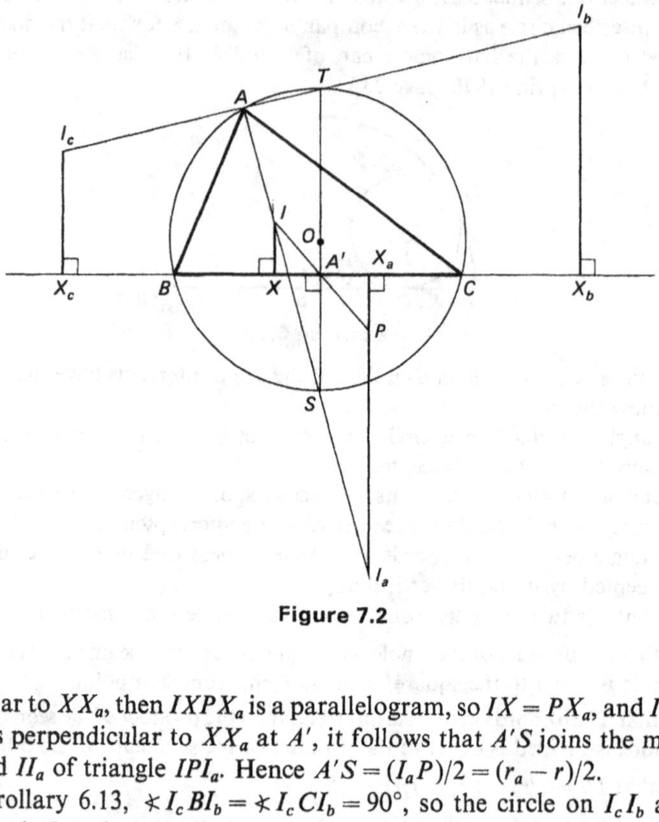

Figure 7.2

perpendicular to XX_a, then $IXPX_a$ is a parallelogram, so $IX = PX_a$, and $I_aP = r_a - r$. Since $A'S$ is perpendicular to XX_a at A', it follows that $A'S$ joins the midpoints of sides IP and II_a of triangle IPI_a. Hence $A'S = (I_aP)/2 = (r_a - r)/2$.

By Corollary 6.13, $\measuredangle I_cBI_b = \measuredangle I_cCI_b = 90°$, so the circle on I_cI_b as diameter passes through B and C. It follows that T is the midpoint of I_cI_b. From trapezoid $I_cX_cX_bI_b$, we have

$$TA' = \tfrac{1}{2}(I_cX_c + I_bX_b) = \tfrac{1}{2}(r_c + r_b).$$

Now we have

$$2R = TS = TA' + A'S = \tfrac{1}{2}(r_c + r_b) + \tfrac{1}{2}(r_a - r),$$

from which the theorem follows. □

▶**7.3 Corollary** The circle on any two excenters as diameter passes through the two vertices of the triangle that are not collinear with the excenters. (See Fig. 7.2.)

▶**7.4 Corollary** The midpoint of a side of a triangle is the midpoint of the segment cut off from that side by the points of tangency of the two excircles not named for that side. (See Fig. 7.2.)

▶**7.5 Corollary** The segment joining two excenters of a triangle is bisected by the circumcircle of the triangle. If the triangle is not isosceles when viewed from the vertex collinear with the equicenters, then that midpoint is not a vertex. (See Fig. 7.2.)

7.6 Theorem The altitudes of a triangle are the angle bisectors of the orthic triangle.

The circle on side BC of triangle ABC as diameter passes through the feet E and F of the altitudes BE and CF, since $\angle BFC = \angle BEC = 90°$. (See Fig. 7.6.) Thus

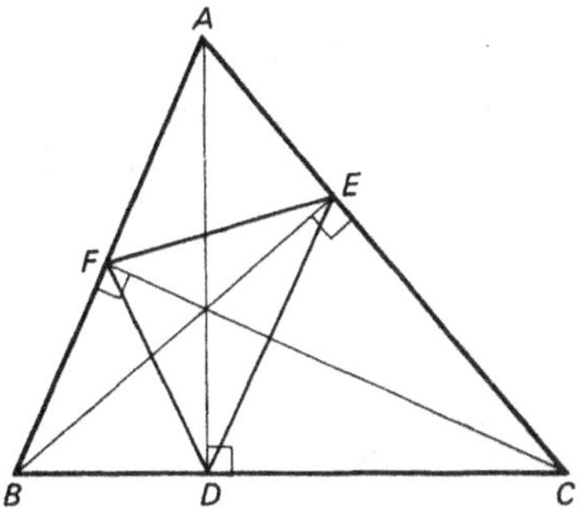

Figure 7.6

$BCEF$ is a cyclic quadrilateral; that is, it can be inscribed in a circle, so $\angle FEC + \angle B = 180°$. Thus $\angle FEC = \angle DEA$, since $BDEA$ is also cyclic, and since BE is perpendicular to AC, then BE bisects angle DEF. Similarly the other two altitudes bisect the other angles of the orthic triangle. ▯

7.7 Corollary The circle on the side of a triangle as diameter passes through the opposite vertices of the orthic triangle.

7.8 Corollary The sides of the orthic triangle form with those of the given triangle three triangles similar to the given triangle.

In Fig. 7.6, $\angle B = 180° - \angle FEC = \angle FEA$. Hence $\triangle ABC \sim \triangle AEF$ by AA (two angles of one triangle congruent to the corresponding two angles of the second triangle). Similarly $\triangle ABC \sim \triangle DBF \sim \triangle DEC$. ▯

7.9 Theorem The tangent to the circumcircle of a triangle at a vertex of the triangle is parallel to the opposite side of the orthic triangle.

This theorem follows from the proof of Theorem 4.5 and from Corollary 7.8. ▯

7.10 Corollary The angle made at an endpoint of one side of the orthic triangle with the side of the given triangle on which it terminates is congruent to the angle between the other two sides of the triangle.

The result follows from Theorem 7.8. ▯

7.11 Corollary The altitudes of a triangle concur.

By Theorem 6.10 applied to the orthic triangle. ▯

▶**7.12 Theorem** The product of the segments into which the orthocenter divides each altitude is a constant for a given triangle.

Any two altitudes, say AD and BE (see Fig. 7.6), are chords of the same circle, by Corollary 7.7. Hence the products of the segments into which they divide each other are equal, by Exercise 6.19. That is, $AH \cdot HD = BH \cdot HE$. By symmetry, it follows that this common value is also equal to $CH \cdot HF$. ▯

7.13 Theorem The feet of the altitudes of a triangle lie on its ninepoint circle.

Referring to Fig. 7.13, $AC' \cong C'D$, since C' is the midpoint of the hypotenuse of right triangle ABD. Thus triangle $C'DA$ is isosceles, and since $C'B'A'B$ is a parellelogram, it follows that $C'B'A'D$ is an isosceles trapezoid, so the circle through C', B' and A' also passes through D. Similarly E and F lie on this circle, the ninepoint circle. ▯

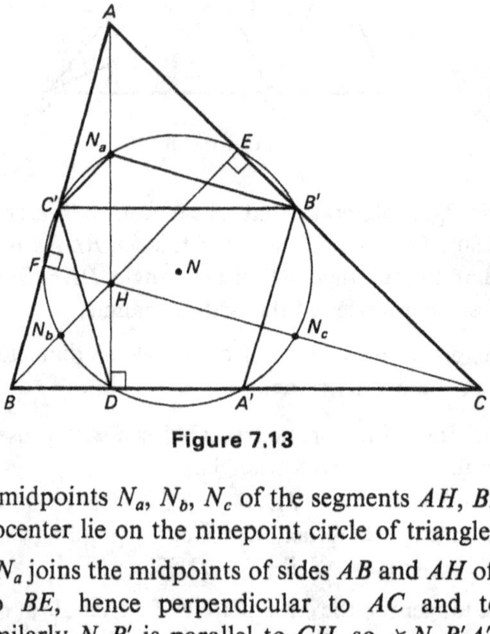

Figure 7.13

7.14 Theorem The midpoints N_a, N_b, N_c of the segments AH, BH, CH joining the vertices to the orthocenter lie on the ninepoint circle of triangle ABC.

In Fig. 7.13, $C'N_a$ joins the midpoints of sides AB and AH of triangle ABH, so $C'N_a$ is parallel to BE, hence perpendicular to AC and to $A'C'$. That is, $\measuredangle N_aC'A' = 90°$. Similarly N_aB' is parallel to CH, so $\measuredangle N_aB'A' = 90°$. Hence the circle on N_aA' as diameter passes through B' and C', so it is the ninepoint circle. Similarly the ninepoint circle passes through N_b and N_c, too. ▯

7.15 We see that the nine points on the ninepoint circle are the midpoints of the sides, the feet of the altitudes, and the midpoints of the segments joining the orthocenter to each of the vertices.

7.16 Corollary The ninepoint center lies midway between the orthocenter and the circumcenter. In fact, any segment joining the orthocenter to a point on the circumcircle is bisected by the ninepoint circle.

Since the ninepoint center N lies on the perpendicular bisectors of DA' and of $B'E$, the first part of this corollary follows. (See Fig. 7.16.) ▯

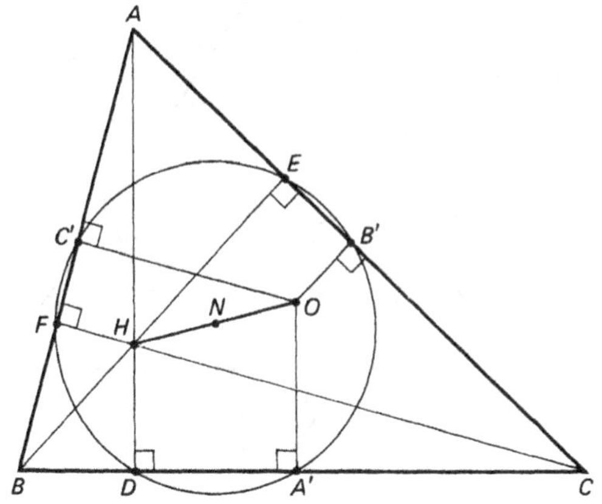

Figure 7.16

7.17 Theorem Let H be the orthocenter for triangle ABC. Then A, B, C are the orthocenters for triangles HBC, AHC, ABH, respectively.

7.18 Definition Four points such that each is the orthocenter of the triangle formed by the other three are said to form an *orthocentric quadrangle.*

7.19 Theorem The four triangles of an orthocentric quadrangle have the same orthic triangle, the same ninepoint circle, and equal circumradii. (See Figs. 7.6 and 5.5.)

▶**7.20 Theorem** The circumcenters of the four triangles of an orthocentric quadrangle form another orthocentric quadrangle having the same ninepoint circle. Furthermore, the four points of either quadrangle are the circumcenters of the triangles of the other quadrangle.

Let the circumcenters of triangles ABC, HBC, AHC, ABH be denoted by O, O_a, O_b, O_c as in Fig. 7.20. Since O_bO_c is the perpendicular bisector of AH, then it is parallel to BC. Since OO_a is perpendicular to BC, O lies on the altitude to vertex O_a of triangle $O_aO_bO_c$. It follows that O is the orthocenter of this triangle. Hence the four circumcenters form an orthocentric quadrangle.

Since N bisects HO (by Corollary 7.16), then N also bisects AO_a, the corresponding segment for triangle HBC. It follows that quadrangle $OO_aO_bO_c$ is symmetric to quadrangle $HABC$ in point N. Now, since O_bO_c is the perpendicular bisector of AH, then BC is the perpendicular bisector of O_aO. Thus A', B', C' are the midpoints of the segments joining the vertices O_a, O_b, O_c to the orthocenter O of triangle $O_aO_bO_c$. The theorem follows. ▯

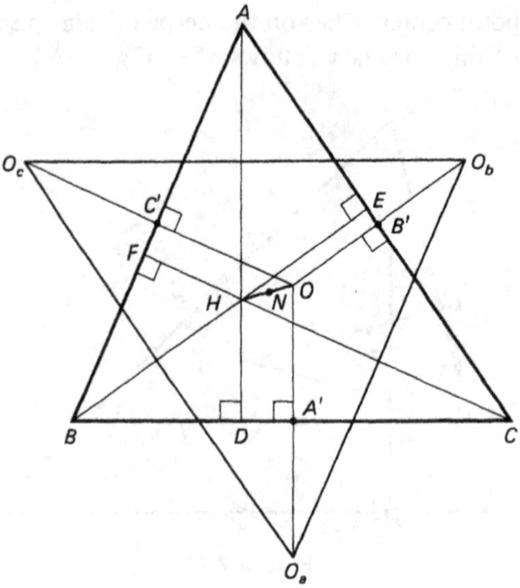

Figure 7.20

7.21 Here we find eight triangles all having the same ninepoint circle. To show just how rich such a figure is, we comment without proof that the nineteenth-century German mathematician Karl Wilhelm Feuerbach proved that the ninepoint circle is tangent to each of the four equicircles of the given triangle (see Theorem 44.6). From Theorem 7.19, the four triangles of an orthocentric quadrangle provide a total of 16 equicircles, all tangent to the ninepoint circle of the quadrangle. (Try drawing all these circles.) By Theorem 7.20, its circumcenters provide four more triangles having this same ninepoint circle, so they in turn provide another 16 equicircles tangent to it. Hence from one triangle we obtain a grand total of 32 equicircles, all tangent to the ninepoint circle!

7.22 We conclude this section with a faulty "proof" of a false "theorem." Recall the theorems we have proved in this and the preceding sections to locate the fallacy in the argument. It should help to draw a most accurate figure, or several accurate figures.

▶**7.23 False "Theorem"** All triangles are isosceles.

If $AB \cong AC$, we are done. So suppose it is not given that $AB \cong AC$. Let the bisector of angle A and the perpendicular bisector of BC meet at P. (See Fig. 7.23.) Draw BP and CP and drop perpendiculars PQ and PR to sides CA and AB. Then $\triangle APR \cong \triangle APQ$ by AAS, so $AR \cong AQ$ and $PR \cong PQ$. Also $\triangle BPA' \cong \triangle CPA'$ by SAS, so $BP \cong CP$. Now $\triangle BPR \cong \triangle CPQ$ by HL, so $RB \cong QC$. Then

$$AB = AR + RB = AQ + QC = AC,$$

so triangle ABC is isosceles. □(?)

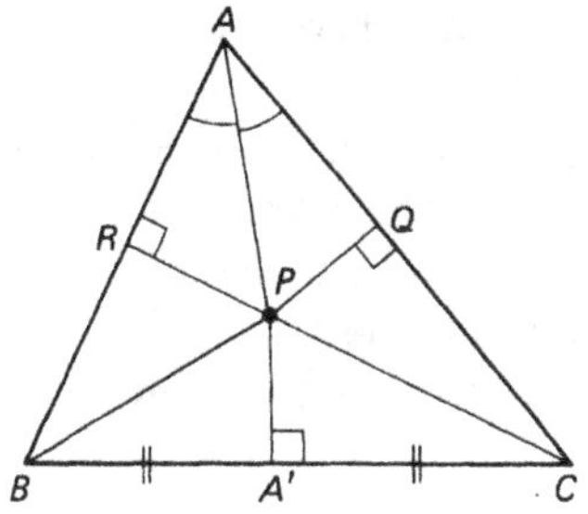

Figure 7.23

Exercise Set 7

1. Prove Corollary 7.3.
2. Prove Corollary 7.4.
3. Prove Corollary 7.5.
4. Prove that the opposite angles of a convex cyclic quadrilateral are supplementary.
5. Prove Corollary 7.7.
6. Prove Corollary 7.11.
7. Prove that the midpoint of the hypotenuse of a right triangle is equidistant from its three vertices.
8. Prove the second part of Corollary 7.16.
9. Prove Theorem 7.17.
10. Prove Theorem 7.19.
11. Find the fallacy in the "proof" of "Theorem" 7.23.
12. Draw the diameters PA and PB in two circles intersecting at P, as shown in the accompanying figure. Let AB cut the two circles in R and S as shown. Then $\measuredangle PRA = 90°$ and $\measuredangle PSB = 90°$, since each angle is inscribed in a semicircle. Thus PR and PS are *two* perpendiculars from a point P to a line AB. Find the fallacy in this argument.

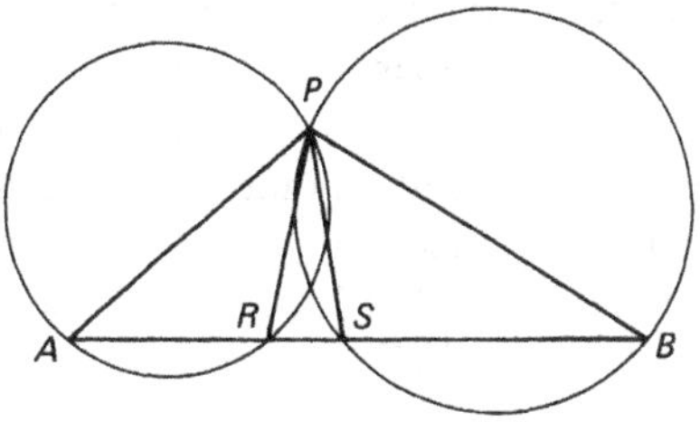

Exercise 7.12

13. Prove that ON_a and AA' in triangle ABC bisect each other.
14. Prove that $(AH)^2 + (BC)^2 = 4(AO)^2$ in triangle ABC.
15. Prove that, in triangle ABC, HA' and AO concur on the circumcircle.
16. Prove that the four centroids of the triangles of an orthocentric quadrangle form a similar orthocentric quadrangle.

17. Prove that the four vertices of an orthocentric quadrangle are the centroids of another similar orthocentric quadrangle.
18. Prove that the equicenters of a triangle form an orthocentric quadrangle whose nine-point circle is the circumcircle of the given triangle.
19. Prove that the radii of the circumcircle to the vertices of a triangle are perpendicular to the opposite sides of the orthic triangle.
20. Prove Theorem 7.2 algebraically using Theorems 6.15 through 6.21.

SECTION 8 | GEOMETRIC CONSTRUCTIONS

8.1 One of the oldest games in the world is the game of Euclidean constructions. Much energy has been expended discovering just what quantities can be constructed using only a compass and straightedge (an unmarked single-edged ruler).

8.2 Equally large amounts of effort have shown that certain quantities cannot be constructed—that it is *impossible* to construct them. No construction of a finite number of steps will ever be found, using just Euclidean tools, that will enable one to trisect every general angle (*trisect an angle*), to draw a square having exactly the same area as a given circle (*square a circle*), or to construct the edge of a cube having just twice the volume of a cube whose edge is given (*duplicate a cube*). These are the three so-called classical construction problems.

Amateur geometers not familiar with the content of such impossibility proofs still invent constructions that appear to solve these ancient problems. A thorough examination of each such construction will uncover errors. Any such construction is either only an approximation (and some are excellent approximations) or else it is incorrect in that it uses the Euclidean tools improperly or makes use of other tools.

8.3 It can be shown that, given any set of numbers (line lengths), one can construct the sum, difference, product, and quotient of any two numbers in the set, hence any rational combination of these numbers. Also one can construct square roots of numbers (see 8.11). And these are the only constructible magnitudes. Hence cube roots or fifth roots or transcendental numbers cannot be constructed generally. Each of the classical construction problems involves finding such an inconstructible number. The exercises pursue this matter further, and a complete discussion is found in the references listed in Exercise 40.7.

8.4 Notation We denote the circle with center A and radius $r = BC$ by $A(r)$ or $A(BC)$. The circle with center A and passing through point P is denoted by $A(P)$.

8.5 We assume that the reader is familiar with the basic construction of the sum and difference of given lengths, of angles congruent to given angles, and of perpendiculars and parallels to given lines. The details of such basic constructions appear in Appendix B.

8.6 Our first construction involves another ancient problem. The first three postulates of Euclid's *Elements* describe the uses of the straightedge and compass. They state:

1. A line segment can be drawn between any two points.
2. A line segment can be extended indefinitely.
3. A circle can be drawn having any given point as center and passing through any other given point.

The first two postulates indicate how the straightedge is to be used and the third describes the use of the compass. (A complete list of Euclid's postulates appears in Appendix A.)

8.7 According to the third postulate, Euclid's compass could not be used to transfer distances. His compass could be set on two points and an arc swung, but the lifting of either point from the paper would cause the compass to lose its setting, to collapse. The modern compass can be used to transfer distances. That is, one may draw the circle $A(BC)$ with a modern compass by setting its points on B and C, then lifting the compass and moving it to point A.

The question arises whether modern compasses can do more than collapsing Euclidean compasses. Certainly they can do no less. To show that the two compasses are indeed equivalent, we present the following theorem.

8.8 Theorem The modern and Euclidean compasses are equivalent.

We shall construct the circle $A(BC)$ using a Euclidean compass. Note that the construction does not require the aid of a straightedge.

Given points A, B, C, as in Fig. 8.8, draw circles $A(B)$ and $B(A)$ to intersect at P and Q. Now draw circles $P(C)$ and $Q(C)$ to meet again at D. Circle $A(D)$ is the desired circle $A(BC)$.

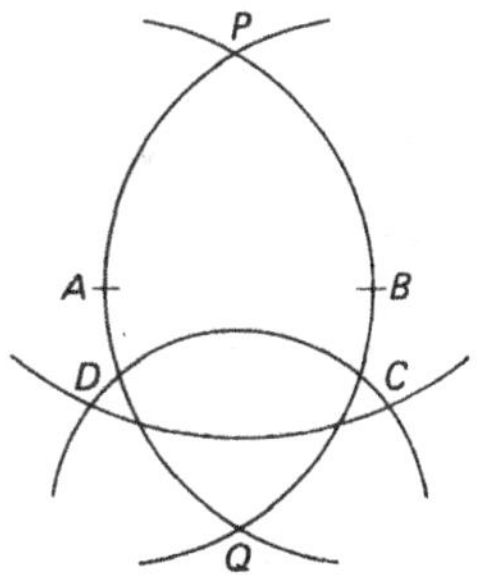

Figure 8.8

The construction is proved by observing that P and Q lie on the perpendicular bisectors of segments AB and CD. Thus A and D are the reflections of B and C in line PQ as mirror, so $AD \cong BC$. ▯

8.9 As shown in the proof of Theorem 8.8, a construction consists of two parts; first, the explanation of the steps one takes to find the desired object; and second, a proof that the constructed object is precisely that which was desired. In the remainder of this section the proofs are left for the reader to supply.

8.10 Problem Construct the mean proportional between two segments.

We are given two segments of lengths a and b and we are required to find their *mean proportional*; that is, we need to find a length x so that $a/x = x/b$; that is, $x = \sqrt{ab}$. On a line (see Fig. 8.10) mark $AP = a$ and $PB = b$. Draw a semicircle on AB as diameter and let the perpendicular to AB at P meet this semicircle at point X. Now $PX = x$. ▯

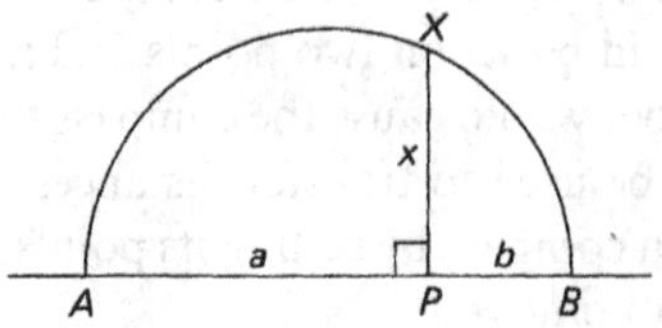

Figure 8.10

8.11 To construct $\sqrt{a}$, we use the construction of Problem 8.10, taking $b = 1$. Hence it is necessary to have a unit segment available when taking square roots of lengths. A unit segment is required also for multiplication and for division, as Problem 8.12 shows.

8.12 Problem Find the product of two segments.

Given segments of lengths a, b, and 1, we shall use proportion in similar triangles to construct a length $x = ab$. On a line, we mark $PU = 1$ and $UB = b$. On any other ray through P, we mark $PA = a$ (see Fig. 8.12). We let the parallel to AU through B cut PA at X. Then AX has length $x = ab$. ▯

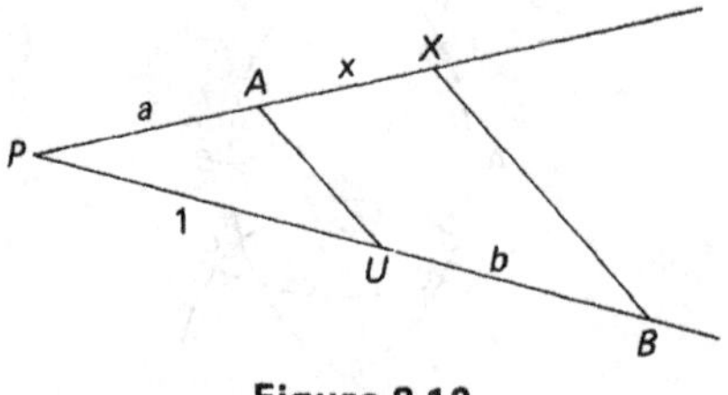

Figure 8.12

8.13 Problem Inscribe a square in a given semicircle.

From the solved problem as shown in Fig. 8.13a, let the square $PQRS$ have side s and the semicircle AOB have radius r. By symmetry, $OQ = s/2$. Since $OR = r$ and $QR = s$, then $(s/2)^2 + s^2 = r^2$, so $s = 2r/\sqrt{5}$. Hence we must construct $2r/\sqrt{5}$.

Given the semicircle AOB, draw right triangle EFG having leg EG twice the length of leg FG. (See Fig. 8.13b.) On hypotenuse EF, mark $EH = r$. Then $EJ = s$, where J is the foot of the perpendicular dropped from H to EG. Now mark length $s/2 = m(HJ)$ on either side of point O on line AB to get points P and Q. The square is easily completed. □

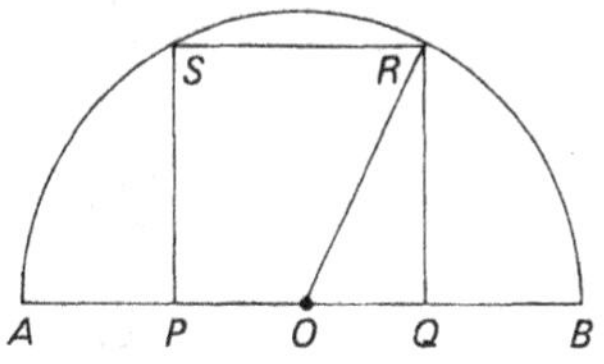

Figure 8.13a

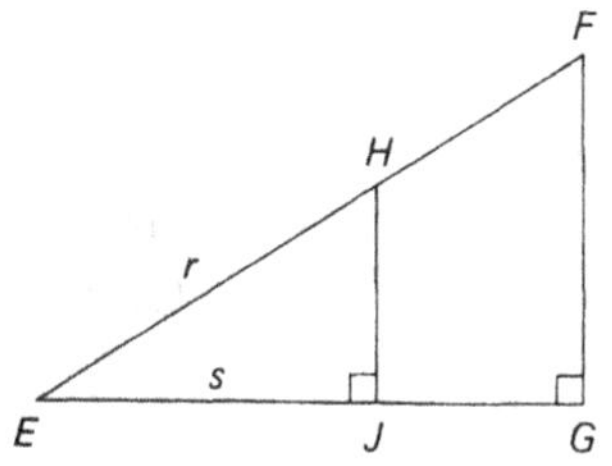

Figure 8.13b

8.14 Problem On a given segment as hypotenuse, construct a right triangle equal (in area) to a given triangle.

Let the segment be PQ and the triangle ABC, as in Fig. 8.14a. Draw altitude AD. Then construct $x = AD \cdot BC/PQ$, as shown in Fig. 8.14b. Let the parallel to PQ, x units distant from PQ, cut in points R and R' the semicircle on PQ as diameter. Then PQR is the desired triangle. □

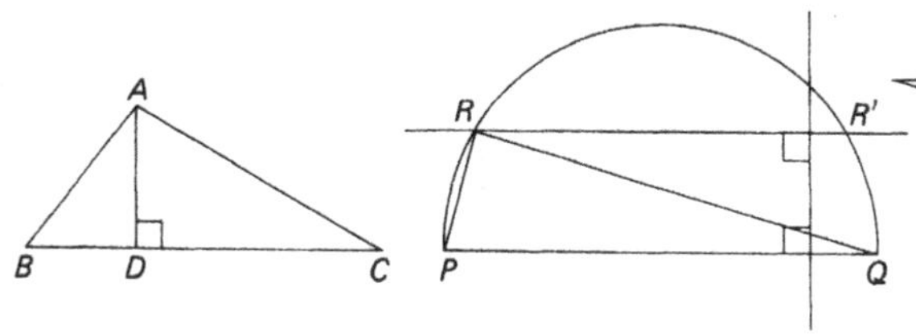

Figure 8.14a

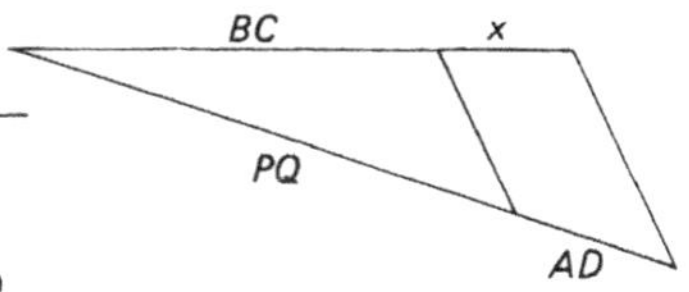

Figure 8.14b

8.15 Problem Construct triangle ABC, given t_a, h_b, c.

At a point E on a base line m, erect a perpendicular h_b units in length to point B, as in Fig. 8.15. Draw circle $B(c)$ to cut m at A. Bisect both angles at A and on these bisectors mark AU_1 and AU_2, each of length t_a. Let BU_1 and BU_2 cut m at C_1 and C_2. Then each of triangles ABC_1 and ABC_2 is a solution. ▯

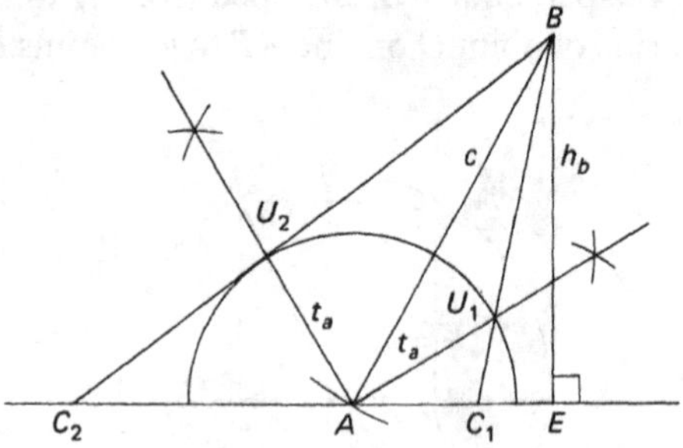

Figure 8.15

▶**8.16 Definition** A *rusty compass* is a compass whose opening is fixed and cannot be altered.

8.17 Many interesting problems occur when the permissible uses of the Euclidean tools are altered slightly. Consider the problem of performing various constructions with a compass whose opening cannot be changed (perhaps because it has rusted). These so-called *rusty compass* problems can be quite challenging. It has been shown that every construction possible with compass and straightedge can be done with rusty compass and straightedge (insofar as the desired objects are points and lines). For a complete discussion, see Eves' *Survey of Geometry* (Vol. 1, Section 4.5).

▶**8.18 Problem** Construct a perpendicular at a point on a line using a rusty compass.

Given point P on line m and letting the fixed compass opening be r, draw circle $P(r)$ to cut m at A and B. Draw circles $A(r)$ and $B(r)$ to cut semicircle $P(r)$ at C and D, as shown in Fig. 8.18. Now let circles $C(r)$ and $D(r)$ meet at Q (they also meet at P). Then PQ is the desired perpendicular. ▯

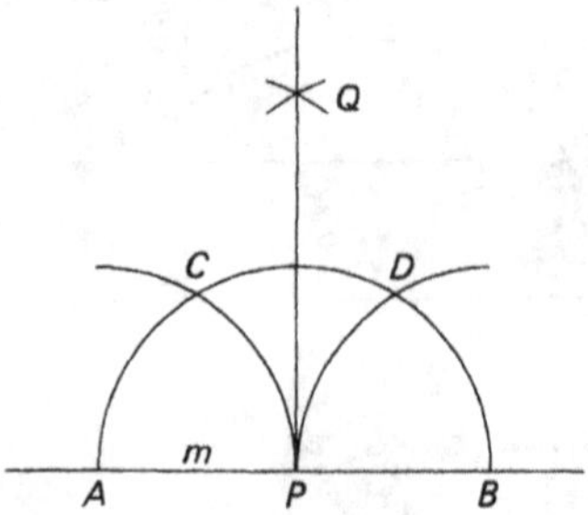

Figure 8.18

▶**8.19 Problem** Using a rusty compass, construct a segment of a given length AB at point P on a line m.

Draw AP. Construct a line n through P parallel to AB as follows (see Fig. 8.19). Erect a perpendicular p to line AB so that p passes within a compass length of P. Then drop a perpendicular n from P to line p. Similarly draw a parallel to AP through B to meet line n at C. Now $PC \cong AB$. Draw line r bisecting the angle between lines m and n, and let the perpendicular from C to r cut m at a point Q. Then $PQ \cong AB$. ▯

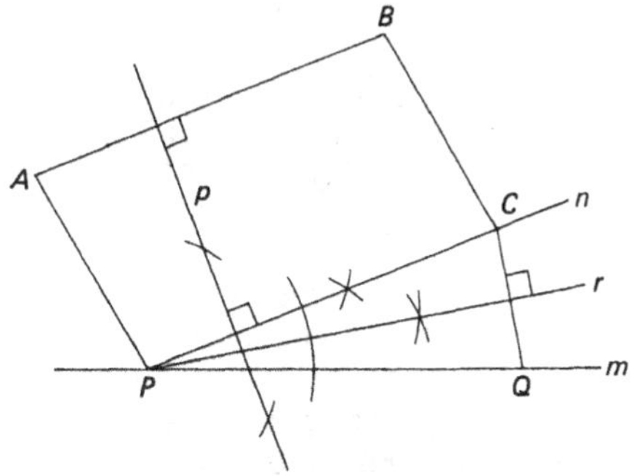

Figure 8.19

Exercise Set 8

1. Complete each problem by proving that the given construction is correct.
   a) Problem 8.10 b) Problem 8.12
   c) Problem 8.13 d) Problem 8.14
   e) Problem 8.15 f) Problem 8.18
   g) Problem 8.19
2. Construct the quotient of two segments.
3. Show how to bisect an angle greater than 120° using a rusty compass.
4. At a given point on a given line construct an angle congruent to a given angle using straightedge and rusty compass.
5. Show how to find square roots with a straightedge and a rusty compass.
6. Construct a triangle similar to a given triangle and four-ninths as large (in area).
7. Construct a trapezoid similar to a given trapezoid and two-thirds as large.
8. On a given segment as base, construct a rectangle equal to a given parallelogram.
9. Construct an isosceles triangle on a given line segment as base and equal to a given triangle.
10. Construct an isosceles triangle equal to a given parallelogram, given the length of its equal sides. Is the construction always possible?
11. Construct triangle ABC, given
    a) B, t_b, h_b b) h_a, m_a, B
    c) b, h_b, c d) A, C, t_c
    e) b, m_a, C f) C, A, h_c
12. "Given a circle, take one-fourth of the circumference as the side of a square. Since the square and the circle now have equal perimeters, they have equal areas. This squares the circle." What is wrong with this argument?

13. Given angle AOB, extend BO and draw any circle with center O cutting the sides of the angle at A and B. (See the accompanying figure.) On a straightedge, mark segment CD equal to OA. Slide the straightedge along point A with point D lying on line BO until point C lies on the circle. Draw this line. Show that angle ADB trisects angle AOB. Does this construction solve the ancient trisection problem?

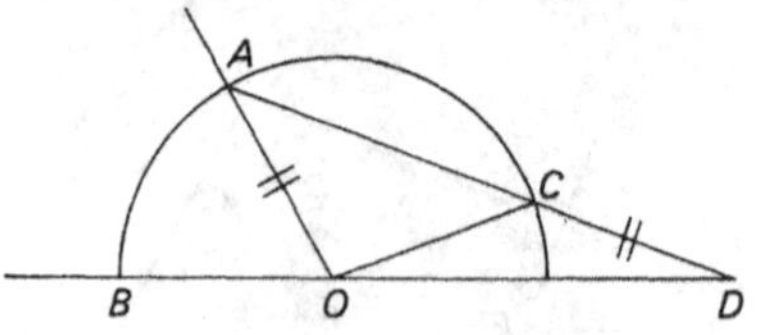

Exercise 8.13

14. An "angle trisection" appears here (see figure), simplified somewhat from its original form in the February 1966 issue of *Mechanix Illustrated*. Find the fallacy in the argument. Draw a circle with center O to cut the sides of the given angle in points A and B. Extend AO and BO to cut the circle again at C and D as in the accompanying figure. Let the bisector of angle AOB cut chord CD at G and the circle at R, as shown. Mark H on OR so that $HR = RG$. Let circle $G(H)$ cut BC and AD at C' and D'. Then $\measuredangle C'OD' = (\measuredangle AOB)/3$.

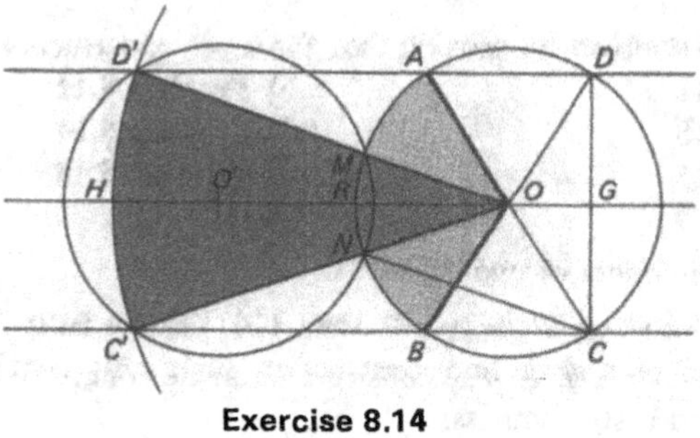

Exercise 8.14

Proof. The circle with radius equal to OA and passing through C' and D' as shown cuts the first circle at points M and N where OD' and OC' meet that circle. Then $\measuredangle RON = \measuredangle NC'C = \measuredangle NCC' = (\measuredangle NOB)/2$, showing that $\measuredangle RON$ trisects $\measuredangle ROB$. The theorem follows. ▯ (?)

15. In the "trisection" of Exercise 8.14, take $\measuredangle AOB = 120°$ and find $\measuredangle C'OD'$ using trigonometry.

16. Prove the statements listed below, from which it follows that $\sqrt[3]{2}$ cannot be constructed, so that the duplication of a cube is impossible. Since π is transcendental, then $\sqrt{\pi}$ is also transcendental, so the circle cannot be squared. Finally, since $\sin 20°$ is a root of the equation $8x^3 - 6x - 1 = 0$, which has no rational roots, then $\sin 20°$ cannot be constructed, so a 60° angle cannot be trisected.
    a) Show how to construct the sum, difference, product, and quotient of two given lengths, and the square root of a given length, in the presence of a unit segment.
    b) Given the unit segment OU, in which $O(0, 0)$ and $U(1, 0)$ are points in the Cartesian plane, show how to construct any point having rational coordinates.

c) Show that the equation of a line through two points with rational coordinates has rational coefficients.
d) Show that the point of intersection of two lines whose equations have rational coefficients has rational coordinates.
e) Show that, given a straightedge and points with rational coordinates, then one can construct only points with rational coordinates.
f) Show that the equation of circle $A(B)$, where A and B are points with rational coordinates, has rational coefficients.
g) Show that the points of intersection of two circles or of a circle and a line, whose equations have rational coefficients, have coordinates that are rational or that involve only square roots of rational numbers.
h) Show that a straightedge and compass can construct only points whose coordinates can be obtained by finitely many applications of the operations of addition, subtraction, multiplication, division, and square root applied to the coordinates of given points.
i) Show that the comments at the beginning of this exercise now follow.

2 ISOMETRIES IN THE PLANE

SECTION 9 THE AMAZING GREEKS

9.1 About 600 B.C. marked the start of a new era in mathematics. People started to ask "Why?" and to demand logical answers. More leisure time and a society which placed increasing emphasis on logical reasoning caused the thinking about mathematics to change from the practical to the more theoretical. The result was the so-called "material axiomatics" which dominated mathematical reasoning for more than 2000 years. In material axiomatics one considers a body of applied mathematics, such as the geometry of the real world. Certain obvious facts about this study are assumed true without proof as axioms or postulates. Then all other facts are logically deduced from the postulates.

9.2 The method of material axiomatics reached full maturity in just 300 years, blossoming forth brilliantly in Euclid's *Elements* about 300 B.C.

9.3 The first man to prove simple theorems was the merchant Thales (640–546 B.C.) of the island of Meletus. In his travels about the Mediterranean Sea, he learned of the mathematics of Egypt and other Eastern countries. Then he *proved* such simple theorems as, for example, that a diameter of a circle bisects the circle and that the base angles of an isosceles triangle are congruent.

9.4 While residing in Egypt, Thales is said to have amazed King Amasis by calculating the height of a pyramid by measuring its shadow and the shadow of a vertical stick of known height and then using simple proportion. Very likely Thales did calculate the height of the pyramid using shadows and proportions, but it is absurd to imagine that he could use the method described, for one could not possibly measure directly the length of the pyramid's shadow.

By predicting a solar eclipse in 585 B.C. he attained great fame. It is said that he fell into a ditch one night while studying the stars during a stroll. An old woman who helped him out protested, "How can you possibly see anything in the heavens when you cannot even behold what is at your feet?"

9.5 Clothed in much mysticism and clouded with many legends, Pythagoras (572?–500? B.C.), of the Aegean island of Samos, and his followers contributed greatly to mathematics. The Pythagorean Society which he founded, a secret fraternity which valued highly both brotherhood and mathematics, credited all discoveries to Pythagoras, so it is not easy to tell what he himself developed as against what his followers deduced.

9.6 It is felt, nonetheless, that Pythagoras did prove the theorem that bears his name. It is said that he was so delighted with this discovery that he sacrificed a hundred oxen. Undoubtedly this story is not true, since the Pythagoreans believed in transmigration of the soul. In fact, one day Pythagoras was reported to have come upon a man beating a dog. He cried for the man to stop, since he could recognize the voice of a departed friend in the dog's howls. And in the next life, he told the man, the tables might be reversed, so that he might be the dumb animal. So effective were these words that the man fell to his knees and begged forgiveness from the dog.

9.7 One must not overlook the great Academy of Plato (429–348 B.C.), over whose door was the motto, "Let no one unversed in geometry enter here." When asked what occupied the Deity, Plato immediately replied, "God geometrizes continually." Eudoxus (408–355 B.C.), the most brilliant of the early mathematicians, invented a theory of proportion that took irrational numbers into account. Two thousand years later Richard Dedekind (1831–1916) showed just how accurate Eudoxus had been in his theory. Aristotle (384–322 B.C.), although known primarily for his systematization of deductive logic, improved some geometric definitions and discussed continuity.

9.8 The most famous scientific treatise of all antiquity, which has been published in more editions than any other book except the Bible, is the *Elements* by Euclid (365?–300? B.C.), written about 325 B.C. This basic mathematics textbook so overshadowed all earlier writings that no trace of its predecessors remains. Indeed it is difficult to find much information about any mathematician prior to Euclid. The thirteen books of the *Elements* embrace not only basic geometry but also number theory and elementary algebra. How much of the material was Euclid's own work is difficult to ascertain, but it is felt that he contributed at least many new proofs.

9.9 Little is known of Euclid's life, but he is reported to be the first professor of mathematics at the great University of Alexandria. When asked by Ptolemy if there were an easier way to master geometry, he replied, "There is no royal road to geometry." Perhaps geometry treated by algebraic means—using analytic geometry, isometries, and similarities—is actually that "royal road" that did not exist in Euclid's day.

9.10 The greatest mathematician of all antiquity was Archimedes (272–212 B.C.) of Syracuse. His accomplishments in geometry and mechanics were legion. He located π between $3\frac{10}{71}$ and $3\frac{1}{7}$ by inscribing and circumscribing polygons of 96 sides in and about a circle. His wondrous war machines forestalled the fall of Syracuse to the

Romans. When the city was finally conquered during a festival when its guard was down, a Roman soldier came upon the aged Archimedes working at a geometric figure in a sand tray; his admonition not to disturb the work enraged the man of war and he ran Archimedes through with his sword. It is worthy of note here that no ancient Roman ever achieved fame in mathematics.

9.11 Many lesser mathematicians followed, each contributing his bit, but none bringing back the great glory that reached its zenith in Archimedes. Eratosthenes (*ca.* 230 B.C.), Apollonius (262?–190? B.C.), Hipparchus (*ca.* 140 B.C.), Heron (*ca.* 110 B.C.), Menelaus (*ca.* 100 B.C.), Claudius Ptolemy (85?–165?), Pappus (*ca.* 340), Theon of Alexandria (*ca.* 390), and his daughter Hypatia (375–415), the first recorded woman mathematician, are but a few.

9.12 A woman of outstanding beauty and an excellent teacher, Hypatia was also a devout and outspoken pagan in this early Christian era. So much so that one day in March, 415, a band of these gentle Christians dragged her from her chariot, stoned her to death with clam shells, dismembered her, and then burned her body, just to make sure the job was well done with proper Christian affection.

9.13 The famous University of Alexandria, founded by Ptolemy about 300 B.C. and the center of all knowledge for more than 900 years, now began to fade as Roman society started to break up, its glory becoming just a shadow of its former self. Finally, in 641, the Arabs conquered Alexandria, burning what had not been destroyed by the Christians. Now Greece was dead. And the Dark Ages overtook Europe.

Exercise Set 9

1. Explain why Thales could not have calculated the height of a pyramid as indicated in 9.4, and show how he could have calculated its height by simple proportion, using two shadow observations at different times of day.
2. Find several different proofs of the Pythagorean theorem.
3. a) Show that when x is the side of a regular polygon inscribed in a circle of radius r, then
$$y = (2r^2 - r(4r^2 - x^2)^{1/2})^{1/2}$$
is the length of the side of a regular polygon having twice the number of sides and inscribed in the same circle.
   b) Show that the perimeter of a regular polygon of 96 sides inscribed in a circle of diameter 1 leads to the formula
$$\pi > 48(2 - \{2 + [2 + (2 + \sqrt{3})^{1/2}]^{1/2}\}^{1/2})^{1/2}.$$
4. Find the formulas corresponding to those of Exercise 9.3 for circumscribed polygons.

SECTION 10 | INTRODUCTION TO TRANSLATIONS, ROTATIONS, AND REFLECTIONS

10.1 If a first triangle ABC is congruent to a second triangle $A'B'C'$, then it is possible to move the first triangle and perhaps turn it over so that it will coincide with the second triangle. Such a motion is called an isometry. That is, an *isometry* is a rigid motion (or map or transformation) of the points of the plane. The defining property of an isometry is that it preserves distances. If the isometry α maps P and Q to P' and Q', then $PQ \cong P'Q'$. It follows that each point maps to just one point (its *image*) and each point P' is the image of just one point P (the *preimage* of P'). We write $\alpha(P) = P'$ to denote that α maps P to P'.

10.2 If the isometry α maps triangle ABC to congruent triangle $A'B'C'$ and if the lengths of the segments AA', BB', CC' are equal and these segments are parallel to one another and similarly directed, then α is called a *translation* through vector* $\overrightarrow{AA'}$. (See Fig. 10.2a.) Our point of view is that a translation moves *all* the points of the

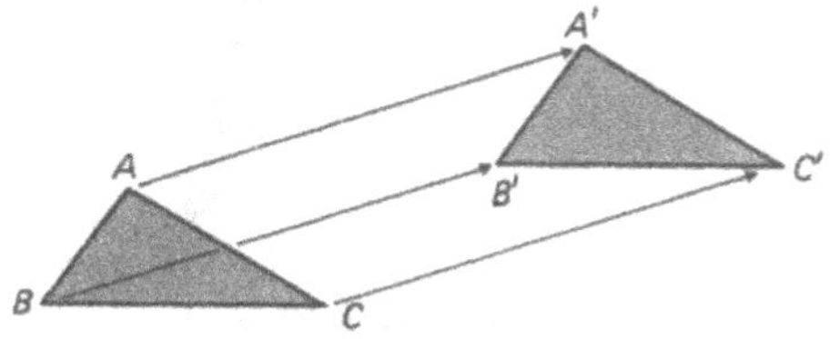

Figure 10.2a

plane. That is, a translation through vector $\overrightarrow{AA'}$ maps each point P in the plane to a point P' such that $\overrightarrow{AA'}$ and $\overrightarrow{PP'}$ are parallel, equal in length, and have the same direction; that is, they are equal vectors (see Fig. 10.2b). Thus, if lines AA' and PP'

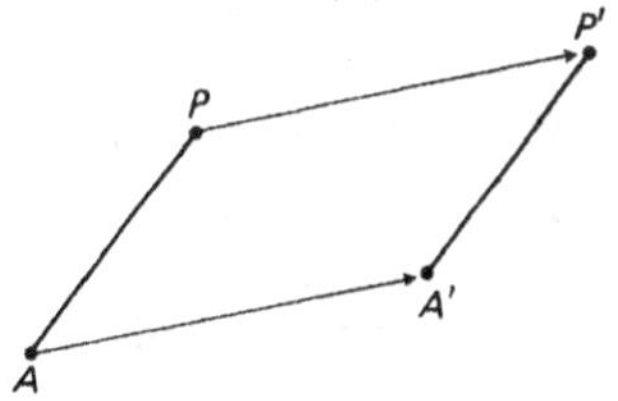

Figure 10.2b

do not coincide, then $AA'P'P$ is a parallelogram. In the event that these lines do coincide, we shall still call the figure they form a (degenerate) parallelogram. Observe that triangle ABC translates to triangle $A'B'C'$ iff AB, CA, BC are similarly oriented to and parallel to or lie on the same line as $A'B'$, $C'A'$, $B'C'$, respectively.

10.3 Let O be a fixed point in the plane and let θ be a fixed directed angle. A *rotation about point O* (called the *center* of the rotation) *through angle* θ is an isometry that maps each point P into a point P' so that $OP \cong OP'$ and $\measuredangle \mathbf{POP'} = \theta$, measured coun-

* We take a vector to be a directed distance, but not a specific directed segment. Thus the vectors $\overrightarrow{AA'}$, $\overrightarrow{BB'}$, $\overrightarrow{CC'}$ of Fig. 10.2a are considered equal to each other. A more detailed discussion of the properties of vectors is to be found in Sections 33 and 34.

terclockwise. Figure 10.3 shows a rotation applied to a triangle ABC. Point O is its own image. Any point that is its own image under a given isometry is called a *fixed point* (or an *invariant point*) for that isometry. Thus the center of a rotation is a fixed

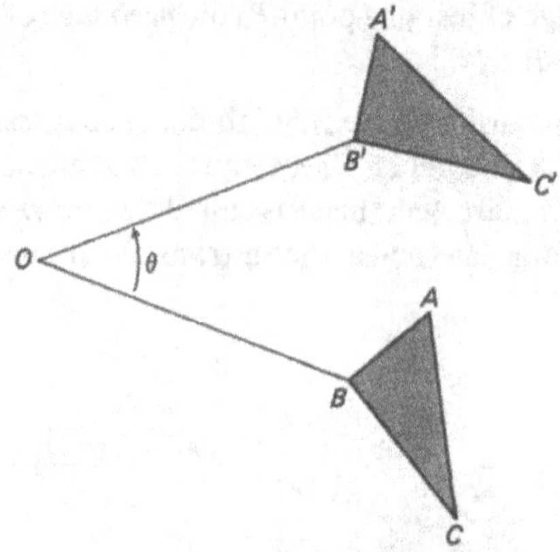

Figure 10.3

point, and no other point is fixed by a rotation through an angle not a multiple of 360°. A translation through a nonzero vector has no fixed points. Note that, although a rotation preserves distances, it is not true that vectors $\vec{AB}$, $\vec{BC}$, $\vec{CA}$ are each equal to vectors $\overrightarrow{A'B'}$, $\overrightarrow{B'C'}$, $\overrightarrow{C'A'}$ when triangle ABC is rotated into triangle $A'B'C'$ through an angle not a multiple of 360°. That is, each side of the triangle, as a vector, is not equal to the corresponding side of its image.

10.4 The third isometry we consider in this section is perhaps the most basic and most important isometry of all. A *reflection in a line m* maps each point P of the plane into its "mirror image" P' with line m as *mirror*. Figure 10.4 shows this reflection, along with a triangle ABC and its image under the reflection. Thus P' is the reflection of P in line m iff m is the perpendicular bisector of PP' or P and P' coincide on line m.

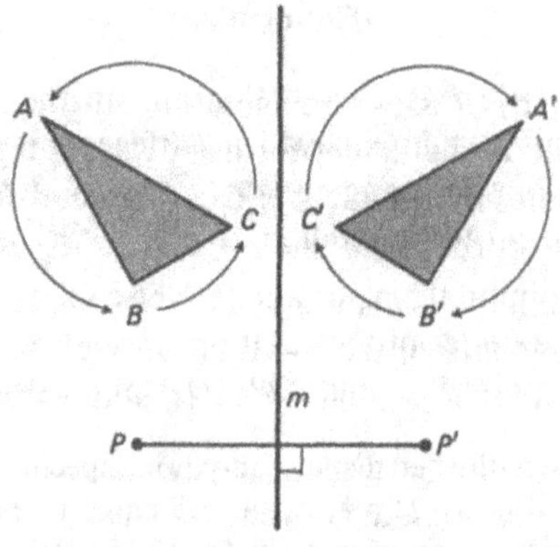

Figure 10.4

Hence all points of m are fixed points. If one reads the vertices of triangle ABC (in Fig. 10.4) in cyclic order, that is, "A–B–C–A," one is traveling around the triangle in the counterclockwise direction. On the other hand, triangle $A'B'C'$ is clockwise. Thus a reflection reverses the sense of a triangle. Any isometry that reverses the sense of a triangle is called *opposite*. A reflection is an opposite isometry. An isometry that preserves the sense of a triangle is called *direct*. Translations and rotations are direct isometries.

10.5 The reflection is the basic building block of isometries. Each translation and each rotation can be written as a product of two reflections. Denoting the reflection in line m by σ_m, if σ_m maps P to P' and then σ_n maps P' to P'', then the *product* of the reflections in lines m and n, in that order, maps P to P'', and we write

$$(\sigma_n \sigma_m)(P) = \sigma_n(\sigma_m(P)) = \sigma_n(P') = P''.$$

Note carefully that $\sigma_n \sigma_m$ means first perform σ_m, then perform σ_n on the result. This definition of "product" holds for transformations in general; it is not restricted just to reflections.

10.6 Let m and n be two parallel lines and let $\mathbf{v}$ denote the vector from line m to line n (measured perpendicular to these lines). Let P be any point with $\sigma_m(P) = P'$ and $\sigma_n(P') = P''$, let $2\mathbf{x}$ denote the vector from P to P' and $2\mathbf{y}$ the vector from P' to P''. (See Fig. 10.6.) Then $\mathbf{x} + \mathbf{y} = \mathbf{v}$, so $2\mathbf{x} + 2\mathbf{y} = 2\mathbf{v}$. That is, $\sigma_n \sigma_m$ translates each point P

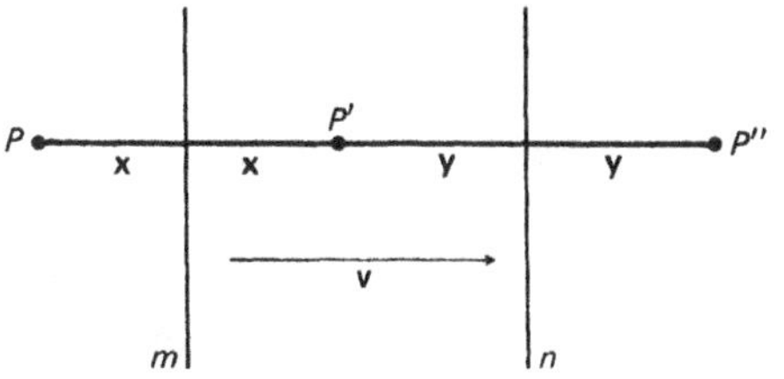

Figure 10.6

through vector $2\mathbf{v}$. It follows that the product of two reflections $\sigma_n \sigma_m$ in parallel lines m and n is a translation through twice the vector from the first line to the second line. Conversely, each translation through a vector $2\mathbf{v}$ can be factored into a product of two reflections in lines m and n, line m being chosen arbitrarily as any line perpendicular to the direction of $\mathbf{v}$, and then line n is that line parallel to m so that the directed distance from m to n is $\mathbf{v}$.

10.7 It deserves repeating that we write $\beta\alpha$ to denote the product of the transformations α and β in that order. In this text, products of transformations are performed from right to left. Since some authors prefer to write products from left to right, be sure to observe the notation conventions of the particular book you are using.

10.8 Now let m and n be two lines intersecting at a point O so that the directed angle from line m to line n is θ. Let P be any point, let $\sigma_m(P) = P'$ and $\sigma_n(P') = P''$. Let the directed angles POP' and $P'OP''$ be $2x$ and $2y$. (See Fig. 10.8.) Then $x + y = \theta$, so the

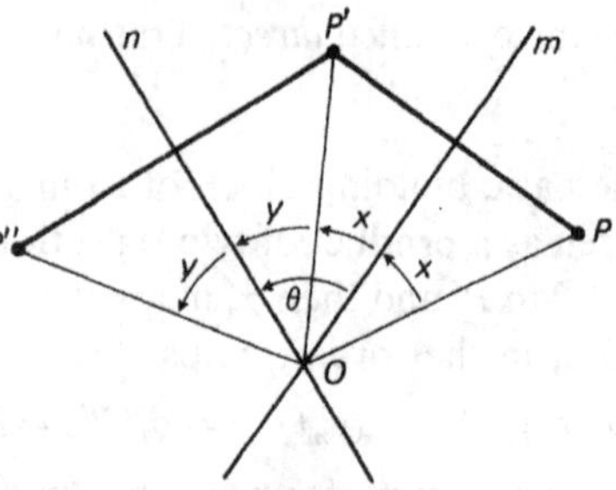

Figure 10.8

directed angle POP'' is $2x + 2y = 2\theta$. Furthermore, $OP \cong OP' \cong OP''$, so $\sigma_n \sigma_m$ is a rotation about point O in angle 2θ. Conversely, every rotation about a point O in an angle 2θ can be factored into a product of reflections in two lines, the first one m being chosen arbitrarily through O, then the second one n is that line through O such that the directed angle from line m to line n is θ.

Exercise Set 10

1. By definition, an isometry preserves lengths. Show that an isometry also preserves angles.
2. If $ABB'A'$ is a rectangle, which two isometries map segment AB to $A'B'$?
3. In the Cartesian plane, if one effects a translation by moving the coordinate axes instead of the points of the plane, how should the axes be moved for the translation through vector $\overrightarrow{AB}$?
4. Show that the product of two translations is a translation.
5. Show that any two directly congruent triangles are related to each other either by a translation or a rotation.
6. Given two directly congruent triangles, find the vector of translation or the center and angle of rotation that maps one to the other.
7. Display two congruent triangles with corresponding sides parallel that are related by a rotation and not a translation. What is the angle of rotation?
8. Use a reflection to prove that the base angles of an isosceles triangle are congruent.
9. Show that a product of reflections in three parallel lines is equal to a reflection in another line of this pencil.
10. Show that a product of reflections in three concurrent lines is equal to a reflection in another line of this pencil.

SECTION 11 | INTRODUCTION TO ISOMETRIES

11.1 An isometry α is completely determined by any three points that form a triangle and their images, because any point P in the plane can be located by giving its distances $m(AP)$, $m(BP)$, $m(CP)$ from these three points. Hence its image P' is determined since the isometry preserves these distances. Figure 13.6 shows this idea.

11.2 It follows that each isometry in the plane is the product of at most three reflections, for it would take at most three reflections to map a triangle ABC to a congruent triangle $A'B'C'$, essentially one reflection per vertex, as shown in Fig. 13.13. Thus each direct isometry, being a product of an even number of reflections, is a product of two reflections, hence it is a translation or a rotation. Each opposite isometry is either a reflection or a product of three reflections.

11.3 Let ι denote that isometry that leaves each point fixed; that is, $\iota(P) = P$ for every point P. We call ι (lower-case Greek iota) the *identity map*. If m is any line, then $\sigma_m \sigma_m = \iota$, for Fig. 10.4 shows that whenever a line m reflects a point P to a point P', then it reflects P' to P; that is, the reflection of the reflection of a point is the point itself. Furthermore, if m and n are any two lines, then

$$(\sigma_m \sigma_n)(\sigma_n \sigma_m) = \sigma_m(\sigma_n \sigma_n)\sigma_m = \sigma_m \iota \sigma_m = \sigma_m \sigma_m = \iota,$$

since transformation multiplication is associative. It is clear that ι is a direct isometry, since it can be written as a product of an even number of reflections.

11.4 If α and β are transformations such that $\alpha\beta = \beta\alpha = \iota$, then β is called the *inverse transformation* to α. If $\alpha(P) = Q$, then $\beta(Q) = P$. Such an inverse always exists for transformations and is unique, so we write α^{-1} for the inverse to α. Since $\sigma_m \sigma_m = \iota$, then a reflection is its own inverse; $\sigma_m^{-1} = \sigma_m$. The letter σ is reserved for self-inverse isometries. Writing α^2 for $\alpha\alpha$, a transformation α, other than the identity ι, that is self-inverse ($\alpha^2 = \iota$) is called *involutoric*. Rotations are generally not involutoric, and translations through nonzero vectors are never involutoric.

11.5 The inverse of the rotation $\alpha = \sigma_n \sigma_m$ is given by $\alpha^{-1} = \sigma_m \sigma_n$ and is a rotation about the same point through the opposite angle. The inverse of the translation $\alpha = \sigma_n \sigma_m$ is again given by $\alpha^{-1} = \sigma_m \sigma_n$ and is a translation in the opposite direction through the same distance. Figures 10.6 and 10.8 show that if the order of reflection is reversed from $\sigma_n \sigma_m$ to $\sigma_m \sigma_n$, then point P'' is mapped first to P', and then to P, just reversing the translation or rotation, hence producing its inverse.

11.6 There is just one rotation that is involutoric, the rotation about a point A through 180° (or through an odd multiple of 180°), called a *halfturn about* (or a *reflection in*) *point A* and denoted by σ_A. Figure 11.6 shows triangle PQR and its image triangle $P'Q'R'$ in a halfturn about point A. Note that involutoric isometries with lower-case subscripts (as in σ_m) represent reflections in lines, whereas those with upper-case subscripts (as in σ_A) represent halfturns about points. In fact, throughout this book, unless otherwise stated, we shall always denote points by upper-case italic letters, lines by lower-case italic letters, and transformations by lower-case

Greek letters. Later, when needed, we shall denote planes by upper-case Greek letters. This convention is subjected to some minor violation in the chapter on complex numbers, but that need not concern us here.

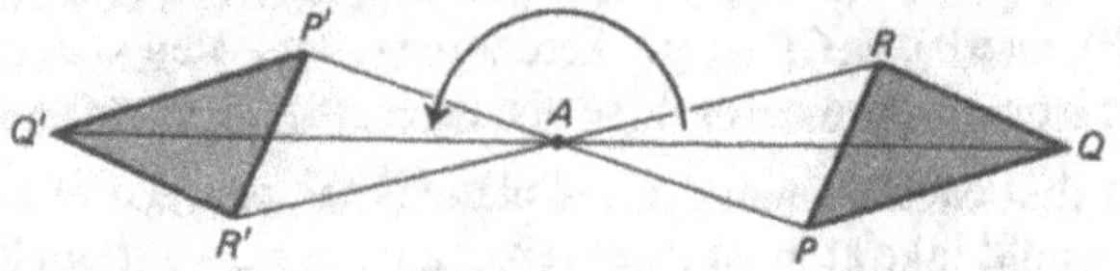

Figure 11.6

11.7 A halfturn about point A is the product of two reflections in any two perpendicular lines m and n through A (see Fig. 11.7a). Thus $\sigma_A = \sigma_n \sigma_m$. Since also n and m are two perpendicular lines through A, it follows that $\sigma_A = \sigma_m \sigma_n$, too. Equating these two expressions for σ_A, we obtain $\sigma_n \sigma_m = \sigma_m \sigma_n$. In fact, $\sigma_s \sigma_r = \sigma_r \sigma_s$ iff $r = s$ or r and s are perpendicular.

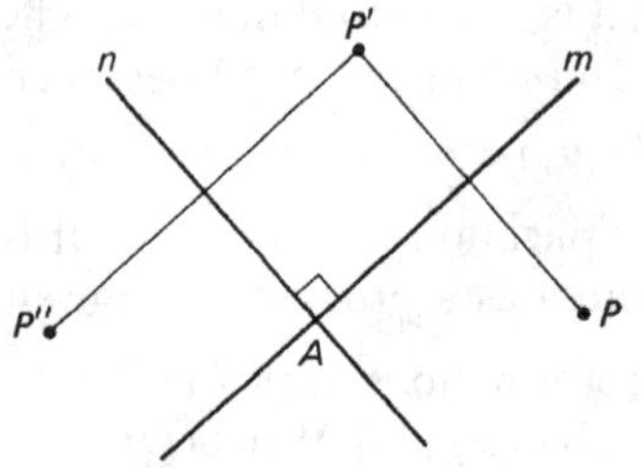

Figure 11.7a

The product $\sigma_B \sigma_A$ of two halfturns is a translation through vector $2\vec{AB}$. Figure 11.7b shows that if $\sigma_A(P) = P'$ and $\sigma_B(P') = P''$, then A and B are midpoints of sides PP' and $P'P''$ of triangle $P'PP''$. It follows that PP'' is parallel to and twice as long as AB. Now for given points A, B, and C, take point D so that $ABCD$ is a parallelogram

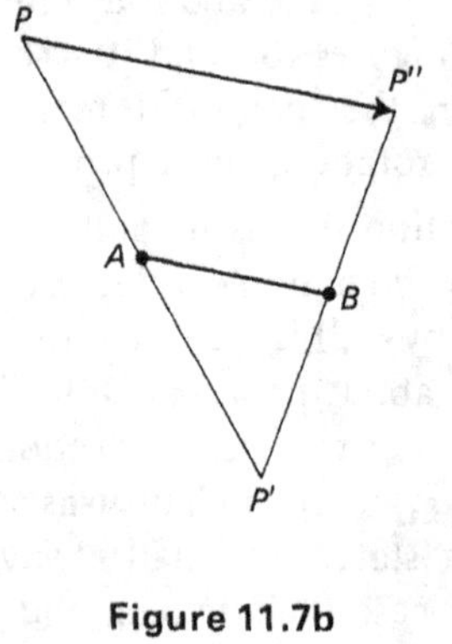

Figure 11.7b

(Fig. 11.7c). Then $\sigma_D\sigma_C\sigma_B\sigma_A = \iota$, since $\sigma_D\sigma_C$ and $\sigma_B\sigma_A$ are inverse translations, so

$$\sigma_D = \sigma_D\,\iota = \sigma_D(\sigma_D\sigma_C\sigma_B\sigma_A) = (\sigma_D\sigma_D)\sigma_C\sigma_B\sigma_A = \iota\sigma_C\sigma_B\sigma_A = \sigma_C\sigma_B\sigma_A;$$

that is, the product of three halfturns ($\sigma_C\sigma_B\sigma_A$) is a halfturn (σ_D) in the fourth vertex of the parallelogram the first three vertices of which are the first three points in their given order.

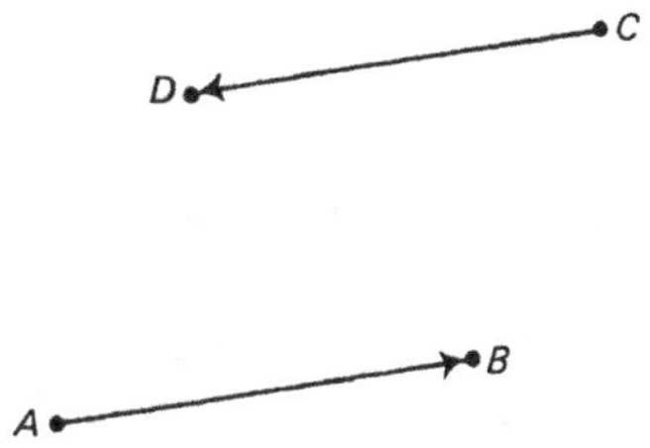

Figure 11.7c

11.8 We next turn our attention to products of reflections in lines m, n, and p. If m, n, p all concur in an ordinary point A, let r be the line through A so that the rotations $\sigma_n\sigma_m$ and $\sigma_p\sigma_r$ are equal (see Fig. 11.8a). Then

$$\sigma_p\sigma_n\sigma_m = \sigma_p(\sigma_n\sigma_m) = \sigma_p(\sigma_p\sigma_r) = (\sigma_p\sigma_p)\sigma_r = \sigma_r.$$

Similarly, if m, n, p are parallel, let r be a line parallel to them so that the translations $\sigma_n\sigma_m$ and $\sigma_p\sigma_r$ are equal (see Fig. 11.8b). Here again we have

$$\sigma_p\sigma_n\sigma_m = \sigma_r.$$

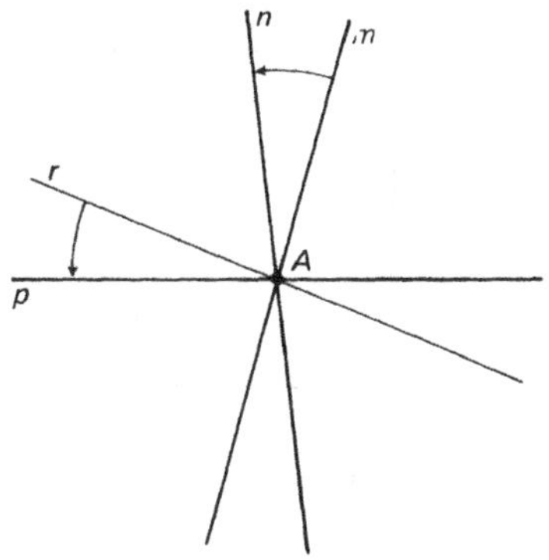

Figure 11.8a

So, if the three lines m, n, p form a pencil, then the product $\sigma_p\sigma_n\sigma_m$ of reflections in these lines reduces to a reflection in some line of the same pencil.

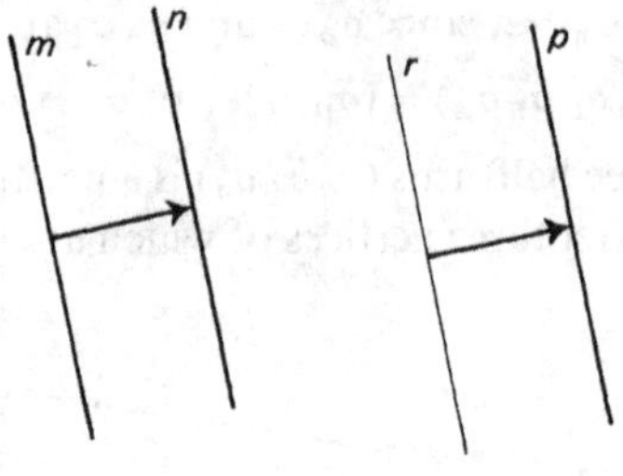

Figure 11.8b

11.9 Any other opposite isometry α can be factored into a product of three reflections $\alpha = \sigma_p \sigma_n \sigma_m$ with m parallel to n, and p perpendicular to both m and n (see Theorem 16.4). This isometry is called a *glide-reflection*; and is the product of the translation $\sigma_n \sigma_m$ followed by the reflection σ_p in line p parallel to the direction of the translation. Note that

$$\alpha = \sigma_p \sigma_n \sigma_m = \sigma_n \sigma_p \sigma_m = \sigma_n \sigma_m \sigma_p$$

since p is perpendicular to lines m and n. The inverse of this glide-reflection $\alpha = \sigma_p \sigma_n \sigma_m$ can be written as

$$\alpha^{-1} = \sigma_m \sigma_n \sigma_p = \sigma_m \sigma_p \sigma_n = \sigma_p \sigma_m \sigma_n,$$

and in its last form α^{-1} is seen to be another glide-reflection, namely the translation inverse to the given glide, and the reflection equal to the given reflection. Thus, in Fig. 11.9, $\alpha = \sigma_p \sigma_n \sigma_m$ carries point A to point A' while α^{-1} carries A' back (dotted lines) to point A.

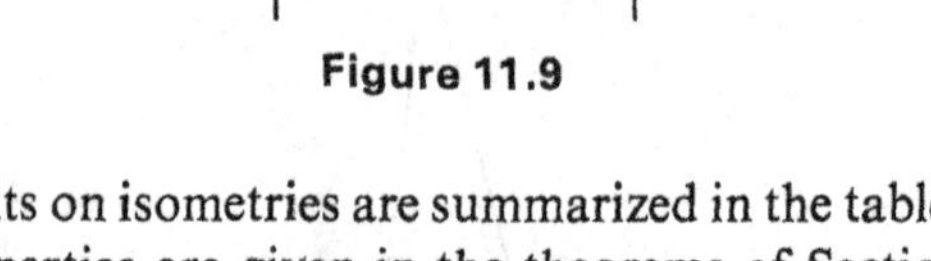

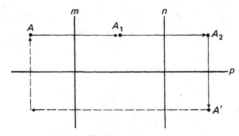

Figure 11.9

11.10 Principal results on isometries are summarized in the table of Theorem 16.14. Other geometric properties are given in the theorems of Section 17.

Exercise Set 11

1. For the given triangle ABC in the coordinate plane with $A(0, 0)$, $B(1, 0)$, and $C(0, 2)$, find coordinates for the vertices of the image triangle $A'B'C'$ under each isometry:
   a) a translation of 5 units in the positive x-direction
   b) a translation of $2\sqrt{2}$ units at 45° into the first quadrant

c) a rotation of 90° about the point $P(3, 0)$
d) a rotation of −45° about the point $P(3, 0)$
e) a rotation of −90° about the point $Q(2, -3)$
f) a reflection in the mirror $y = 0$
g) a reflection in the mirror $y = x$
h) a reflection in the mirror $x + y = 4$
i) a halfturn about $P(3, 0)$
j) a halfturn about $Q(2, -3)$
k) a glide-reflection of 5 units in the positive x-direction with mirror $y = 2$
l) a glide-reflection of $2\sqrt{2}$ units at 45° into the first quadrant with mirror $x - y = 4$
m) the identity ι

2. Write each isometry of Exercise 11.1 as a reflection or as a product of reflections.
3. No translation or glide-reflection (with nonzero vector) can map a rectangle into itself. Find all rotations, halfturns, and reflections that map a given rectangle into itself.
4. Repeat Exercise 11.3 for the following:
   a) square
   b) parallelogram
   c) rhombus
   d) equilateral triangle
   e) regular hexagon
   f) regular pentagon
5. Write the inverse of each isometry in Exercise 11.1.
6. Decide which of the isometries in Exercise 11.1 are involutoric.
7. Decide for each isometry α of Exercise 11.1 whether there is a positive integer n such that $\alpha^n = \iota$. If so, find the smallest such n. We call that n the *order* of the isometry. Involutoric isometries have order 2, ι has order 1. Translations and glide-reflections through nonzero vectors and many rotations have no such positive n, so the *order* of such isometries is defined to be zero.
8. Given the lines m, n, p with equations $y = 0$, $y = 2x$, $x = 0$, find line q so that
   a) $\sigma_q = \sigma_p \sigma_n \sigma_m$
   b) $\sigma_q = \sigma_m \sigma_n \sigma_p$
   c) $\sigma_q = \sigma_m \sigma_p \sigma_n$
   d) $\sigma_q = \sigma_p \sigma_m \sigma_n$
9. Repeat Exercise 11.8 taking $y = 0$, $y = 2$, $y = 3$ as equations for the lines m, n, p.
10. Show that a translation can be written as a product of two rotations. Furthermore, show that one center of rotation and one angle of rotation ($\neq 0°$) may be chosen arbitrarily.
11. Show that the midpoint of the hypotenuse of a right triangle is equidistant from the three vertices; do so by using a halfturn about that midpoint.

SECTION 12 | TRANSFORMATION THEORY

12.1 In this section one will find a formal treatment of the transformation theory required for a formal treatment of isometries. It is assumed that the reader has read the overview given in Sections 10 and 11, so that he is familiar with the direction this development is to take.

12.2 Definition A *transformation of the plane*, or more simply, a *transformation*, α is a one-to-one map or function of the points of the plane onto themselves. That is, α maps each point of the plane to a unique image point, and each point in the plane is the image of exactly one point.

12.3 Definition If α and β are transformations, we define their *product* (or *composition*) $\beta\alpha$ by $(\beta\alpha)(P) = \beta(\alpha(P))$ for each point P. We also write α^2 for $\alpha\alpha$, α^3 for $\alpha\alpha^2$, etc.

12.4 Since a transformation is a function, it follows that two transformations α and β are equal iff they are identical, that is, iff for each point P in the plane, $\alpha(P) = \beta(P)$.

12.5 Theorem If α and β are transformations, then $\beta\alpha$ is a transformation.

12.6 Theorem Transformation multiplication is associative; that is, if α, β, γ are transformations, then $(\gamma\beta)\alpha = \gamma(\beta\alpha)$.

For a given point P, let $\alpha(P) = Q$, $\beta(Q) = R$, and $\gamma(R) = S$. Then

$$((\gamma\beta)\alpha)(P) = (\gamma\beta)(\alpha(P)) = (\gamma\beta)(Q) = \gamma(\beta(Q)) = \gamma(R) = S,$$

and

$$(\gamma(\beta\alpha))(P) = \gamma((\beta\alpha)(P)) = \gamma(\beta(\alpha(P))) = \gamma(\beta(Q)) = \gamma(R) = S.$$

Hence $(\gamma\beta)\alpha = \gamma(\beta\alpha)$. ▯

12.7 Definition For transformation α and point P, if $\alpha(P) = P$, then P is called a *fixed point* (or an *invariant point* or *double point*) for transformation α.

12.8 Definition The transformation ι, defined by $\iota(P) = P$ for each point P in the plane, is called the *identity transformation* or the *identity map*. That is, ι is that transformation that leaves each point fixed.

12.9 Theorem If α is a transformation, then $\alpha\iota = \iota\alpha = \alpha$.

12.10 Definition If α and β are transformations and $\beta\alpha = \iota$, then β is said to be *inverse* to α.

12.11 Theorem If β is inverse to α, then α is inverse to β.

Let $\beta(P) = Q$ and $\alpha(Q) = R$ for a given point P. Since $\beta\alpha = \iota$, then

$$Q = (\beta\alpha)(Q) = \beta(\alpha(Q)) = \beta(R).$$

Since β is a transformation, it is one-to-one; that is, since $\beta(R) = \beta(P)$, then $R = P$. Now

$$(\alpha\beta)(P) = \alpha(\beta(P)) = \alpha(Q) = R = P,$$

so $\alpha\beta = \iota$; that is, α is inverse to β. ▯

12.12 Theorem If β and γ are both inverse to α, then $\beta = \gamma$. That is, a transformation has at most one inverse.

For $\beta = \beta\iota = \beta(\alpha\gamma) = (\beta\alpha)\gamma = \iota\gamma = \gamma$ by Theorems 12.6, 12.9, and 12.11. ▯

12.13 Theorem Each transformation has an inverse transformation.

Given transformation α and any point P in the plane, let $\alpha(Q) = P$. This is always possible since α is an onto map. Now define β by $\beta(P) = Q$. Then β is a transformation because α is a transformation, and $\beta\alpha = \iota$. ▯

12.14 Definition If α is a transformation, denote its unique inverse by α^{-1}.

12.15 Definition *A* set G, along with a binary operation $*$, is called a *group* iff the following four postulates are satisfied:

$G1$: whenever $a, b \in G$, then $a*b \in G$,
$G2$: if $a, b, c \in G$, then $(a*b)*c = a*(b*c)$,
$G3$: there is an element i in G such that, if $a \in G$, then $a*i = i*a = a$, and
$G4$: for each a in G, there is an element a^{-1} in G such that $a*a^{-1} = a^{-1}*a = i$.

12.16 Definition A group is called *abelian* iff it satisfies the commutative law:

$G5$: for each $a, b \in G$, $a*b = b*a$.

12.17 Theorem The set T of all transformations of the plane, along with transformation multiplication, is a group.

By Theorems 12.5, 12.6, 12.9, and 12.13. ▯

12.18 That this group of Theorem 12.17 is not abelian will be shown later (see Exercise 12.3).

12.19 Geometry can be defined by means of its transformation groups. Euclidean geometry is the study of those properties (congruence of figures, equality of lengths, parallelism, etc.) which are invariant under the transformations of the transformation group containing all reflections and products of reflections (translations, rotations and glide-reflections).

12.20 High school geometry courses also study similar figures. This study is called *plane equiform geometry*, the study of properties invariant under the group of all similarities, namely reflections, products of reflections, homotheties (or uniform stretches of the plane), and all products of these transformations. These transformations are studied in the next chapter.

12.21 If, in the transformation group, one includes all projections of the points of the plane (projections from one plane onto another), the resulting study is *projective geometry*. Every theorem of projective geometry also holds in equiform and Euclidean geometries since every property preserved by all the transformations of projective geometry will certainly be preserved by any subset of these transformations. Similarly, every theorem of equiform geometry is true in Euclidean geometry.

12.22 In 1872 Felix Klein (1849–1925) conceived the definition of a *geometry* as the study of those properties of a set of points that are invariant under the transformations of some transformation group. He had just accepted a chair on the faculty at the University of Erlangen, and this definition was presented in his inaugural address, now known as his *Erlanger programm*. Thus, when studying the Euclidean geometry determined by isometries, we are concerned with congruence, lengths, measures of angles, similarity, concurrence of lines, collinearity of points, etc. In plane equiform geometry, congruence and lengths are no longer preserved. In projective geometry, only collinearity of points and concurrence of lines are left invariant in the list above. Each time the group of transformations is enlarged, the field of study is narrowed, for fewer properties are preserved when more transformations are considered. Conversely, all properties preserved under a given group of transformations will certainly be preserved under any subgroup of these transformations.

12.23 Euclid assumed (see Appendix A) that "things which coincide with one another are equal." It was his intent to "pick up and move" a figure so that it could be superimposed upon another figure as a test for congruence. Since the logical foundations of this *method of superposition* are not made clear, modern treatments assume the SAS condition for congruence of triangles. Where superposition is hazy (how can you pick up a genuine line, and what happens to it if you do manage to pick it up?), the SAS postulate is precise. Now, an isometry is exactly the motion Euclid had in mind, and we therefore develop the theory of isometries to clarify Euclidean transformations. So a modern treatment of superposition would state that "two triangles are congruent if a reflection or a product of reflections maps one triangle into the other," thereby avoiding the vague "pick up and move" idea. Students beginning high school geometry have the background to base their sophomore geometry course on transformations, provided the material is well written for that level. The text that you use in such a course should do more than pay mere lip service to transformations, but the properties of triangles and circles, etc., integrated with the corresponding properties of prisms and spheres, etc., must receive full coverage.

Exercise Set 12

1. Prove Theorem 12.5.
2. Prove Theorem 12.9.
3. Find transformations α and β such that $\alpha\beta \neq \beta\alpha$.
4. Prove that ι is unique.
5. Prove that $(\alpha^{-1})^{-1} = \alpha$.
6. Prove that $(\beta\alpha)^{-1} = \alpha^{-1}\beta^{-1}$.
7. If $\alpha^2 = \alpha$, then transformation α is said to be *idempotent*.
   a) Show that if α is idempotent, then $\alpha = \iota$.
   b) Show that there are transformations other than ι such that $\alpha^3 = \alpha$.
8. For given transformations α and β, show that there are unique transformations γ and δ such that $\gamma\alpha = \beta$ and $\alpha\delta = \beta$. Find expressions for γ and δ in terms of α and β.
9. Show that if $\alpha\gamma = \beta\gamma$, or if $\gamma\alpha = \gamma\beta$, for transformations α, β, γ, then $\alpha = \beta$.
10. Show that if $\alpha^2\beta^2 = (\alpha\beta)^2$ for transformations α and β, then $\alpha\beta = \beta\alpha$.
11. a) Find transformations α, β, γ such that $\gamma\alpha = \beta\gamma$ and $\alpha \neq \beta$.
    b) If α and γ are given, find an expression for β.
    c) Can such β always be found for given α and γ?

12. Show that any set S of transformations forms a group under transformation multiplication when:
    1) S is nonempty,
    2) whenever $\alpha \in S$, then $\alpha^{-1} \in S$, and
    3) whenever $\alpha, \beta \in S$, then $\beta\alpha \in S$.
13. Show that the three conditions of Exercise 12.12 can be replaced by the two conditions:
    1) S is nonempty, and
    2*) if $\alpha, \beta \in S$, then $\beta^{-1}\alpha \in S$.
14. Show that each set of isometries found in Exercises 11.3 and 11.4 is a group.

SECTION 13 | ISOMETRIES AS PRODUCTS OF REFLECTIONS

13.1 To give us a starting point, we shall postulate that two triangles are congruent if they satisfy the SAS condition; that is, if two sides and the included angle of one triangle are each congruent to the corresponding parts of the second triangle. Other sufficient conditions for congruence of triangles will be proved later in this chapter.

13.2 Definition An *isometry* is a map of the points of the plane that preserves distance; that is, α is an isometry iff for each pair of points P and Q in the plane, if $\alpha(P) = P'$ and $\alpha(Q) = Q'$, then $PQ \cong P'Q'$.

13.3 Theorem An isometry maps line segments into congruent line segments.

Let α be an isometry and PQ a segment. Let $\alpha(P) = P'$ and $\alpha(Q) = Q'$. Then $PQ \cong P'Q'$. Take any point A on segment PQ. Then $\mathbf{PA} + \mathbf{AQ} = \mathbf{PQ}$. So, if $\alpha(A) = A'$, then $\mathbf{P'A'} + \mathbf{A'Q'} = \mathbf{P'Q'}$ since α preserves distances. Hence A' lies on segment $P'Q'$, and the theorem follows. ▯

13.4 Corollary An isometry maps lines into lines and circles into circles.

13.5 Definition If S is a point set and α is a transformation, let $\alpha(S)$ denote the set of all images of points of S under the map α; that is, $P \in S$ iff $\alpha(P) \in \alpha(S)$.

13.6 Theorem There is at most one point X at given distances from three given noncollinear points P, Q, R.

There is a circle (which may reduce to a single point) of points Y such that $PX \cong PY$ and a circle of points Y such that $RX \cong RY$. These two circles meet in at most two points Y_1 and Y_2. Since PQR is a triangle, Q does not lie on PR, the perpendicular bisector of Y_1Y_2. Hence $QY_1 \not\cong QY_2$, so at most one of these segments can be congruent to QX. Thus there is at most one point X at given distances from three noncollinear points P, Q, R. (See Fig. 13.6.) ▯

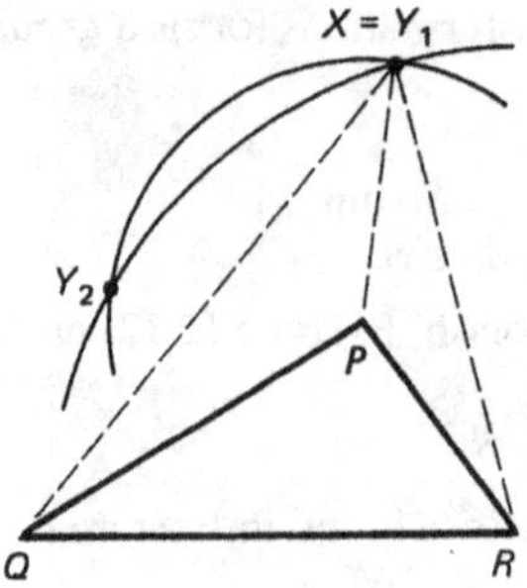

Figure 13.6

13.7 Corollary If a map of the points of the plane to the plane preserves distance, then it is one-to-one and onto; that is, an isometry is a transformation of the plane.

13.8 Theorem If PQR is a triangle and α and β are isometries such that $\alpha(P) = \beta(P)$, $\alpha(Q) = \beta(Q)$, and $\alpha(R) = \beta(R)$, then $\alpha = \beta$.

By Theorem 13.6, for each point X in the plane, $\alpha(X) = \beta(X)$ since α and β preserve distance. So $\alpha = \beta$. ▯

13.9 Definition A plane map α is called a *reflection in line m*, or a *reflection*, iff whenever $B = \alpha(A)$ for points A and B, then m is the perpendicular bisector of segment AB, or else $A = B$ and $A \in m$. The reflection in line m is denoted by σ_m and line m is called its *mirror*.

13.10 Theorem A reflection is an isometry.

Suppose A and B are any two points, and let $A' = \sigma_m(A)$ and $B' = \sigma_m(B)$. Let AA' and BB' cut m in points F and G, and suppose A, B, and A' are not collinear, as shown in Fig. 13.10. Then triangles AGF and $A'GF$ are congruent by SAS since

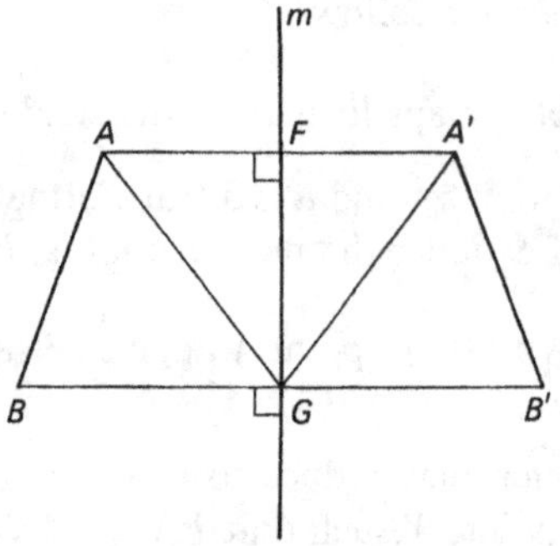

Figure 13.10

$AF \cong A'F$, $FG = FG$, and $\measuredangle AFG \cong \measuredangle A'FG = 90°$. Thus $AG \cong A'G$. Also $\measuredangle AGB \cong \measuredangle A'GB'$ by subtraction and $BG \cong B'G$. Hence triangles ABG and $A'B'G$ are congruent by SAS. Now $AB \cong A'B'$. The case when A, B, and A' are collinear is not difficult. Thus σ_m preserves distances, and every point has an image. Thus σ_m is a map that preserves distances. Hence it is an isometry. ▯

13.11 Corollary Every product of reflections is an isometry.

13.12 Theorem For each reflection σ_m, $\sigma_m^{-1} = \sigma_m$.

13.13 Theorem Each isometry is a product of at most three reflections.

Let the isometry α map triangle ABC to $A'B'C'$.

Case 1. If $A = A'$, $B = B'$, and $C = C'$, then $\alpha = \iota$, the square of any given reflection σ_m.

Case 2. If $A = A'$ and $B = B'$, but $C \neq C'$ as in Fig. 13.13a, then AB is the perpendicular bisector of CC' since C and C' are equidistant from A and from B. Hence a reflection in line AB maps triangle ABC to $A'B'C'$.

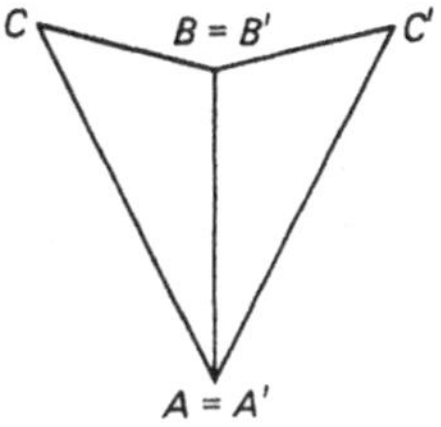

Figure 13.13a

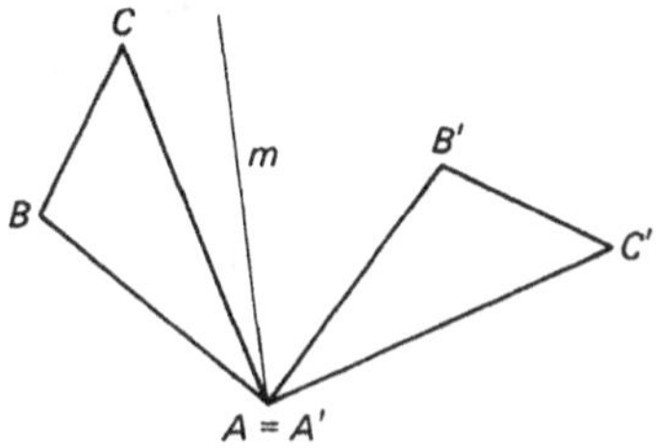

Figure 13.13b

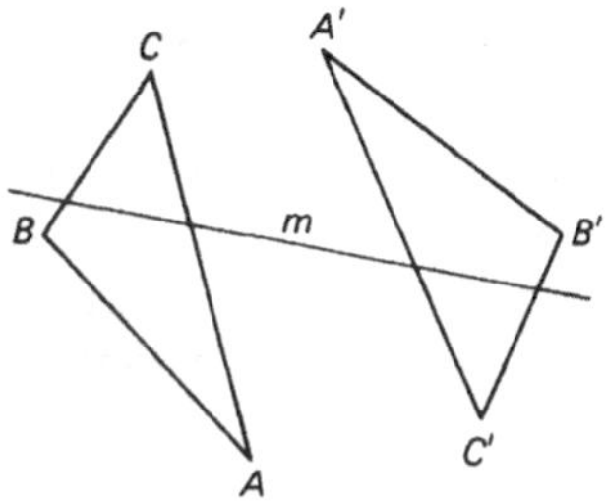

Figure 13.13c

Case 3. If $A = A'$ but $B \neq B'$ as in Fig. 13.13b, then A lies on m, the perpendicular bisector of BB'. Then σ_m maps B to B' and leaves A fixed, reducing this case to one of the two cases above.

Case 4. If $A \neq A'$ (see Fig. 13.13c), reflect triangle ABC in line m, the perpendicular bisector of AA', so that $\sigma_m(A) = A'$. Now this case reduces to one of the three cases above. ▯

13.14 Definition Each isometry that is a product of an even number of reflections is called *direct*. An *opposite* isometry is the product of an odd number of reflections.

13.15 Definition The *sense* of a triangle ABC is *clockwise* iff one travels in the clockwise direction when reading the vertices in cyclic order A-B-C-A; it is *counterclockwise* iff one goes in the counterclockwise direction when reading A-B-C-A. When triangles ABC and $A'B'C'$ are congruent and both are read in the same sense, they are said to be *directly* congruent; if their senses are not the same, they are *oppositely* congruent (Fig. 13.15).

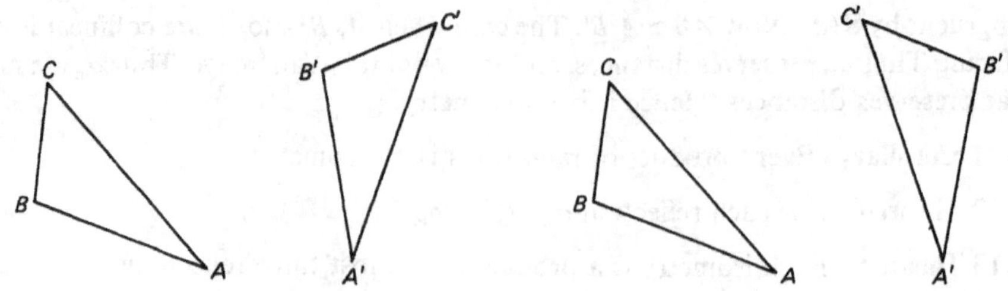

Directly congruent triangles *Oppositely congruent triangles*

Figure 13.15

13.16 Theorem A reflection maps a triangle into a congruent triangle. Furthermore, the congruence is opposite.

We must first prove, assuming only the SAS condition for congruence of triangles, that if $\sigma_m(ABC) = A'B'C'$, then $\triangle ABC \cong \triangle A'B'C'$. We do this in three cases.

Case 1. Suppose A and B both lie on the mirror m, and let CC' cut m at R (see Fig. 13.16a). Now any two corresponding segments are congruent since σ_m is an isometry. Since CC' is perpendicular to m, then $\triangle ACR \cong \triangle AC'R$ and $\triangle BCR \cong \triangle BC'R$ by SAS. At least one of these two triangles is not degenerate, so at least one of the two statements $\measuredangle BAC \cong \measuredangle BAC'$ and $\measuredangle ABC \cong \measuredangle ABC'$ must be true. Now triangles ABC and ABC' are congruent by SAS.

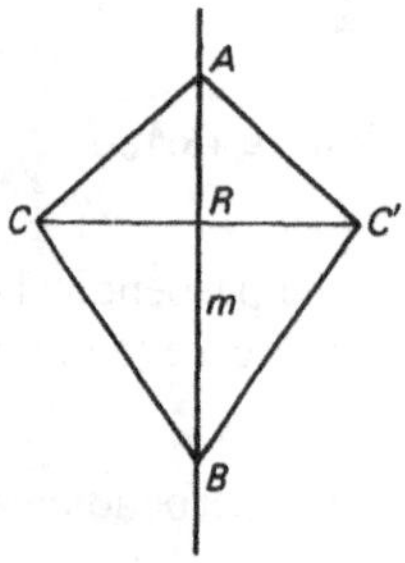

Figure 13.16a

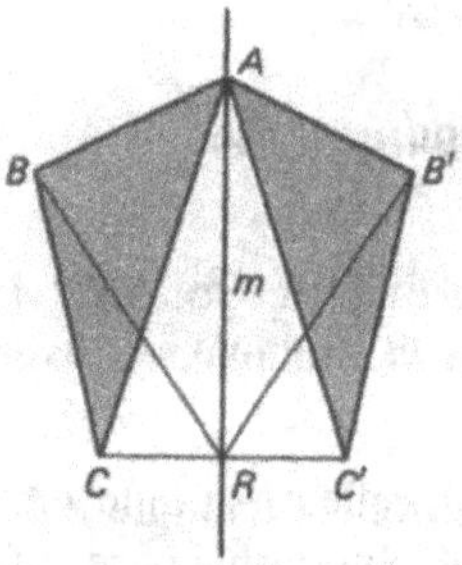

Figure 13.16b

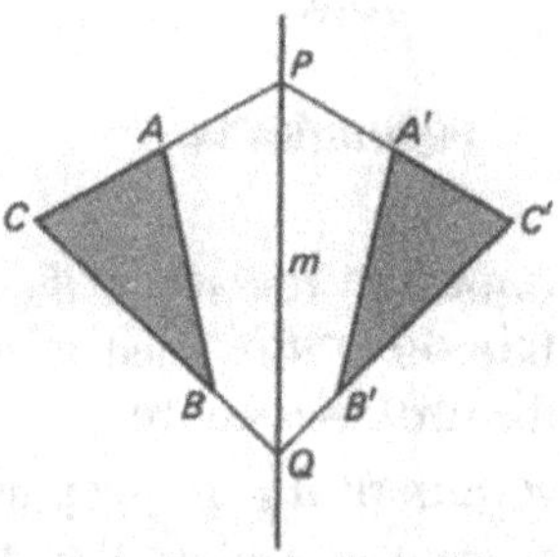

Figure 13.16c

Case 2. If A lies on the mirror m, but B and C do not lie on m, then A cannot lie on both BB' and CC'. Suppose A does not lie on CC', and let CC' cut m at R as in Fig. 13.16b. By Case 1, $\triangle CAR \cong \triangle C'AR$ and $\triangle BAR \cong \triangle B'AR$, so $\measuredangle CAR \cong \measuredangle C'AR$ and $\measuredangle BAR \cong \measuredangle B'AR$. Then $\measuredangle BAC \cong \measuredangle B'AC'$ by subtraction. Hence $\triangle ABC \cong \triangle AB'C'$ by SAS.

Case 3. In any other case, at most one side, say AB, of triangle ABC can be parallel to the mirror m, so suppose CA cuts m at P and BC cuts m at Q. By Case 1, $\triangle PCQ \cong \triangle PC'Q$, so $\measuredangle ACB \cong \measuredangle A'C'B'$. Now $\triangle ABC \cong \triangle A'B'C'$ by SAS.

Figure 10.4 gives the reader a basis for a proof of the second sentence of this theorem. ▯

13.17 Corollary An opposite isometry maps a triangle into an oppositely congruent triangle.

13.18 Corollary A direct isometry maps a triangle into a directly congruent triangle.

13.19 Theorem No isometry is both direct and opposite, but each isometry is either direct or opposite.

13.20 Theorem There are exactly two isometries, one direct and one opposite, that carry a given segment AB into a congruent segment $A'B'$ and map A to A' and B to B'.

Let C be any other point so that ABC is a triangle. There are on segment $A'B'$ exactly two triangles $A'B'C'$ and $A'B'C''$ congruent to triangle ABC. Furthermore, $A'B'$ is the perpendicular bisector of $C'C''$, so triangle $A'B'C''$ is the reflection in line $A'B'$ of triangle $A'B'C'$. Thus these two triangles are oppositely congruent, so one of them must be directly congruent and the other oppositely congruent to triangle ABC. Thus the isometries that map triangle ABC into these two triangles will be direct and opposite, respectively. ▯

Exercise Set 13

1. Prove Corollary 13.4.
2. Prove Corollary 13.7.
3. Prove Theorem 13.10 for the case when A, B, and A' are collinear.
4. Use mathematical induction to prove Corollary 13.11.
5. Prove Theorem 13.12.
6. Indicate the numbers of reflections possible for the cases of Theorem 13.13.
7. Complete the proof of Theorem 13.16.
8. Prove Corollary 13.17.
9. Prove Corollary 13.18.
10. Prove Theorem 13.19.
11. Use mathematical induction and Corollaries 13.4, 13.17, and 13.18 to prove that an isometry maps any polygon into a congruent polygon.
12. Let $P(a, 0)$, $Q(b, 0)$, $R(0, c)$ be Cartesian points with $a \neq b$ and $c \neq 0$.
    a) Show that there are just two points X such that $PX = s$ and $QX = t$, where $|s - t| < PQ < s + t$. Given that one of these points X has coordinates (u, v), find the coordinates for the other point X.
    b) Find $m(RX)$ for both points X, and show that the two of these distances are never equal.
13. Find images for each of the points $A(0, 0)$, $B(1, 0)$, $C(0, 3)$, $D(1, 1)$, $E(2, 3)$, $F(15, 12)$, $G(-3, -5)$ in the reflection in the mirror:
    a) $y = 0$
    b) $x = 0$
    c) $y = 2$
    d) $x = -1$
    e) $y = x$
    f) $x + y = 2$

14. Draw two congruent triangles with no corresponding vertices coinciding and construct the reflections that carry one to the other. Find such a pair of triangles that require exactly:
 a) one reflection b) two reflections
 c) three reflections d) four reflections

SECTION 14 | TRANSLATIONS AND ROTATIONS

14.1 Definition A plane map α is called a *translation through vector* $\overrightarrow{PQ}$ iff for each point A, $\alpha(A) = B$ where $\overrightarrow{AB} = \overrightarrow{PQ}$; that is, AB and PQ are parallel, congruent, similarly-directed segments.

14.2 Theorem If lines m and n are parallel and if the normal vector from m to n is $\mathbf{v}$, then $\sigma_n \sigma_m$ is a translation through vector $2\mathbf{v}$.

14.3 In Theorem 14.2, note that vector $2\mathbf{v}$ is perpendicular to m and to n, is directed from m towards n, and is in length *twice* the distance between m and n. (See Fig. 10.6.)

14.4 Theorem Every translation is an isometry and can be factored into a product of two reflections in parallel mirrors m and n, one of these lines being arbitrarily chosen perpendicular to the vector $2\mathbf{v}$ of translation, then the other mirror is parallel to the first and placed so that the directed distance from m to n is $\mathbf{v}$.

14.5 Theorem The inverse of the translation through vector $\overrightarrow{PQ}$ is the translation through vector $\overrightarrow{QP}$. If the translation can be written as $\sigma_n \sigma_m$, then its inverse is $\sigma_m \sigma_n$.

14.6 Theorem Translations commute.

By vector methods, $2\mathbf{v} + 2\mathbf{w} = 2\mathbf{w} + 2\mathbf{v}$, since in either case this vector is a diagonal of the parallelogram determined by the vectors $2\mathbf{v}$ and $2\mathbf{w}$. (See Fig. 14.6.) ▯

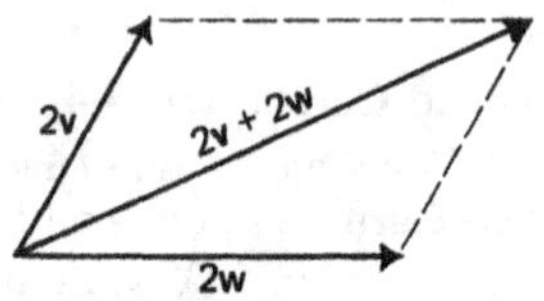

Figure 14.6

14.7 Theorem 14.6 can also be proved algebraically. Such a proof is left for the reader in Exercise 15.11.

14.8 Definition Let O be a point and θ the measure of a directed angle. A plane map α is called a *rotation about point O through angle* θ iff for each point A, $\alpha(A) = B$ where $A = B = O$ or $OB \cong OA$ and $d(\measuredangle AOB) = \theta$. Point O is called the *center* of the rotation.

14.9 Theorem If lines m and n intersect at point O with the directed angle from m to n of measure θ, then $\sigma_n \sigma_m$ is a rotation about point O through angle 2θ.

Let $\sigma_m(P) = P'$ and $\sigma_n(P') = Q$. If $P \notin m$, then OPP' is an isosceles triangle with m the bisector of its vertex angle POP'. Similarly, if $P' \notin n$, then n is the bisector of the vertex angle $P'OQ$ of the isosceles triangle $OP'Q$. (See Fig. 14.9.) It follows that

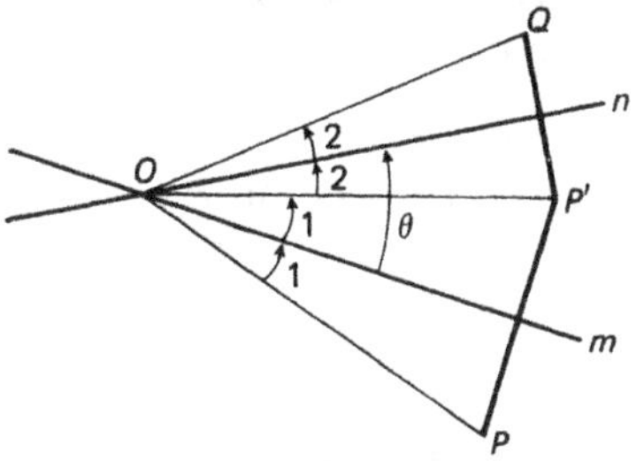

Figure 14.9

$\measuredangle POQ = \measuredangle POP' + \measuredangle P'OQ$, so $\measuredangle POQ$ is twice the angle from line m to line n. Also, since triangles OPP' and $OP'Q$ are both isosceles, then $OP \cong OQ$.

The case when $P \in m$ or when $P' \in n$ is simpler. The theorem follows. □

14.10 Theorem Each rotation is an isometry and can be factored into a product of two reflections in intersecting lines m and n. Furthermore, one of the mirrors may be arbitrarily chosen through the center O of rotation, then the other mirror is that line through O, such that the directed angle from m to n is half the angle of rotation.

14.11 Theorem Each direct isometry is a rotation or a translation.

14.12 Theorem The product of two translations is a translation.

Geometrically, the proof is trivial. The vector of the product of the two translations is the vector sum of the vectors of the two given translations. Hence the product is a translation.

Alternatively, let us give a proof utilizing Theorems 14.2 and 14.4. Let the translations be α and β and let $\alpha = \sigma_n \sigma_m$ and $\beta = \sigma_q \sigma_p$. If m, n, p, q are all parallel or coincident, then let r be another line of this pencil such that $\sigma_p \sigma_r = \sigma_n \sigma_m$. (See Fig. 14.11a.) Now

$$\beta\alpha = \sigma_q \sigma_p \sigma_n \sigma_m = \sigma_q \sigma_p \sigma_p \sigma_r = \sigma_q \iota \sigma_r = \sigma_q \sigma_r,$$

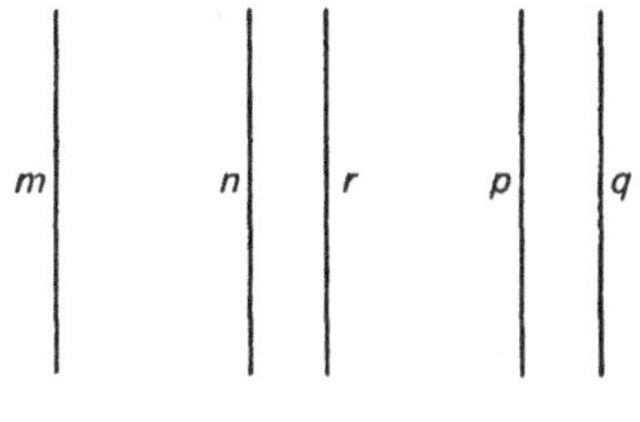

Figure 14.11a

a translation since it is the product of two reflections in parallel mirrors.

If n and p are not parallel or coincident, let them meet at A. (See Fig. 14.11b.) Let lines n' and p' pass through A so that n' is perpendicular to m and $\sigma_{p'}\sigma_{n'} = \sigma_p\sigma_n$. Then p' is perpendicular to q by Theorems 14.9 and 14.10. (See Fig. 14.11c.) Let m and n' meet at B, and p' and q at C. Let r denote line BC and let m' and n' be the perpendiculars to r at B and C, respectively. (See Fig. 14.11d.) By Theorems 14.9 and 14.10 again, $\sigma_q\sigma_{p'} = \sigma_{q'}\sigma_r$ and $\sigma_r\sigma_{m'} = \sigma_{n'}\sigma_m$. Then

$$\beta\alpha = \sigma_q\sigma_p\sigma_n\sigma_m = \sigma_q\sigma_{p'}\sigma_{n'}\sigma_m = \sigma_{q'}\sigma_r\sigma_r\sigma_{m'} = \sigma_{q'}\sigma_{m'},$$

a translation since m' and q' are parallel mirrors. ▯

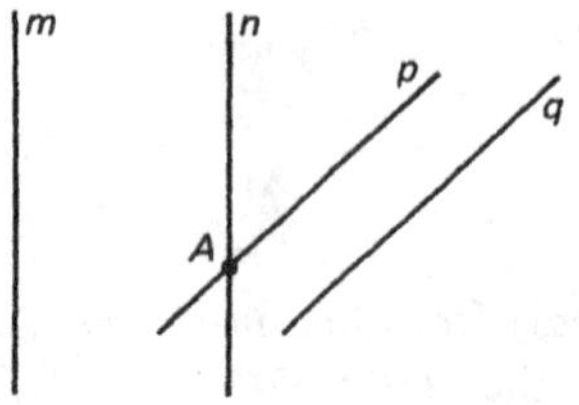

Figure 14.11b

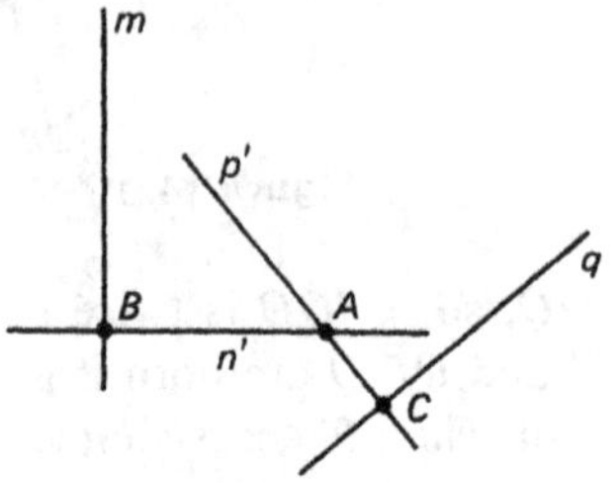

Figure 14.11c

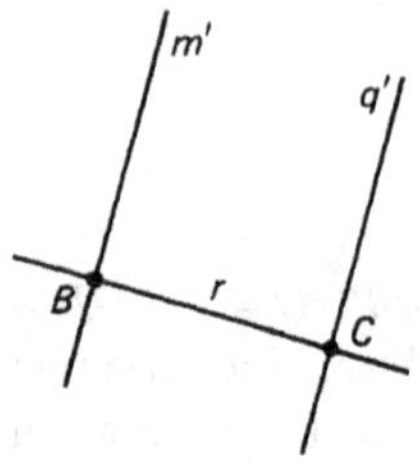

Figure 14.11d

14.13 Theorem The inverse of a rotation about O through angle θ is the rotation about O through angle $-\theta$. If the rotation can be written as $\sigma_n\sigma_m$, then its inverse is $\sigma_m\sigma_n$ (see Fig. 14.9).

14.14 Theorem The product of two rotations about the same center O is a rotation about O, and the product commutes.

Let α and β be the rotations with angles θ_1 and θ_2. By Theorem 14.10, write $\alpha = \sigma_n\sigma_m$ and $\beta = \sigma_p\sigma_n$ where m, n, p are three properly chosen lines through O (see Fig. 14.14). Then

$$\beta\alpha = \sigma_p\sigma_n\sigma_n\sigma_m = \sigma_p\,\iota\sigma_m = \sigma_p\sigma_m,$$

a rotation about O with angle $\theta_1 + \theta_2$. Since $\theta_1 + \theta_2 = \theta_2 + \theta_1$, it follows that the rotations commute. ▯

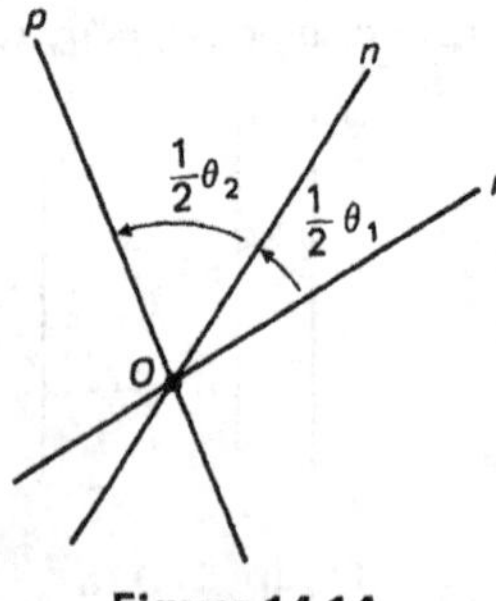

Figure 14.14

14.15 Theorem The product of two rotations is a rotation or a translation and is, in general, not commutative.

If the centers of the rotations are the same, then Theorem 14.14 applies. So assume the centers A and B are not the same. Let p denote the line AB. (See Fig. 14.15.) By Theorem 14.10, there are lines m and n through A and B respectively so that the two rotations α and β are given by

$$\alpha = \sigma_p \sigma_m \qquad \text{and} \qquad \beta = \sigma_n \sigma_p.$$

Now we have

$$\beta\alpha = \sigma_n \sigma_p \sigma_p \sigma_m = \sigma_n \sigma_m,$$

a translation if m is parallel to n, or a rotation if m and n meet in an ordinary point. ▯

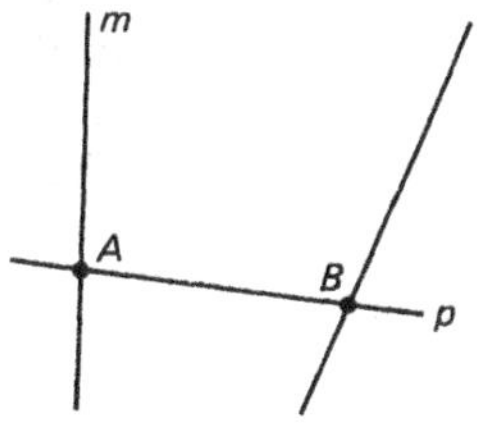

Figure 14.15

14.16 Definition A *glide-reflection* is the product of a translation and a reflection in a mirror parallel to the direction of the translation (see Fig. 11.9).

14.17 The glide-reflection completes our list of opposite isometries. In the next section we shall show that every opposite isometry is either a glide-reflection or, as a special glide-reflection, a reflection. Since each direct isometry is either a rotation or a translation, we have introduced enough basic isometries. Nonetheless, in the next section we shall introduce—for convenience—one more direct isometry to complete our list of isometries.

14.18 Theorem The translation and reflection of a glide-reflection commute.

Exercise Set 14

1. Prove Theorem 14.2.
2. Prove Theorem 14.4
3. Prove Theorem 14.5.
4. Prove Theorem 14.10.
5. Prove Theorem 14.11.
6. Prove Theorem 14.13.
7. Show that two rotations about different centers do not, in general, commute by examining the nature of the products $\alpha\beta$ and $\beta\alpha$ where α and β are 90° rotations about points $A(-1, 0)$ and $B(1, 0)$.
8. What must be true of two rotations α and β if their product $\beta\alpha$ is a translation?

9. If the product $\beta\alpha$ of two rotations α and β is a translation, what can be said about the product $\alpha\beta$?
10. Prove Theorem 14.18.
11. Explain how a reflection is a special glide-reflection. Is it true that a translation is also a special glide-reflection?
12. Prove that a glide-reflection is an isometry. Is it direct or opposite?
13. Find the inverse of a glide-reflection.
14. What sort of isometry might the product of two glide-reflections be? Could it be a glide-reflection? Explain.
15. Is the product of two reflections ever a reflection? Explain.
16. Find images for each of the points $A(0, 0)$, $B(1, 0)$, $C(0, 3)$, $D(1, 1)$, $E(2, 3)$, $F(15, 12)$, $G(-3, -5)$ under a rotation with the indicated center and the indicated angle:
    a) $(0, 0)$, $90°$
    b) $(0, 0)$, $180°$
    c) $(0, 0)$, $-90°$
    d) $(1, 0)$, $90°$
    e) $(1, 1)$, $90°$
    f) $(2, 3)$, $180°$
    g) $(-1, -3)$, $-90°$
    h) $(1, -2)$, $45°$
17. Find images for the points listed in Exercise 12.16 under each translation with vector $\overrightarrow{PQ}$:
    a) $P(0, 0)$ and $Q(5, 0)$
    b) $P(0, 0)$ and $Q(2, -3)$
    c) $P(2, 5)$ and $Q(1, 7)$
    d) $P(3, -2)$ and $Q(3, -5)$
18. Prove that the product of a rotation and a translation is a rotation.
19. The center of the rotation $\beta\alpha$ of Theorem 14.15 is at the intersection of the lines m and n. Locate the center of the rotation $\alpha\beta$. When is $\alpha\beta$ a translation?

SECTION 15 | HALFTURNS

15.1 Theorem An isometry preserves angles.

From Theorem 13.16, a reflection preserves angles; that is, a reflection maps an angle into a congruent angle. The theorem then follows from Theorem 13.13. □

15.2 It follows that an isometry is a congruence transformation, and that every congruence transformation is an isometry. Thus isometries are the very transformations we seek for Euclidean superposition.

15.3 Definition An *involutoric* isometry is any nonidentity isometry the square of which is the identity map.

15.4 Notice that it is customary *not* to call ι involutoric even though $\iota^2 = \iota$. Exercise 11.7 helps to clarify this idea.

15.5 Definition A *halfturn about point P* (or a *reflection in point P*) is a rotation about P through an angle of $180°$. It is denoted by σ_P (See Fig. 11.6.)

15.6 The term "reflection in point P" is quite indicative of how this isometry maps the points of the plane. In this chapter, however, we shall strictly reserve the term "reflection" for a reflection in a line. Of course, the term "halfturn" also indicates quite well the idea of this isometry. Although a halfturn is just a (special) rotation, it has been given its own name (*halfturn*) because of its importance and its unique properties. It is the only involutoric rotation and, in fact, the only involutoric direct isometry. Theorem 15.7 establishes that halfturns and reflections are the only involutoric isometries.

15.7 Theorem Each involutoric isometry is either a reflection or a halfturn.

Let σ be the involutoric isometry and let $\sigma(A) = B$ with $A \neq B$. Since $A = \sigma^2(A)$, then $\sigma(B) = \sigma(\sigma(A)) = \sigma^2(A) = A$. Choose P equidistant from A and B so that PAB is an isosceles triangle. Let $\sigma(P) = P'$ so that $\sigma(\triangle PAB) = \triangle P'BA$. Now P' has exactly two possible locations: so that $P = P'$ or else $PAP'B$ is a rhombus. In the first case, σ is a reflection in the perpendicular bisector of AB, and in the second case σ is a halfturn about the midpoint of AB by Theorem 13.8. ▯

15.8 Theorem If σ_P is a halfturn about point P, and a and b are any two perpendicular lines through P, then $\sigma_P = \sigma_b \sigma_a = \sigma_a \sigma_b$.

15.9 Corollary Reflections in two perpendicular lines commute.

15.10 Only involutoric isometries will be denoted by σ (lower-case Greek sigma); that is, only reflections and halfturns. A proof that a halfturn is involutoric is left for the exercises. Which involutoric isometry is intended is made clear by the subscript on the sigma; upper-case italic letters shall always refer to points, and lower-case italic letters to lines. Hence σ_a is a reflection, and σ_A is a halfturn.

15.11 Theorem The product $\sigma_B \sigma_A$ is a translation through vector $2\overrightarrow{AB}$.

Let $\sigma_A(P) = P'$ and $\sigma_B(P') = P''$ (see Fig. 11.7b). In triangle $PP'P''$, segment AB joins the midpoints of two sides, so AB is parallel and congruent to half the third side PP''. Hence $\overrightarrow{PP''} = 2\overrightarrow{AB}$, and the theorem follows. ▯

15.12 Theorem The product $\sigma_C \sigma_B \sigma_A$ is a halfturn about point D, the fourth vertex of parallelogram $ABCD$.

The theorem can be proved easily as a corollary to Theorem 15.11. ▯

It is also quite instructive to prove Theorems 15.11 and 15.12 by writing each given halfturn as a product of two reflections in perpendicular lines, one of which is chosen through (or parallel to) AB in each case. ▯

Exercise Set 15

1. Prove Theorem 15.1.
2. Explain how a halfturn can be considered to be a "reflection in a point."
3. Prove that a reflection is involutoric.
4. Prove that a halfturn is involutoric.
5. Prove Theorem 15.8.

6. Prove Theorem 15.12 as a corollary to Theorem 15.11.
7. Prove Theorem 15.12 by writing each halfturn as a product of reflections as suggested in the text.
8. Find images for each of the points $A(0, 0)$, $B(1, 0)$, $C(0, 3)$, $D(1, 1)$, $E(2, 3)$, $F(15, 12)$, $G(-3, -5)$ under a glide-reflection with vector PQ and with the given equation for mirror.
   a) $P(0, 0)$, $Q(1, 0)$, and $y = 0$
   b) $P(0, 0)$, $Q(1, 0)$, and $y = 5$
   c) $P(2, 5)$, $Q(2, -1)$, and $x = -4$
   d) $P(0, 0)$, $Q(1, 1)$, and $y = x$
9. Repeat Exercise 15.8 for a halfturn about the given point:
   a) $(0, 0)$ b) $(5, 0)$
   c) $(2, 3)$ d) $(-3, 7)$
10. Prove Theorem 15.11 by factoring σ_B and σ_A into products of reflections having a common mirror $m = AB$.
11. Use Theorem 15.12 and a converse to Theorem 15.11 to prove that two translations commute (Theorem 14.6).

SECTION 16 | PRODUCTS OF REFLECTIONS

16.1 Theorem If lines a, b, c form a pencil, then there is a line d in that pencil such that $\sigma_c\sigma_b\sigma_a = \sigma_d$. Conversely, if $\sigma_c\sigma_b\sigma_a = \sigma_d$, then lines a, b, c, d belong to a pencil.

Given that lines a, b, c belong to a pencil, then find line d so that $\sigma_b\sigma_a = \sigma_c\sigma_d$ by one of Theorems 14.4 and 14.10 (see Figs. 11.8a and 11.8b). Then $\sigma_c\sigma_b\sigma_a = \sigma_d$. The argument reverses to establish the converse. ▯

16.2 Corollary If lines a, b, c form a pencil, then $\sigma_c\sigma_b\sigma_a = \sigma_a\sigma_b\sigma_c$.

16.3 Theorem If $\sigma_g\sigma_b\sigma_a$ is a glide-reflection with both a and b perpendicular to g, then, letting $B = g \cap b$ and $A = g \cap a$, we have

$$\sigma_g\,\sigma_b\,\sigma_a = \sigma_b\,\sigma_g\,\sigma_a = \sigma_b\,\sigma_a\,\sigma_g = \sigma_B\,\sigma_a = \sigma_b\,\sigma_A.$$

16.4 Theorem Each product $\sigma_c\sigma_b\sigma_a$ is a glide-reflection. (A reflection is a special glide-reflection.)

The theorem is trivial if a, b, c form a pencil. So assume that (for example) only a and b intersect at P (see Fig. 16.4a). Rotate a and b about P into lines a' and b' so that $\sigma_{b'}\sigma_{a'} = \sigma_b\sigma_a$ and b' is perpendicular to c at a point Q (Fig. 16.4b). Similarly rotate b' and c about Q into lines b'' and c' so that $\sigma_{c'}\sigma_{b''} = \sigma_c\sigma_{b'}$ and a' is perpendicular to b'' (Fig. 16.4c). Now we have

$$\sigma_c\sigma_b\sigma_a = \sigma_c\sigma_{b'}\sigma_{a'} = \sigma_{c'}\sigma_{b''}\sigma_{a'} = \sigma_{b''}\sigma_{c'}\sigma_{a'},$$

a glide-reflection, since a' and c' are both perpendicular to b''. ▯

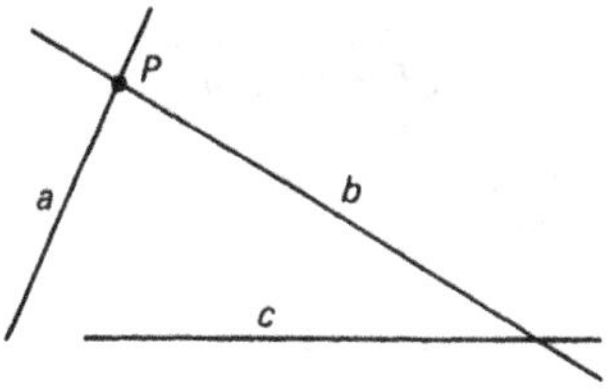

Figure 16.4a

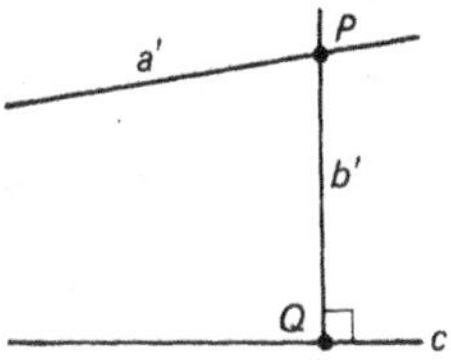

Figure 16.4b

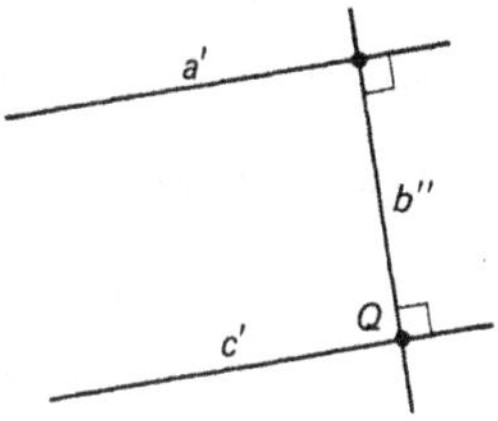

Figure 16.4c

16.5 Theorem Each isometry can be written as a product of the form $\sigma_b \sigma_a$ or of the form $\sigma_B \sigma_a$.

16.6 Theorem All reflections and products of reflections, that is, all isometries, form a subgroup of the transformation group of Theorem 12.17.

▶**16.7 Theorem** All rotations and translations form a subgroup of the group of Theorem 16.6.

▶**16.8 Theorem** All halfturns and translations form a subgroup of the group of Theorem 16.7.

▶**16.9 Theorem** All translations form a subgroup of the group of Theorem 16.8.

▶**16.10 Theorem** All translations along a fixed line form a subgroup of the group of Theorem 16.9.

16.11 Theorem An isometry with no invariant point is either a translation or a glide-reflection according to whether it is direct or opposite.

Such an isometry could not be a reflection or a rotation since each of these isometries has at least one invariant point. The center of a rotation is invariant and each point on the mirror of a reflection is fixed. □

16.12 Theorem An isometry with at least one invariant point is either a rotation or a reflection according to whether it is direct or opposite.

16.13 Theorem An isometry with more than one invariant point is either the identity or a reflection according to whether it is direct or opposite.

16.14 Theorem The chart in Fig. 16.14 summarizes the relations between each isometry and its representation in terms of reflections.

16.15 The theorems of this section tie together the algebraic properties of reflections and products of reflections with the geometric properties of lines and points. This interplay between algebra and geometry is sufficiently important to warrant one more section (Section 17) to collect, classify, and emphasize theorems that show this relationship.

Isometry	*Sense*	*Fixed points*	*Fixed lines*	*Minimum number of reflections*
Reflection in line m	Opposite	Points on line m	m and all lines perpendicular to m	1 in line m
Identity map	Direct	All	All	2 in any one line
Rotation about O in angle $\theta \neq 180°$	Direct	O (only)	None	2 in lines intersecting at O in angle $\theta/2$
Halfturn about O	Direct	O (only)	All lines through O	2 in lines perpendicular at O
Translation with vector $\overrightarrow{PQ}$	Direct	None	All lines parallel to $\overrightarrow{PQ}$	2 in lines perpendicular to $\overrightarrow{PQ}$ and half the length of $\overrightarrow{PQ}$ apart
Glide-reflection along m and in m	Opposite	None	m (only)	3, in m and two lines perpendicular to m

Figure 16.14

Exercise Set 16

1. Fill in the details of the proof of Theorem 16.1.
2. Prove Theorem 16.3.
3. Prove Theorem 16.5.
4. Prove Theorem 16.6.
5. Prove Theorems 16.7 through 16.10.
6. Prove Theorem 16.12.
7. Prove Theorem 16.13.
8. Justify the statements about fixed points and fixed lines in Theorem 16.14.
9. Show that a rotation that is not a halfturn is not involutoric.
10. Show that neither a translation nor a glide-reflection with a nonzero vector is involutoric.
11. Let lines m and n meet at P. Show that when both m and n are fixed lines under an isometry, then P is a fixed point.
12. Given point A and line m, find point B and line n so that $\sigma_A \sigma_m = \sigma_n \sigma_B$.
13. Decide whether each set of transformations forms a subgroup of the group of all isometries.
    a) All reflections in mirrors through a given point P and all rotations about P (remember that the identity can be considered a rotation).
    b) All rotations.
    c) All rotations about a fixed point P.
    d) All reflections in mirrors parallel to a given direction and all translations having vectors perpendicular to the mirrors.
    e) All halfturns.

SECTION 17 | PROPERTIES OF ISOMETRIES; A SUMMARY

17.1 The theorems of this section summarize the algebraic and geometric properties of isometries and especially the relations between the algebra and geometry of isometries. Remember that lower case letters represent lines and upper case letters represent points. Although all the theorems of this section are of interest, perhaps special emphasis should be laid upon theorems 17.3, 17.4, 17.8, 17.13, and 17.15. These results will prove quite useful in the applications that follow in the next three sections. Drawing a picture to illustrate each theorem and relating each part of the theorem to the figure should make the theorem easier to remember. Be sure to draw such an illustration for each theorem.

17.2 Theorem The following conditions are equivalent to one another:

1) $A \in b$,
2) $\sigma_A \sigma_b = \sigma_b \sigma_A$,
3) $A = \sigma_b(A)$,
4) $b = \sigma_A(b)$,
5) $\sigma_b \sigma_A$ (or $\sigma_A \sigma_b$) is involutoric, and
6) $\sigma_b \sigma_A$ is a reflection in the mirror through A perpendicular to b.

17.3 Theorem The following conditions are equivalent:

1) $a = b$ or a and b are perpendicular,
2) $\sigma_b \sigma_a = \sigma_a \sigma_b$,
3) $a = \sigma_b(a)$,
4) $(\sigma_b \sigma_a)^2 = \iota$, and
5) $\sigma_b \sigma_a$ is either the identity or a halfturn.

17.4 Theorem The following conditions are equivalent:

1) a, b, c, d belong to a pencil and the directed angle or directed distance from a to b is equal to that from c to d,
2) $\sigma_b \sigma_a = \sigma_d \sigma_c$,
3) $\sigma_d \sigma_b \sigma_a$ is involutoric (and hence it is the reflection σ_c),
4) $\sigma_d \sigma_b \sigma_a$ is a reflection (namely σ_c),
5) $\sigma_c \sigma_d \sigma_b \sigma_a = \iota$, and
6) $\sigma_d \sigma_b \sigma_a = \sigma_a \sigma_b \sigma_d = \sigma_c$.

17.5 Theorem If $a \neq c$, then the following conditions are equivalent:

1) b lies midway between a and c or b bisects the angle between a and c,
2) $\sigma_c \sigma_b = \sigma_b \sigma_a$,
3) $c = \sigma_b(a)$, and
4) $\sigma_c = \sigma_b \sigma_a \sigma_b$.

17.6 Theorem If $A \neq C$, then the following conditions are equivalent:

1) b is the perpendicular bisector of AC,
2) $\sigma_C \sigma_b = \sigma_b \sigma_A$,

3) $C = \sigma_b(A)$, and
4) $\sigma_C = \sigma_b \sigma_A \sigma_b$.

17.7 Theorem If $a \neq c$, then the following conditions are equivalent:

1) a is parallel to c and B lies midway between them,
2) $\sigma_c \sigma_B = \sigma_B \sigma_a$,
3) $c = \sigma_B(a)$, and
4) $\sigma_c = \sigma_B \sigma_a \sigma_B$.

17.8 Theorem The following conditions are equivalent:

1) $ABCD$ is a parallelogram,
2) $\sigma_B \sigma_A = \sigma_C \sigma_D$,
3) $\sigma_D = \sigma_C \sigma_B \sigma_A$, and
4) $\sigma_D \sigma_C \sigma_B \sigma_A = \iota$.

17.9 Theorem If $A \neq C$, then the following conditions are equivalent:

1) B is the midpoint of AC,
2) $\sigma_C \sigma_B = \sigma_B \sigma_A$,
3) $C = \sigma_B(A)$, and
4) $\sigma_C = \sigma_B \sigma_A \sigma_B$.

17.10 Theorem The following conditions are equivalent:

1) $A = B$,
2) $A = \sigma_B(A)$,
3) $\sigma_B \sigma_A = \sigma_A \sigma_B$, and
4) $\sigma_B \sigma_A = \iota$.

17.11 Theorem If $\sigma_C \sigma_b \sigma_A$ is involutoric, then it is a reflection.

17.12 Theorem If $\sigma_c \sigma_B \sigma_a$ is involutoric, then it is a halfturn.

17.13 Theorem $\sigma_C \sigma_B \sigma_A$ is always involutoric, so $\sigma_C \sigma_B \sigma_A = \sigma_A \sigma_B \sigma_C$.

17.14 Theorem σ_a and σ_A are always involutoric.

17.15 Theorem $(\sigma_c \sigma_b \sigma_a)^2$ is a translation.

The theorem states that the square of each glide-reflection is a translation. Theorem 16.3 implies that the glide and the reflection of a glide-reflection commute. The square of the glide-reflection, then, reduces to the square of its translation since the reflections can be arranged so as to cancel. □

17.16 To conclude this section we show an interesting relation, a product of 22 reflections that is equal to the identity! The importance of this result lies not in its being a vital theorem that every student must be able to recall instantly. Neither is it likely that Thomsen's relation would be an appropriate topic of conversation at a dinner party. It is of interest to us for the utter simplicity of its proof. Before reading the proof given in the text, try to imagine how you would prove this theorem.

▶**17.17 Theorem** Thomsen's relation. For any three lines a, b, c,

$$\sigma_a\sigma_b\sigma_c\sigma_a\sigma_b\sigma_c\sigma_b\sigma_c\sigma_a\sigma_b\sigma_c\sigma_a\sigma_c\sigma_b\sigma_a\sigma_c\sigma_b\sigma_c\sigma_b\sigma_a\sigma_c\sigma_b = \iota.$$

Since $(\sigma_a\sigma_b\sigma_c)^2$ and $(\sigma_b\sigma_c\sigma_a)^2$ are translations by Theorem 17.15, then they commute. Furthermore,

$$((\sigma_a\sigma_b\sigma_c)^2)^{-1} = (\sigma_c\sigma_b\sigma_a)^2 \qquad \text{and} \qquad ((\sigma_b\sigma_c\sigma_a)^2)^{-1} = (\sigma_a\sigma_c\sigma_b)^2.$$

Now the desired relation is simply the identity

$$(\sigma_a\sigma_b\sigma_c)^2(\sigma_b\sigma_c\sigma_a)^2(\sigma_c\sigma_b\sigma_a)^2(\sigma_a\sigma_c\sigma_b)^2 = \iota. \square$$

Exercise Set 17

1. Theorems 17.3 to 17.12 can be paired so that the second theorem says for lines and points approximately what the first theorem of the pair says for points and lines. The two theorems of such a pair are called *duals* of each other. For example, Theorems 17.6 and 17.7 are duals.
   a) Arrange Theorems 17.3 to 17.12 in dual pairs.
   b) Find duals for Theorems 17.2 and 17.14.
   c) What can be said about a dual for Theorem 17.13?

2 through 14. Prove Theorems 17.2 to 17.14.

15. For any three lines a, b, c, prove that
$$\sigma_a\sigma_b\sigma_c(\sigma_a\sigma_b)^2(\sigma_c\sigma_a)^2\sigma_b\sigma_c\sigma_a\sigma_c\sigma_b\sigma_a(\sigma_c\sigma_b)^2(\sigma_a\sigma_c)^2\sigma_b\sigma_a\sigma_c = \iota.$$

16. Show that, for any four parallel lines a, b, c, d,
$$\sigma_a\sigma_b\sigma_c\sigma_d\sigma_a\sigma_b\sigma_c\sigma_b\sigma_c\sigma_a\sigma_d\sigma_b\sigma_c\sigma_a\sigma_d\sigma_c\sigma_b\sigma_a\sigma_d\sigma_c\sigma_b\sigma_c\sigma_b\sigma_d\sigma_a\sigma_c\sigma_b\sigma_d = \iota.$$

17. Show that Thomsen's relation and the formulas of Exercises 17.15 and 17.16 are still true if each reflection σ_x is replaced by a halfturn σ_X. Explain how the proofs must be changed.

SECTION 18 | APPLICATIONS OF ISOMETRIES TO ELEMENTARY GEOMETRY

18.1 We are now ready to apply the theory of isometries developed in the earlier sections of this chapter to problems in Euclidean geometry. We begin with very basic properties and progress toward more advanced ideas through Sections 19 and 20. Many of the theorems and proofs presented here can be used in high school classes, if isometries are presented sufficiently early. Indeed, most of the theorems in Sections 18 and 19 appear in high school geometry texts (with different proofs). The high school student who understands the workings of isometries should find the methods of transformations quite logical and understandable.

The procedure we are using might best be called *transform-solve-transform*, for when we are given a problem, we *transform* it into a new problem through the use of isometries, *solve* the new problem, then *transform* the solution back to the original problem. In Theorem 18.4, for example, the given problem is transformed into one in which the two triangles share a common side, then the problem is solved (the theorem is proved) for this case, and finally this solution is applied to the given problem of proving any two triangles that satisfy the SSS condition congruent.

Remember that we have assumed the SAS condition for congruence as a postulate, so that other congruence conditions must be proved.

18.2 Theorem Vertical angles are congruent.

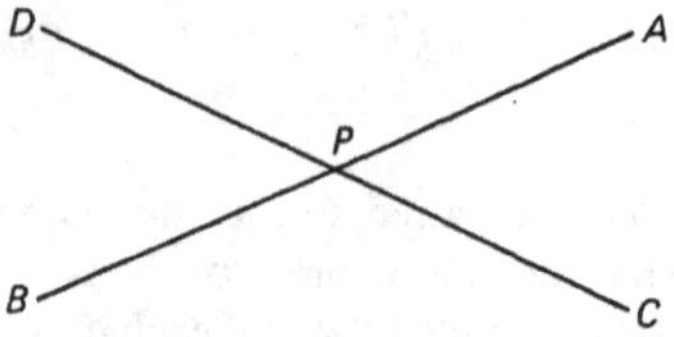

Figure 18.1

Let AB and CD meet at P as in Fig. 18.1. Then $\sigma_P(\sphericalangle APC) = \sphericalangle BPD$ and $\sigma_P(\sphericalangle APD) = \sphericalangle BPC$, since lines through P are fixed under the map σ_P. ▯

18.3 Theorem If a point is equidistant from two points, then it lies on the perpendicular bisector of the segment between the two points.

Let A be equidistant from P and Q as in Fig. 18.3. Let m be the bisector of angle PAQ. Then σ_m maps line AP into line AQ, and since $AP \cong AQ$, $\sigma_m(P) = Q$. Hence m is the perpendicular bisector of PQ. ▯

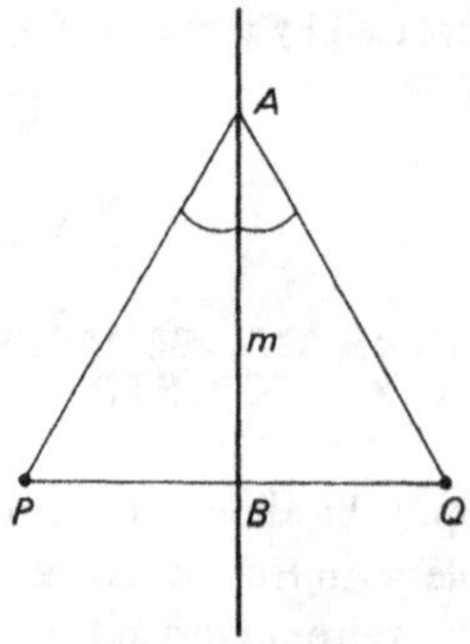

Figure 18.3

18.4 Theorem Two triangles are congruent if they satisfy the SSS condition; that is, if the three sides of the first triangle are congruent respectively to the corresponding sides of the second triangle.

Let the sides of triangle ABC be congruent to the corresponding sides of triangle $A'B'C'$. Use an isometry to map triangle $A'B'C'$ to triangle ABC'' with C and C'' on opposite sides of line AB (Fig. 18.4). Since A and B are both equidistant from C and from C'', then line AB is the perpendicular bisector of CC''. Hence the reflection in line AB carries triangle ABC'' into triangle ABC. Now, since an isometry is a congruence transformation, the theorem follows; that is, triangles ABC and $A'B'C'$ are congruent. ▯

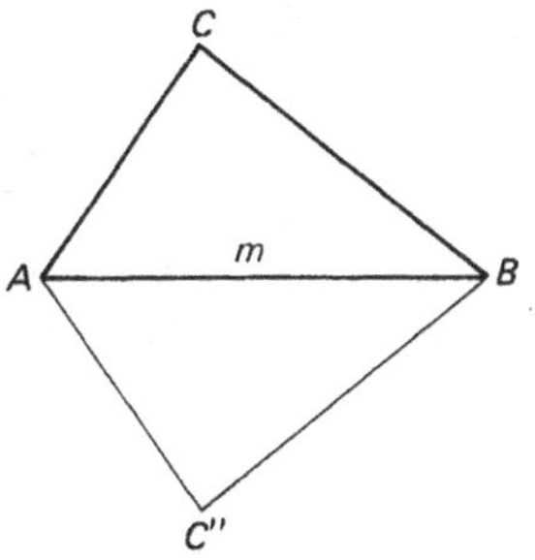

Figure 18.4

18.5 Theorem Two triangles are congruent if they satisfy the ASA condition; that is, if two angles and the included side of the first triangle are congruent respectively to the corresponding parts of the second triangle.

18.6 Theorem Two triangles are congruent if they satisfy the AAS condition; that is, if two angles and a side opposite one of them in the first triangle are congruent respectively to the corresponding parts in the second triangle.

18.7 Theorem Parallel lines cut by a transversal form congruent corresponding and alternate angles.

Let transversal p cut the parallel lines m and n at points A and B (Fig. 18.7). A translation through vector AB carries the angles at A into those at B (since lines m and n are parallel), establishing the theorem. ▯

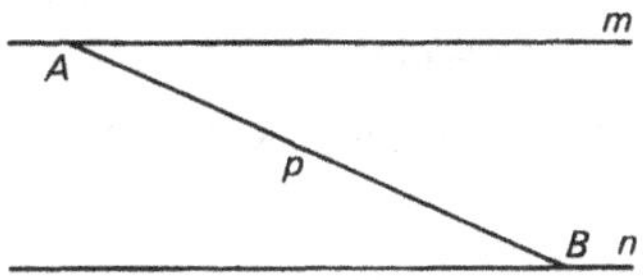

Figure 18.7

18.8 Theorem The base angles of an isosceles triangle are congruent.

18.9 Theorem The diagonals of a parallelogram bisect each other.

Let N be the midpoint of diagonal AC of parallelogram $ABCD$. Then $\sigma_A\sigma_N = \sigma_N\sigma_C$. We must show that N is the midpoint of BD; that is, that $\sigma_D\sigma_N = \sigma_N\sigma_B$. ▯

18.10 Corollary A parallelogram is symmetric with respect to the point of intersection of its diagonals.

18.11 Corollary A parallelogram and either diagonal form two congruent triangles.

18.12 Theorem The perpendicular bisector of the (major) base of an isosceles trapezoid is the perpendicular bisector also of the summit (minor base).

Reflect the isosceles trapezoid $ABCD$ in its base AB into the image trapezoid $ABC'D'$ (Fig. 18.12). Let CC' and DD' cut AB at E and F. The angles at E and F are right angles and $DF \cong CE$. Now $AD \cong BC$ is given, and $AD \cong AD'$, $BC \cong BC'$ and $CC' \cong DD'$. So isosceles triangles ADD' and BCC' are congruent. Then their corresponding altitudes AF and BE are congruent. Hence the perpendicular bisector of AB is the perpendicular bisector of EF, hence also of CD, since $CDFE$ is a rectangle. ▯

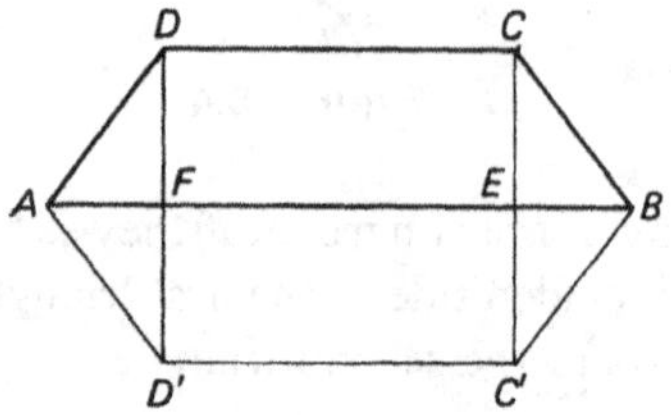

Figure 18.12

18.13 Corollary

a) The base angles of an isosceles trapezoid are congruent.
b) The summit angles of an isosceles trapezoid are congruent.
c) The diagonals of an isosceles trapezoid are congruent.
d) An isosceles trapezoid is symmetric with respect to the perpendicular bisector of its bases.

18.14 Theorem If the base angles (or the summit angles) of a trapezoid are congruent, then the trapezoid is isosceles.

▶**18.15 Theorem** In Fig. 18.15, if $AB \cong BC \cong DE$, $AD \cong BE$, and $BD \cong CE$, then A, B, and C are collinear.

Triangles ABD and BCE are congruent by SSS and $BCED$ is a parallelogram, so BD is parallel to CE. Similarly, AD is parallel to BE. Thus there is a translation mapping triangle ADB into triangle BEC. Since a translation preserves directions of lines, it follows that A, B, and C are collinear. ▯

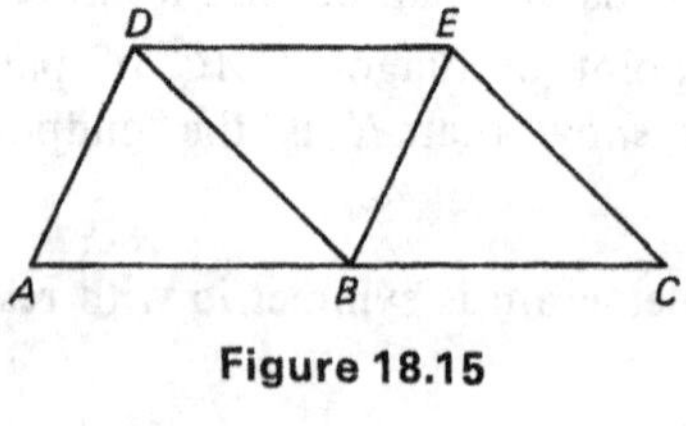

Figure 18.15

Exercise Set 18

1. Prove Theorem 18.5.
2. Prove Theorem 18.6.
3. Prove that two right triangles are congruent if they satisfy the HL condition.
4. Prove Theorem 18.8 as a corollary to Theorem 18.3.
5. Prove Theorem 18.8 by synthetic geometry using as an auxiliary line from the *apex* (the vertex between the two congruent sides) of the isosceles triangle,
   a) the median,
   b) the altitude,
   c) the angle bisector.
6. a) Prove Theorem 18.8 by reflecting in the bisector of the vertex angle.
   b) Prove Theorem 18.8 by showing triangles ABC and ACB congruent by the SAS postulate (where sides AB and AC are given congruent).
7. Prove Theorem 18.9.
8. Prove Corollary 18.10.
9. Prove Corollary 18.11.
10. Prove Corollary 18.13.
11. Prove Theorem 18.14.
12. Prove that (a) the medians, (b) the altitudes, and (c) the angle bisectors drawn from the base angles of an isosceles triangle are congruent.
13. Prove that no median of a scalene triangle is perpendicular to the side it bisects.
14. Prove that the diagonals of a rhombus are perpendicular.
15. Prove that if the diagonals of a parallelogram are perpendicular, then the parallelogram is a rhombus.
16. Prove Theorem 18.12 for a rectangle (assumed true in the proof of that theorem).

SECTION 19 | FURTHER ELEMENTARY APPLICATIONS

19.1 Theorem A circle is symmetric in any diameter.

19.2 Theorem A diameter of a circle perpendicular to a chord of that circle bisects the chord.

19.3 Theorem Two chords of a circle are congruent iff they are equidistant from the center.

Given that the chords are equidistant from the center, then reflect in that diameter that bisects the angle between the radii perpendicular to the chords. Establish the converse by a rotation. ▯

19.4 Corollary Congruent chords of a circle intercept congruent arcs of the circle, and conversely.

19.5 Theorem The angles of intersection of two intersecting circles are congruent.

Each of the centers A and B of the two circles is equidistant from P and Q, the points of intersection of the circles, so A and B lie on the perpendicular bisector of PQ. Reflect the figure in line AB. The circles map into themselves by Theorem 19.1 and P maps into Q. The theorem follows. ▯

19.6 Theorem The area of a parallelogram is equal to that of a rectangle with the same base and altitude; that is, it is the product of its base and its altitude to that base.

In parallelogram $ABCD$ (Fig. 19.6), there is a translation α that carries AD into BC. Assuming, without loss of generality, that angle A is less than a right angle, drop perpendicular DE to side AB. Let $\alpha(\triangle ADE) = \triangle BCF$. Then E, B, and F are collinear, so $DEFC$ is a rectangle, and its area is equal to that of parallelogram $ABCD$. ▯

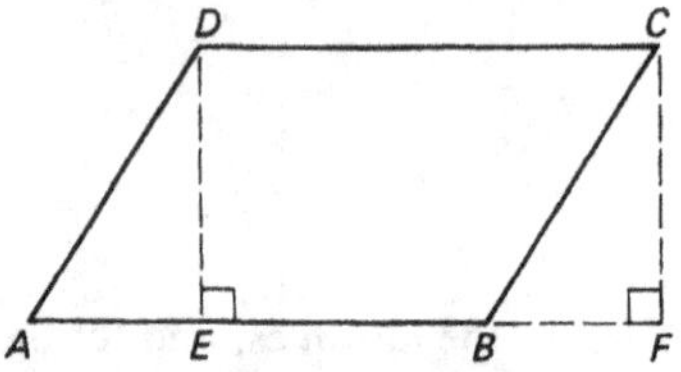

Figure 19.6

19.7 Theorem The area of a trapezoid is equal to the product of the altitude and half the sum of the bases.

Perform a halfturn about the midpoint M of a nonparallel side AD to form a parallelogram $B'C'BC$ as shown in Fig. 19.7. Then apply Theorem 19.6 to establish the theorem. ▯

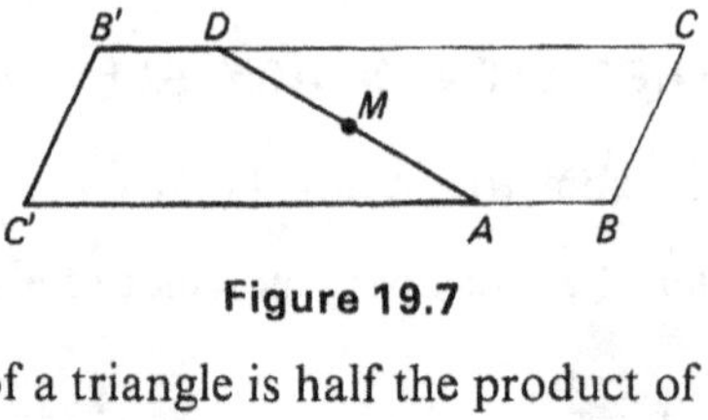

Figure 19.7

19.8 Theorem The area of a triangle is half the product of any side as base and the altitude to that side.

19.9 Theorem The midpoints of the sides of a quadrilateral form a parallelogram.

Let $ABCD$ be a quadrilateral having M, N, O, P as midpoints of its sides as shown in Fig. 19.9. Now $\sigma_P \sigma_O \sigma_N \sigma_M$ is a translation, and a translation through a nonzero vector has no fixed points. But

$$(\sigma_P \sigma_O \sigma_N \sigma_M)(A) = (\sigma_P \sigma_O \sigma_N)(B) = (\sigma_P \sigma_O)(C) = \sigma_P(D) = A,$$

so we must have $\sigma_P \sigma_O \sigma_N \sigma_M = \iota$. Thus $MNOP$ is a parallelogram. ▯

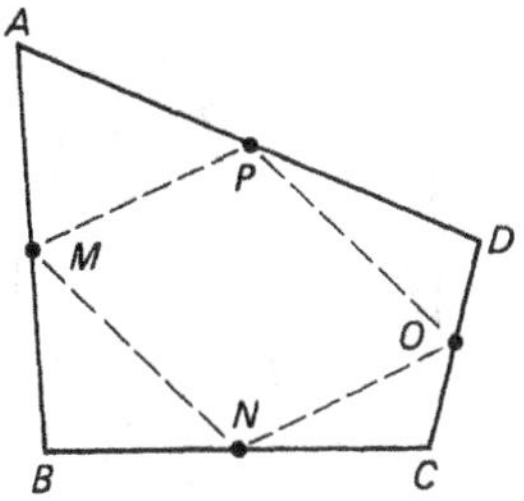

Figure 19.9

▶ **19.10 Theorem** If $ABQP$ and $BCRQ$ are parallelograms, then so also is $ACRP$ a parallelogram.

We have (see Fig. 19.10)

$$\sigma_A \sigma_C \sigma_R \sigma_P = \sigma_A(\sigma_C \sigma_R \sigma_P) = \sigma_A \sigma_P \sigma_R \sigma_C$$
$$= (\sigma_B \sigma_Q \sigma_P \sigma_A)(\sigma_A \sigma_P \sigma_R \sigma_C) = \sigma_B \sigma_Q \sigma_R \sigma_C = \iota,$$

so $ACRP$ is a parallelogram. ▯

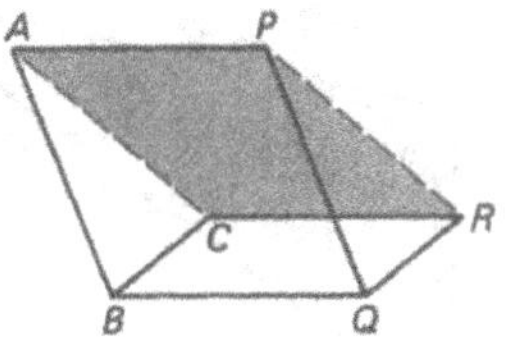

Figure 19.10

▶ **19.11 Theorem** If $ABPQ$ and $BCQR$ are parallelograms, then so also is $ACPR$ a parallelogram (Fig. 19.11).

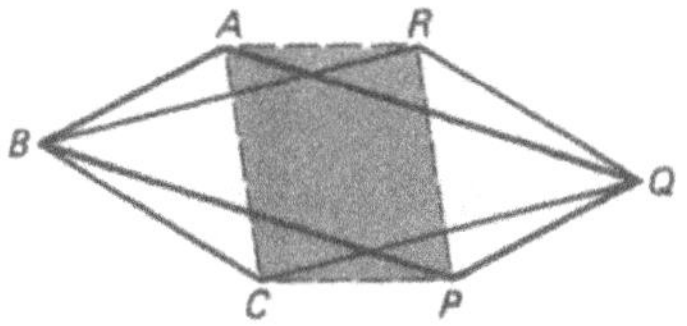

Figure 19.11

▶ **19.12 Theorem** If A, B, C, D, E, F are six points lying on a circle, and lines a, b, c are concurrent, where a is the perpendicular bisector of AB and of DE, b is the perpendicular bisector of BC and EF, and c is that for CD, then c is the perpendicular bisector also of FA (Fig. 19.12).

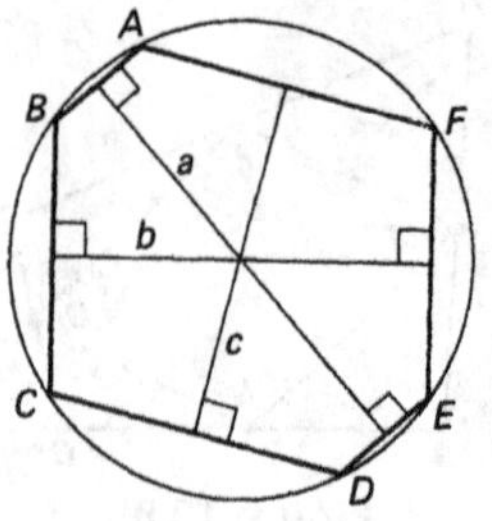

Figure 19.12

▶ **19.13 Theorem** If P is the midpoint of AB and of DE, Q is the midpoint of BC and of EF, and R is the midpoint of CD, then R is the midpoint also of FA.

See Fig. 19.13. Since $(\sigma_R \sigma_Q \sigma_P)^2 = \iota$, then we have

$$\sigma_R(A) = (\sigma_Q \sigma_P \sigma_R \sigma_Q \sigma_P)(A) = F,$$

then R is the midpoint of AF. ▯

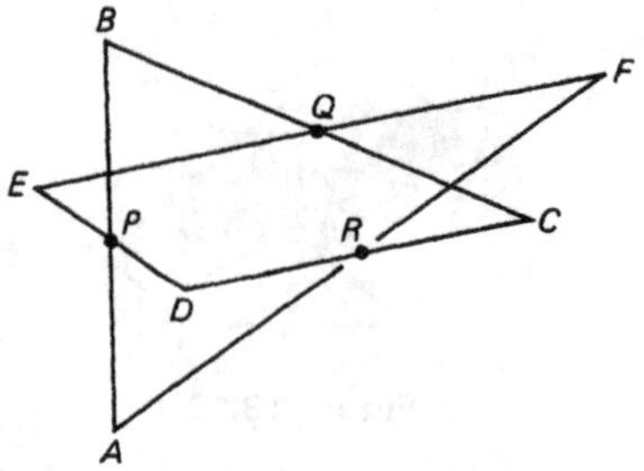

Figure 19.13

Exercise Set 19

1. Prove Theorem 19.1.
2. Prove Theorem 19.2.
3. Prove Theorem 19.3.
4. Prove Corollary 19.4.
5. Prove Theorem 19.8.
6. Prove Theorem 19.11.
7. Prove Theorem 19.12.
8. What figure do the four internal angle bisectors of a quadrilateral form?
9. Prove that the medians of a triangle, as vectors, form a triangle.
10. Prove that the midpoint of the hypotenuse of a right triangle is equidistant from all three vertices.
11. Prove that the two tangents from a point to a circle are congruent.

12. Review the proof of Theorem 8.8 (that the Euclidean and modern compasses are equivalent).
13. Let lines a, b, c form a triangle whose orthic triangle is DEF. Prove that $\sigma_c\sigma_b\sigma_a$ is the glide-reflection whose mirror is the line DF, and whose vector is directed from D toward F and has length equal to the perimeter of the orthic triangle.
14. *The coin game*. Two players alternately place coins one at a time on a rectangular table top, without letting the coins overlap. The winner is the last player who can find a space for a coin. Assuming that you play first, what is a winning strategy? For what shapes of tables does your strategy apply? (For practical purposes one might play the game with quarters or nickels on a 3-by-5 card.)
15. *The polygon game*. A coin is placed on each of the vertices of a regular polygon. Each player in turn removes either two coins from adjacent vertices or one coin from any vertex. The winner is the one who takes the last coin. Assuming that you play second, what is a winning strategy? For what other configurations of coins does your strategy apply?

SECTION 20 | ADVANCED APPLICATIONS

20.1 Theorem The perpendicular bisectors of the sides of a triangle concur.

20.2 *First proof*. Let the perpendicular bisectors of sides a, b, c of triangle ABC be denoted by d, e, f. (See Fig. 20.2.) Then

$$(\sigma_f\sigma_e\sigma_d)(B) = (\sigma_f\sigma_e)(C) = \sigma_f(A) = B,$$

so $\sigma_f\sigma_e\sigma_d$ is an opposite isometry with a fixed point. By Theorem 16.14 it is a reflection in a line; that is, d, e, f concur by Theorem 16.1. ▯

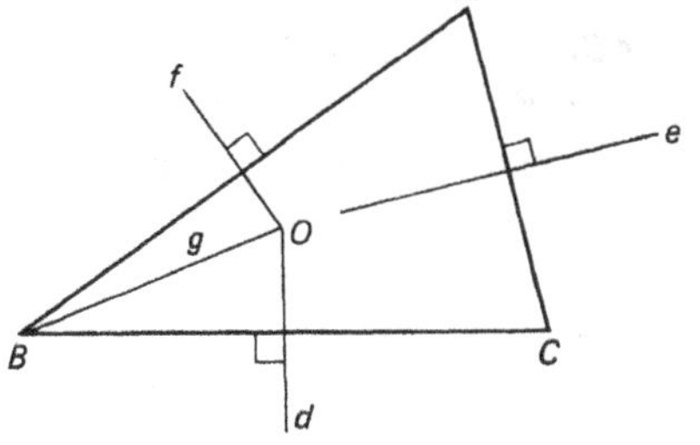

Figure 20.2

20.3 *Second proof*. Using Fig. 20.2 again, let d and f meet at O and denote line BO by g. Then the isometry $\sigma_d\sigma_g\sigma_f$ is a reflection in a line m through O. Since

$$(\sigma_d\sigma_g\sigma_f)(A) = C; \qquad \text{that is,} \qquad \sigma_m(A) = C,$$

it follows that m is the perpendicular bisector of AC. ▯

20.4 Theorem The bisectors of the angles of a triangle concur.

Denote the bisectors of angles B and C in triangle ABC by e and f. Let e and f meet at I (see Fig. 20.4). Then $\sigma_f\sigma_g\sigma_e$, where g is the perpendicular from I to side a, is a reflection in a line m through I by Theorem 16.1. Since $(\sigma_f\sigma_g\sigma_e)(c) = b$, then m is the bisector of angle A by Theorem 17.5. ▯

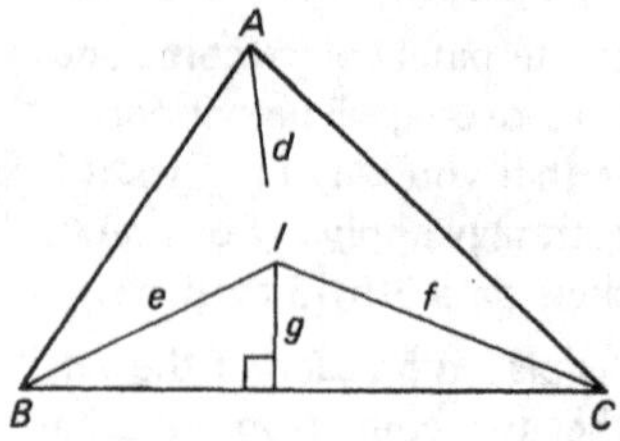

Figure 20.4

20.5 Note that any two of the bisectors in Theorem 20.4 can be taken as external and the third one internal.

20.6 Theorem If $\sigma_C\sigma_A\sigma_B = \sigma_U$, $\sigma_A\sigma_B\sigma_C = \sigma_V$, and $\sigma_B\sigma_C\sigma_A = \sigma_W$, then $\sigma_U\sigma_C = \sigma_C\sigma_V = \sigma_B\sigma_A$, $\sigma_V\sigma_A = \sigma_A\sigma_W = \sigma_C\sigma_B$, and $\sigma_W\sigma_B = \sigma_B\sigma_U = \sigma_A\sigma_C$.

We have both $\sigma_U\sigma_C$ and $\sigma_C\sigma_V$ equal to

$$\sigma_U\sigma_C = \sigma_C\sigma_V = \sigma_C\sigma_A\sigma_B\sigma_C = (\sigma_C\sigma_A\sigma_B)\sigma_C = \sigma_B\sigma_A\sigma_C\sigma_C = \sigma_B\sigma_A.$$

The other equations follow in a similar manner. ▯

20.7 Theorems 20.6 and 20.9 are purely algebraic in statement and in proof, based on only the group properties of halfturns. Geometric interpretations of these algebraic theorems are given in the corollaries that follow the theorems. Results such as these illustrate the close connections between algebra and geometry.

20.8 Corollary If $CABU$, $ABCV$, and $BCAW$ are parallelograms, then ABC is the medial triangle for triangle UVW. Furthermore, the altitudes of triangle ABC concur by Theorem 20.1 since they are the perpendicular bisectors of the sides of triangle UVW (Fig. 20.8).

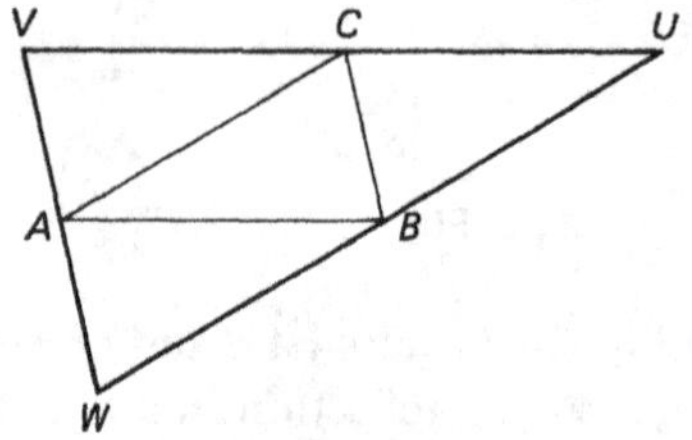

Figure 20.8

20.9 Theorem If $\sigma_C(U) = V$, $\sigma_A(V) = W$, and $\sigma_B(W) = U$, then $\sigma_U = \sigma_B\sigma_A\sigma_C$, $\sigma_V = \sigma_A\sigma_B\sigma_C$, and $\sigma_W = \sigma_B\sigma_C\sigma_A$.

Since the halfturn $\sigma_B \sigma_A \sigma_C$ leaves point U fixed, then it is the halfturn about point U, etc. □

20.10 Corollary The triangle (ABC) formed by joining the midpoints of the sides of a triangle (UVW) has sides parallel to and congruent to halves of the sides of the given triangle (Fig. 20.8).

20.11 Theorem The medians of a triangle meet at a trisection point of each median.

Let G be two-thirds of the way from A to the midpoint A' of the opposite side of triangle ABC (Fig. 20.11). Since vector $\overrightarrow{AG}$ is twice vector $\overrightarrow{GA'}$, then $\sigma_A \sigma_G = \sigma_G \sigma_{A'} \sigma_G \sigma_{A'}$. We must show that $\sigma_C \sigma_G = \sigma_G \sigma_{C'} \sigma_G \sigma_{C'}$, where C' is the midpoint of side AB. By Theorem 17.9, since A' and C' are midpoints of their respective sides, then $\sigma_B = \sigma_{A'} \sigma_C \sigma_{A'}$ and $\sigma_B \sigma_{C'} \sigma_A \sigma_{C'} = \iota$, so $\sigma_{C'} = \sigma_B \sigma_{C'} \sigma_A = \sigma_{A'} \sigma_C \sigma_{A'} \sigma_{C'} \sigma_A$.

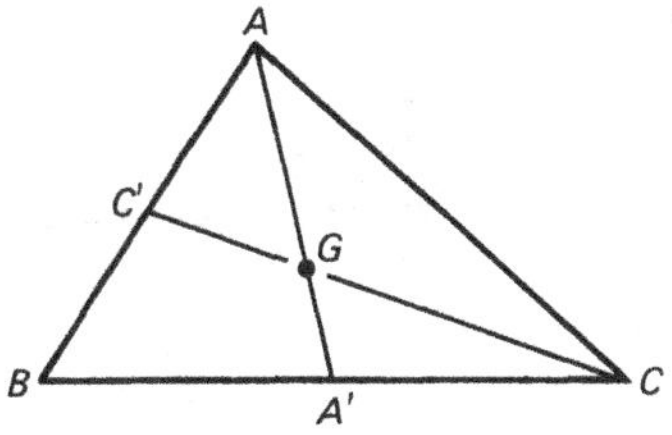

Figure 20.11

Now we have

$$\begin{aligned}\sigma_C \sigma_G \sigma_{C'} \sigma_G \sigma_{C'} \sigma_G &= \sigma_C \sigma_G (\sigma_{A'} \sigma_C \sigma_{A'} \sigma_{C'} \sigma_A) \sigma_G \sigma_{C'} \sigma_G \\ &= \sigma_C (\sigma_G \sigma_{A'} \sigma_C) \sigma_{A'} (\sigma_{C'} \sigma_A \sigma_G) \sigma_{C'} \sigma_G \\ &= \sigma_C \sigma_C \sigma_{A'} \sigma_G \sigma_{A'} \sigma_G \sigma_A \sigma_{C'} \sigma_{C'} \sigma_G \\ &= \sigma_{A'} \sigma_G \sigma_{A'} \sigma_G \sigma_A \sigma_G \\ &= \iota\end{aligned}$$

because G is two-thirds of the way from A to A'. Similarly, G is two-thirds of the way from B to B', and the theorem follows. □

20.12 Theorem If Cevians a', b', c' in triangle ABC concur, then their isogonal conjugates a'', b'', c'' concur also. [*Isogonal conjugate lines* are two lines through a vertex of an angle and symmetric with respect to the bisector of the angle.]

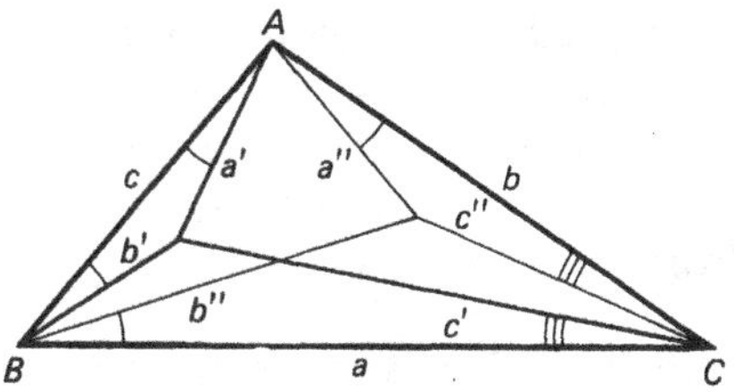

Figure 20.12

We are given (Fig. 20.12) $\sigma_c\sigma_{a'} = \sigma_{a''}\sigma_b$, $\sigma_b\sigma_{c'} = \sigma_{c''}\sigma_a$, $\sigma_a\sigma_{b'} = \sigma_{b''}\sigma_c$, and $(\sigma_{a'}\sigma_{b'}\sigma_{c'})^2 = \iota$. We must prove that $(\sigma_{a''}\sigma_{b''}\sigma_{c''})^2 = \iota$. To that end,

$$\begin{aligned}(\sigma_{a''}\sigma_{b''}\sigma_{c''})^2 &= ((\sigma_c\sigma_{a'}\sigma_b)(\sigma_a\sigma_{b'}\sigma_c)(\sigma_b\sigma_{c'}\sigma_a))^2\\ &= ((\sigma_b\sigma_{a'}\sigma_c)(\sigma_c\sigma_{b'}\sigma_a)(\sigma_a\sigma_{c'}\sigma_b))^2\\ &= (\sigma_b(\sigma_{a'}\sigma_{b'}\sigma_{c'})\sigma_b)^2\\ &= \sigma_b(\sigma_{a'}\sigma_{b'}\sigma_{c'})^2\sigma_b = \sigma_b\sigma_b = \iota. \ \square\end{aligned}$$

20.13 Theorem Let P be any point not on triangle ABC and let D, E, F be the reflections of P in sides a, b, c. Then the circumcenter O of triangle DEF is the isogonal conjugate point of point P in triangle ABC. [Two points are *isogonal conjugate points* with respect to a triangle iff the three cevians determining one point are isogonal conjugate lines of the cevians determining the other point.]

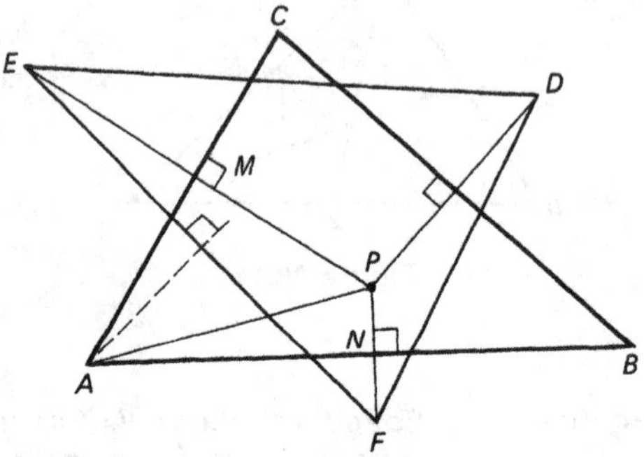

Figure 20.13

By Exercise 20.1 applied to triangle PEF, the perpendicular bisector of EF is the isogonal conjugate of line AP in angle CAB (Fig. 20.13). □

20.14 Theorem The orthocenter and the circumcenter of a triangle are isogonal conjugate points.

In Theorem 20.13 let the given triangle be ABC and its circumcenter be P. Then triangles PEF and ABC have common midpoints M and N for the sides PE and AC, and sides PF and AB (see Fig. 20.13). It follows that EF is parallel and congruent to BC. Similarly for AB and DE and for CA and FD. Hence triangles ABC and DEF are congruent. Now the circumcenter P of triangle ABC is the orthocenter of triangle DEF. Hence the circumcenter of triangle DEF is the orthocenter of triangle ABC. The theorem follows from Theorem 20.13. □

►**20.15 Theorem** In triangle ABC, S is the centroid iff

$$\sigma_S\sigma_C\sigma_S\sigma_B\sigma_S\sigma_A = \iota.$$

Suppose first that we are given $\sigma_S\sigma_C\sigma_S\sigma_B\sigma_S\sigma_A=\iota$. Since A' is the midpoint of side BC, then $\sigma_B=\sigma_{A'}\sigma_C\sigma_{A'}$, by Theorem 17.9. Now

$$\begin{aligned}\iota &= \sigma_S\sigma_C\sigma_S\sigma_B\sigma_S\sigma_A\\ &= \sigma_S\sigma_C\sigma_S\sigma_{A'}\sigma_C\sigma_{A'}\sigma_S\sigma_A\\ &= \sigma_S\sigma_C\sigma_C\sigma_{A'}\sigma_S\sigma_{A'}\sigma_S\sigma_A\\ &= \sigma_S\sigma_{A'}\sigma_S\sigma_{A'}\sigma_S\sigma_A\\ &= (\sigma_S\sigma_{A'})^2(\sigma_S\sigma_A),\end{aligned}$$

which states that S is two-thirds of the way from A to A'. Hence S is the centroid G. (See Fig. 20.15.)

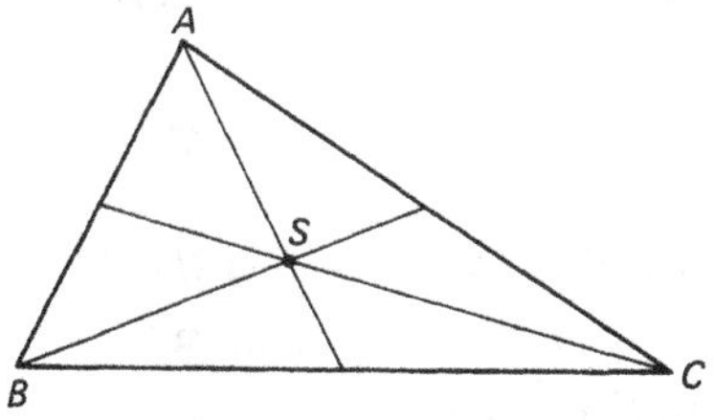

Figure 20.15

Conversely, if S is the centroid, then $(\sigma_S\sigma_C)(\sigma_S\sigma_B)(\sigma_S\sigma_A)=\iota$, since the three medians, as vectors, form a triangle, by Theorem 6.7. See also Exercise 19.9 and Answer 19.9. ▯

▶**20.16 Corollary** The centroid of a triangle is at a trisection point of each median.

▶**20.17 Theorem** *Fagnano's Problem.* In a given acute triangle inscribe a triangle RST having minimum perimeter.

Choose any three points R, S, T on the three sides of the given triangle ABC, as shown in Fig. 20.17. Let $\sigma_b(R)=R'$ and $\sigma_c(R)=R''$. Then the perimeter p of triangle RST is given by

$$p = RS+ST+TR = R'S+ST+TR'',$$

so, for fixed R, p is least when $R'STR''$ is a straight line.

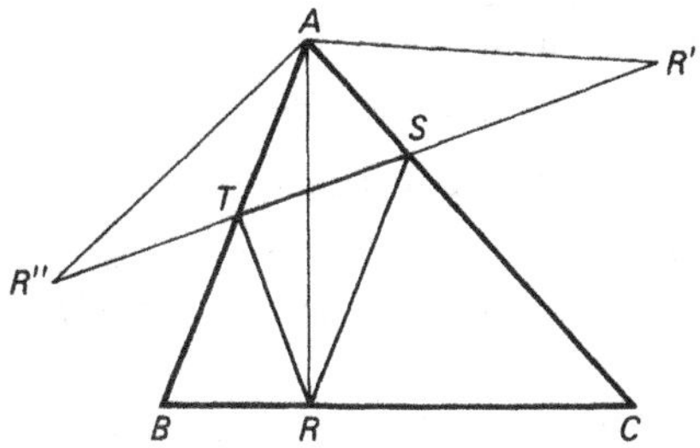

Figure 20.17

Since $AR'' \cong AR \cong AR'$ and $\measuredangle R''AR' \cong 2 \measuredangle A$ by construction of R' and R'', then triangle $AR'R''$ is determined when R is fixed and the length of $R'R''$ varies directly with that of AR. Thus $R'R''$ is a minimum when AR is as short as possible; that is, when AR is perpendicular to BC. Similarly S and T must also be feet of altitudes of triangle ABC. From Theorem 7.6, if RST is the orthic triangle, then $R'STR''$ is a straight line. Hence the orthic triangle is the unique solution. □

Exercise Set 20

1. As a corollary to Theorem 20.1, show that the directed angle from d to g is congruent to that from e to f.
2. Prove Theorem 20.4 for one internal and two external bisectors.
3. Prove Corollary 20.8.
4. Complete the proof of Theorem 20.9.
5. Prove Corollary 20.10.
6. In the proof of Theorem 20.11 show that G is 2/3 of the way along median BB'.
7. Prove Theorem 20.14 as a corollary to Theorem 6.20.
8. In the proof of Theorem 20.14 show that triangle DEF is the image of triangle ABC in a halfturn about the ninepoint center.
9. Prove Corollary 20.16.
10. Prove that a quadrilateral with three right angles has four right angles.
11. Prove that the angle bisectors of a triangle concur, by taking X, Y, Z as the points of contact of the incircle with the sides a, b, c of the triangle, and d, e, f as the bisectors of angles A, B, C. Then show that $(\sigma_f \sigma_e \sigma_d)(Y) = Y$. Complete the proof.
12. Let $ABCD$ be a quadrilateral with right angles at A and C. Let $Q = \sigma_A(D)$ and $R = \sigma_C(D)$. Prove that H, the midpoint of QR, is the orthocenter of triangle ABC.
13. *Hjelmslev's Theorem.* If α is an isometry such that $\alpha(m) = n$, then for each point P in m, there is a point Q in n, such that $\alpha(P) = Q$. Prove that the midpoints of such segments PQ either all coincide or are all distinct.
14. *Eves' problem.* A farmer has his house and barn on the same side of a straight river, at distances h and b from the river, with the barn d units downstream from the house. Toward which point on the river's edge should he head from the house in order to fill a pail with water and carry it to the barn while traveling the shortest total distance?
15. Prove that the perpendiculars dropped from two vertices of a triangle upon the median from the third vertex are congruent.
16. Through a given point P draw a line equidistant from given points A and B.
17. Prove that a trapezoid inscribed in a circle is isosceles.
18. Triangles ABC and DEF have equal areas and congruent bases BC and EF lying on the same straight line. Vertices A and D are on opposite sides of that line. Show that AD is bisected by the line.
19. Medians BB' and CC' of triangle ABC are extended their own lengths to points B'' and C''. Show that A lies on line $B''C''$.

20. From a point inside a circle, more than two congruent segments can be drawn to the circle. Prove that the point is the center of the circle.
21. A tangent intersects at A and B two parallel tangents to a circle with center O. Prove that the circle on AB as diameter passes through O.
22. [From "Dynamic Proofs of Euclidean Theorems" by R. L. Finney, *Mathematics Magazine* **43** (1970) pages 177–185.]
    a) If ABM and CDM are similarly-oriented isosceles right triangles with right angles at M, prove that AC and BD are congruent and perpendicular.
    b) If X and Y are the centers of squares erected externally on sides AB and BC of triangle ABC, prove that XB' and YB' are congruent and perpendicular at B', the midpoint of side AC.
    c) Let W, X, Y, Z be the centers of squares (in order) erected externally on the sides of a quadrilateral. Prove that WY and XZ are congruent and perpendicular segments.
    d) Prove that the segment joining the centers of two squares erected externally on two sides of a triangle is congruent and perpendicular to the segment joining the center of the third such square to the opposite vertex.
    e) Show that the word "externally" can be replaced by "internally" in parts (b), (c), and (d).

▶SECTION 21 ANALYTIC REPRESENTATIONS OF DIRECT ISOMETRIES

21.1 It would seem logical to write first the representation for a reflection, then find a method of multiplying transformations since all other isometries are products of reflections. It is much simpler however, to find representations for translations and rotations directly from their geometric properties. In fact, in the next section we shall utilize the results found here in order to calculate the form of the general reflection!

In contrast to the methods and formulas usually developed in a calculus and analytic geometry course, we shall consider translating, rotating, and reflecting all the points in the plane instead of moving the coordinate axes. In that respect our mappings are just the inverses of the mappings considered in the calculus.

21.2 Suppose that each point $P(x, y)$ is to be translated through vector $\mathbf{v} = \overrightarrow{OA}$ to point $P'(x', y')$ where $A(h, k)$ (Fig. 21.2). Then P is moved h units to the right and k

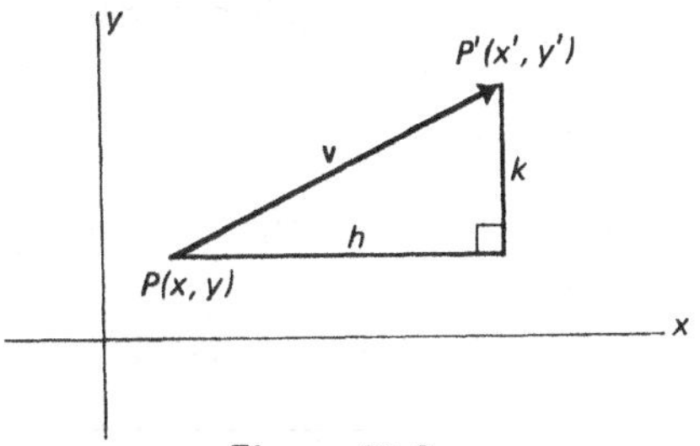

Figure 21.2

units up, so that $x' = x + h$ and $y' = y + k$. These equations define the translation, so we have proved the following theorem.

21.3 Theorem Let $\mathbf{v} = \vec{OA}$ where $O(0, 0)$ and $A(h, k)$. Then the translation that maps each point $P(x, y)$ through vector $\mathbf{v}$ to point $P'(x', y')$ is given by

$$\begin{cases} x' = x + h, \\ y' = y + k. \end{cases}$$

21.4 Definition The vector $\mathbf{v}$ of Theorem 21.3 is said to have *components h* and k and we write $\mathbf{v} = (h, k)$ for this vector or for any vector $\mathbf{v} = \vec{CD}$ where $C(a, b)$ and $D(a + h, b + k)$.

21.5 Consider rotating each point $P(x, y)$ about point $O(0, 0)$ through angle θ to point $P'(x', y')$. Let the directed angle from the positive x-axis to ray OP be ϕ, and let $m(OP) = r$. Then (see Fig. 21.5)

$$x = r\cos\phi \qquad \text{and} \qquad y = r\sin\phi.$$

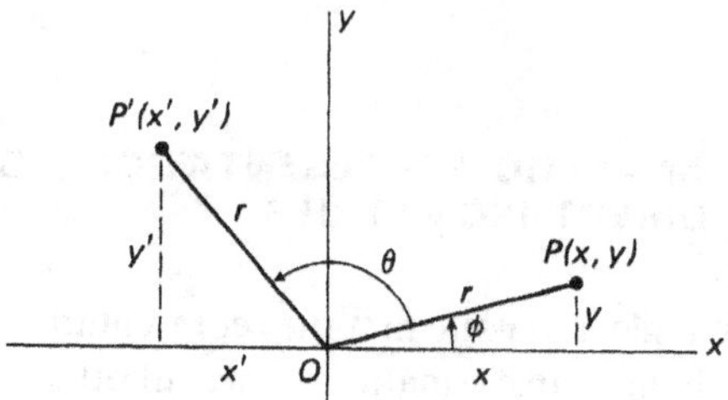

Figure 21.5

The angle from the positive x-axis to ray OP' is then $\theta + \phi$, and we have

$$\begin{aligned} x' &= r\cos(\theta + \phi) \\ &= r\cos\theta\cos\phi - r\sin\theta\sin\phi \\ &= x\cos\theta - y\sin\theta \end{aligned}$$

and

$$\begin{aligned} y' &= r\sin(\theta + \phi) \\ &= r\sin\theta\cos\phi + r\cos\theta\sin\phi \\ &= x\sin\theta + y\cos\theta. \end{aligned}$$

These equations determine the rotation, so we have proved the following theorem.

21.6 Theorem The rotation about the origin through angle θ, mapping each point $P(x, y)$ into $P'(x', y')$, is given by

$$\begin{cases} x' = x\cos\theta - y\sin\theta, \\ y' = x\sin\theta + y\cos\theta. \end{cases}$$

21.7 Let α and β be two transformations of the plane, and let $\alpha(x, y) = (x', y')$ and $\beta(x', y') = (x'', y'')$. Then, by definition of transformation multiplication, we have $(\beta\alpha)(x, y) = (x'', y'')$.

21.8 We may accomplish a rotation about point $C(h, k)$ through angle θ by first translating through vector $\mathbf{v} = (-h, -k)$ so that the point $C(h, k)$ is moved to the origin, second rotating through angle θ about the origin, and third translating back through vector $-\mathbf{v} = (h, k)$. Letting α, β, and α^{-1} denote these isometries, we have

$$\alpha: \begin{cases} x' = x - h, \\ y' = y - k, \end{cases}$$

$$\beta: \begin{cases} x'' = x' \cos\theta - y' \sin\theta, \\ y'' = x' \sin\theta + y' \cos\theta, \end{cases}$$

and

$$\alpha^{-1}: \begin{cases} x''' = x'' + h, \\ y''' = y'' + k. \end{cases}$$

Calculating $\alpha^{-1}\beta\alpha$ by eliminating x', y', x'', and y'' from the equations above, we have the following theorem.

21.9 Theorem The rotation about point $C(h, k)$ through angle θ is accomplished by

$$\begin{cases} x' = (x - h)\cos\theta - (y - k)\sin\theta + h \\ y' = (x - h)\sin\theta + (y - k)\cos\theta + k. \end{cases}$$

21.10 The equations of Theorem 21.9 are sufficiently complicated that the reader is urged to remember the *method* (translate to the origin, rotate about the origin, then translate back again) rather than to memorize the formulas.

21.11 Thus translations and rotations about the origin have relatively simple formulations, so their analytic forms may be used whenever convenient. Rotating about other points probably should be avoided whenever possible when using analytic representations. Of course, lines and other curves whose equations are known may be rotated and translated by making use of the equations given in Theorems 21.3, 21.6, and 21.9. Much more convenient are the implicit forms found by solving these equations for x and y in terms of x' and y'. For a rotation this is accomplished most easily by noting that $P(x, y)$ is found from $P'(x', y')$ by the inverse of the given rotation. The implicit equations for the translation and the rotation about the origin are

$$\begin{cases} x = x' - h \\ y = y' - k \end{cases} \quad \text{and} \quad \begin{cases} x = x' \cos(-\theta) - y' \sin(-\theta) \\ y = x' \sin(-\theta) + y' \cos(-\theta). \end{cases}$$

The equations for the rotation may be written in the simpler form

$$\begin{cases} x = x'\cos\theta + y'\sin\theta \\ y = -x'\sin\theta + y'\cos\theta, \end{cases}$$

although it may be more confusing to try to remember this simpler form in addition to the original rotation form. That is, here again it is probably better to remember the idea that the implicit form comes from a rotation through angle $-\theta$ in mapping P' to P rather than to remember the specific equations.

21.12 Example Rotate the line $2x + 3y = 5$ through 90° about the origin. We have

$$x = x'\cos 90° + y'\sin 90° = y'$$

and

$$y = -x'\sin 90° + y'\cos 90° = -x'.$$

Making these substitutions in the given equation for the line, we obtain $2y' - 3x' = 5$. It is left to the reader to graph both equations to show that the result is correct.

We conclude this section by stating the equations for a halfturn.

21.13 Theorem The halfturn about the point $C(h, k)$ is given by

$$\begin{cases} x' = -x + 2h, \\ y' = -y + 2k. \end{cases}$$

Exercise Set 21

1. For the points O, A, C, D given in Theorem 21.3 and Definition 21.4, calculate analytically the product $\sigma_C\sigma_D\sigma_A\sigma_O$ to show that $OADC$ is a parallelogram.
2. Graph the lines of Example 21.12.
3. Prove Theorem 21.13 from a geometric picture.
4. Prove Theorem 21.13 as a special case of Theorem 21.9.
5. Show analytically that the product of two translations is a translation.
6. Show that the product of two rotations about the origin through angles θ and ϕ is a rotation about the origin through angle $\theta + \phi$.
7. Show that the product of the two rotations about the origin with angle θ and about the point $C(h, k)$ with angle ϕ is a translation iff $\theta + \phi$ is an integral multiple of 360°.
8. Find an analytic representation for the inverse of the general translation as given in Theorem 21.3.
9. Find an analytic representation for the inverse of the rotation about the origin given in Theorem 21.6.
10. Find an analytic representation for the inverse of the general rotation as given in Theorem 21.9.
11. Find an analytic representation for the inverse of the halfturn as given in Theorem 21.13.

12. Show that the rotation about $C(h, k)$ through angle θ can be written in the form below, and find r and s in terms of h, k, and θ:

$$x' = x\cos\theta - y\sin\theta + r \qquad \text{and} \qquad y' = x\sin\theta + y\cos\theta + s.$$

13. Show that the equations of Exercise 21.12 show that every rotation can be written as a rotation about the origin followed by a translation.
14. Given that $a^2 + b^2 = 1$, show that each transformation of the form

$$x' = ax - by + c \qquad \text{and} \qquad y' = bx + ay + d$$

represents the product of a rotation and a translation, and find the angle of the rotation.

15. Show that the isometry of Exercise 21.14 is a translation when $a = 1$; hence show that these equations represent all direct isometries.
16. Rotate $A(1, 0)$ through 120° and through 240° about the origin to locate the vertices of an equilateral triangle. Show that $x = -\frac{1}{2}$ is one of its sides, and rotate it through the same rotations to find equations for the other two sides.

▶SECTION 22 | ANALYTIC REPRESENTATIONS OF OPPOSITE ISOMETRIES

22.1 A reflection in a coordinate axis has a very simple form. The following theorem is obvious.

22.2 Theorem The reflection σ_x in the x-axis and the reflection σ_y in the y-axis are given, respectively, by

$$\sigma_x: \begin{cases} x' = x \\ y' = -y \end{cases} \qquad \text{and} \qquad \sigma_y: \begin{cases} x' = -x \\ y' = y. \end{cases}$$

22.3 Reflections in other lines through the origin are handled without great difficulty. Let m be a line passing through the origin with angle of inclination θ as shown in Fig. 22.3. Let σ_x denote the reflection in the x-axis and let α denote a rotation about the origin in angle 2θ. Since $\sigma_m\sigma_x = \alpha$, we have

$$\sigma_m = \alpha\sigma_x;$$

that is, the reflection in line m is the product of a reflection in the x-axis followed by a rotation about the origin with angle 2θ. Since σ_x and α have the representations

$$\sigma_x: \begin{cases} x' = x \\ y' = -y \end{cases} \qquad \text{and} \qquad \alpha: \begin{cases} x'' = x'\cos 2\theta - y'\sin 2\theta \\ y'' = x'\sin 2\theta + y'\cos 2\theta, \end{cases}$$

then $\sigma_m = \alpha\sigma_x$ has the representation

$$\sigma_m: \begin{cases} x'' = x\cos 2\theta + y\sin 2\theta \\ y'' = x\sin 2\theta - y\cos 2\theta. \end{cases}$$

By setting $\theta = 0$ and $\theta = 90°$ and comparing the results with those given in Theorem 22.2, one can see that the equations for the reflection σ_m as given above are indeed correct for reflections in the x- and y-axes. The results we have obtained are stated as Theorem 22.4.

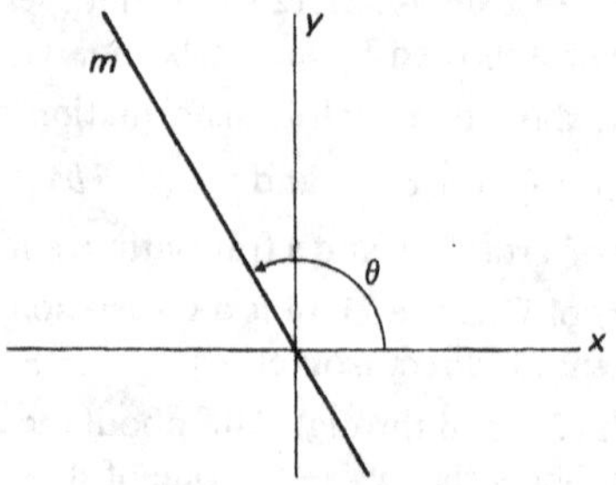

Figure 22.3

22.4 Theorem A reflection σ_m in the line m passing through the origin with angle of inclination θ has the representation

$$\sigma_m: \begin{cases} x' = x\cos 2\theta + y\sin 2\theta \\ y' = x\sin 2\theta - y\cos 2\theta. \end{cases}$$

22.5 When its mirror does not pass through the origin, we may handle the reflection just as we did for a rotation not about the origin. Thus, suppose line m passes through point $A(h, k)$ with inclination θ. First, translate A to the origin, then reflect in line n through the origin and parallel to line m, and finally translate back again. The results are stated in the next theorem.

22.6 Theorem The equations for the reflection σ_m in line m passing through $A(h, k)$ with angle of inclination θ are

$$\sigma_m: \begin{cases} x' = (x-h)\cos 2\theta + (y-k)\sin 2\theta + h \\ y' = (x-h)\sin 2\theta - (y-k)\cos 2\theta + k. \end{cases}$$

22.7 This general form of Theorem 22.6 is again too complicated to be worth memorizing, but it is easily reconstructed when needed.

22.8 Lastly, let us consider a glide-reflection α whose mirror m passes through the origin with inclination θ and whose translation β is r units along m measured positive upward or, for a horizontal line, positive to the right. The translation and the reflection are then given by

$$\beta: \begin{cases} x' = x + r\cos\theta \\ y' = y + r\sin\theta \end{cases} \quad \text{and} \quad \sigma_m: \begin{cases} x' = x\cos 2\theta + y\sin 2\theta \\ y' = x\sin 2\theta - y\cos 2\theta. \end{cases}$$

Now the glide-reflection $\alpha = \sigma_m\beta = \beta\sigma_m$ is given in Theorem 22.9.

22.9 Theorem A glide-reflection whose mirror m passes through the origin with angle of inclination θ and whose translation is along line m through r units, r measured positive from the origin into the first two quadrants or along the positive x-axis, and negative otherwise, is given by

$$\begin{cases} x' = x\cos 2\theta + y\sin 2\theta + r\cos\theta \\ y' = x\sin 2\theta - y\cos 2\theta + r\sin\theta. \end{cases}$$

22.10 When the mirror of the glide-reflection passes through the point $A(h, k)$ instead of the origin, the equations for its representation are readily developed in the same manner as we have done earlier for the reflection and for the rotation. This formulation for the general glide-reflection is left as an exercise.

Exercise Set 22

1. Prove Theorem 22.2.
2. Prove Theorem 22.6.
3. Show analytically that the product of reflections in lines m and n through the origin with inclinations θ and ϕ is a rotation about the origin with angle $2(\phi - \theta)$.
4. Show analytically that the product of reflections in parallel lines m and n of inclination θ with m passing through the origin, and n passing through $A(h, k)$, is a translation. Find its vector.
5. Show that the equations of Theorem 22.9 for a glide-reflection are arrived at whether one analytically uses $\alpha = \sigma_m\beta$ or $\alpha = \beta\sigma_m$, where β and σ_m are the translation and reflection given in 22.8.
6. Find an analytic representation for the inverse of the reflection of Theorem 22.4.
7. Find an analytic representation for the inverse of the reflection of Theorem 22.6.
8. Find an analytic representation for the inverse of the glide-reflection of Theorem 22.9.
9. Formulate the equations for the general glide-reflection.
10. Find an analytic representation for the inverse of the general glide-reflection found in Exercise 22.9.
11. Show that each reflection can be written in the form
$$\begin{cases} x' = ax + by + c \\ y' = bx - ay + d \end{cases} \qquad \text{in which } a^2 + b^2 = 1.$$
12. Show that each glide-reflection also has the form of Exercise 22.11. Hence every opposite isometry has this form.
13. Show that the form of Exercise 22.11 represents a reflection in a line through the origin, followed by a translation in the vector (c, d) not necessarily parallel to the mirror of reflection. Hence all such equations represent opposite isometries.
14. Show that each isometry has the analytic form
$$\begin{cases} x' = ax - by + c \\ y' = \pm(bx + ay) + d \end{cases} \qquad \text{with } a^2 + b^2 = 1,$$

and that it is direct if the plus sign holds and opposite if the minus sign applies. Conversely, show that every transformation given by the equations above is an isometry. Hence these equations characterize isometries analytically.

15. Show that the product of two isometries of the form given in Exercise 22.14 is another isometry of that form.
16. Show that each isometry can be written as the product of either a rotation about the origin or a reflection in a line through the origin, followed by a translation.
17. Find implicit forms for:
    a) The reflections of Theorem 22.2
    b) The reflection of Theorem 22.4
    c) The reflection of Theorem 22.6
    d) The glide-reflection of Theorem 22.9
    e) The general glide-reflection found in Exercise 22.9
18. From the implicit forms for a reflection found in Exercise 22.17, find an equation for the image of a circle $x^2 + y^2 - 4x - 2y = 0$ in a reflection in the line $y = x$. Find any fixed points on the circle and sketch a graph.
19. Show that, in the form given for a glide-reflection, if we let the translation have vector $\overrightarrow{OA}$, take r always positive, and take θ as the directed angle (less than 360°) from the positive x-axis to the *ray* OA, then the formulas given are still correct.
20. Show analytically that a product of three reflections in mirrors through the origin is a reflection in a mirror through the origin.
21. Show analytically that a product of three reflections in parallel mirrors is a reflection in a mirror parallel to the given mirrors.

3 SIMILARITIES IN THE PLANE

SECTION 23 THE REBIRTH OF MATHEMATICAL THINKING

23.1 During the Dark Ages, beginning with the fall of Alexandria in 641, mathematical thinking sank to its nadir. Although minor discoveries appeared from time to time, the next thousand years were essentially barren. Furthermore, absolutely no progress in the mathematical method occurred until the nineteenth century. Thus, from the time of its first zenith about the time of Archimedes, more than 2000 years passed before a second zenith approached.

23.2 The Arabs, by carting away and translating into Arabic some of the library at Alexandria, preserved much of the Greek knowledge, later transmitting it back into Europe. They also adopted the Hindu numeral system and passed it along to Western mathematicians in later years. Thus the primary function of the Arabs was that of custodian of Greek learning.

23.3 At the end of the Dark Ages some knowledge began to filter back into Europe, and the twelfth century became a time of translators, the ancient Greek learning being translated from Arabic into Latin, the language of the emerging scholars. In the thirteenth century universities began to spring up all over the Continent as people began again to delve into the mysteries known to the Greeks.

23.4 The fourteenth century was marked by the Black Death and the Hundred Years' War, to the detriment of mathematics. The fifteenth and sixteenth centuries began to show promise of a bright future with advances in arithmetic, algebra, and trigonometry. Printing was invented, so knowledge could be dispersed much more rapidly and widely than ever before. The stage was set for the great scenes which began to be played in the seventeenth century.

23.5 In 1647 William Oughtred (1574–1660) used π/δ as the symbol for the ratio $\pi = 3.14159\ldots$. He gave a total of more than 150 symbols, mostly in algebra. The straight logarithmic slide rule was his invention. Also he and one of his students each independently invented the circular slide rule. It is said that his wife was so frugal that she would not permit him the use of a candle for his nighttime studies. He worked late into the night, sometimes skipping sleep for two or three nights in a row. And he is supposed to have died "in a transport of joy" when he heard of the restoration of Charles II to the throne. "It should be added, by way of excuse," De Morgan commented, "that he was eighty-six years old."

23.6 Johannes Kepler (1571–1630), a close friend of Galileo Galilei (1564–1642), spent a full 22 years analyzing against observed data the theories he assumed for the motions of the planets. He finally concluded that all the planets travel in elliptical orbits with the sun at one focus, that the radius vector from the sun to a given planet sweeps out equal areas in equal times, and that the square of the period of revolution of a planet is proportional to the cube of the length of its orbit's major axis. By dividing a circle into infinitely many congruent isosceles triangles, he anticipated the calculus.

Kepler's personal life was filled with tragedy. Smallpox at the age of four left him with very poor eyesight. His youth was unhappy and his marriage was miserable. His favorite child died from smallpox and his wife went insane and died. A glutton for punishment, he carefully weighed the good and bad points of eleven girls before choosing the worst one to be his new wife, resulting in a second marriage even more unhappy than his first one. His mother was jailed as a witch, and he nearly suffered the same fate while fighting to free her. She died within a few months of her release. He was fired from one lectureship and his pay was always late. Finally, he died of a fever while on the way to collect some of his back wages.

23.7 Pierre de Fermat (1601–1665) led a quiet personal life, rather unusual for a mathematician. Although his main work was in number theory, his *De maximus et minimis* anticipated the calculus.

23.8 It was commonly held at this time that the infinite could not possibly be comprehensible to man, who had a finite mind. Again equal to the challenge, De Morgan stated that by that reasoning, "who drives fat oxen should himself be fat." Nonetheless, it was to be 200 years before the mathematical infinite was understood.

23.9 René Descartes (1596–1650), a man with a violent temper, was the discoverer of analytic geometry in 1637. Although others before him (Apollonius, Vieta, Oresme, Cavalieri, Roberval, and Fermat) had applied algebra and coordinates to specific curves, Descartes was the first one to introduce a coordinate system to be applied to all geometric curves. He introduced exponents (as in x^3 for xxx) and the rule of signs that bears his name and is used for determining the numbers of positive and negative roots of a polynomial equation.

23.10 Fermat once criticized Descartes, causing his infamous temper to explode. Descartes attacked the former's "method of tangents" with a vengeance. Even though Descartes was dead wrong, he continued the controversy long after it should have been forgotten. He died because of a request from Queen Christina of Sweden: Always frail, he had been permitted as a child to arise each day when he was ready to do so, generally toward noon. In later years he cherished his mornings in bed. Young Christina requested him to come to Sweden as her personal tutor, so great was her thirst for mathematics. After refusing several times, he finally agreed to leave his warm Holland. To his horror, she demanded her lessons at 5 a.m. in a cold, drafty library. Unable to stand the climate and the early hours, he became ill and died just four months later.

It is said that the real reason for his death was this: Christina, having been raised to be quite masculine, was quite an equestrian, riding daily sometime *after* her lesson from Descartes (pronounced day-CART). Had she reversed the order of these two daily events, poor René would undoubtedly have lived much longer. For it is well known that one should never put "Descartes before the horse."

23.11 In 1639 Girard Desargues (1593–1662) introduced projective geometry. His work, coming just two years after Descartes' analytic geometry and written in a very eccentric style, attracted little attention and was soon lost for over two hundred years. He introduced the concept of a pencil of lines or planes, and proved his famous two-triangle theorem.

23.12 Blaise Pascal (1623–1662), not permitted to study mathematics until he had thoroughly mastered Latin and Greek, set down his own axioms and definitions for elementary geometry, and then proved that the sum of the angles of a triangle is equal to two right angles. His father, well versed in mathematics, caught him studying this theorem and was so amazed that he wept for joy. Pascal wrote his treatise on conics when he was only 16 years of age. From his mystic hexagram theorem he himself deduced more than 400 corollaries!

23.13 John Wallis (1616–1703) was the first to treat the conic sections analytically as second-degree polynomials. He introduced the symbol ∞ for infinity and discovered the interesting infinite product

$$\frac{\pi}{2} = \frac{2}{1} \cdot \frac{2}{3} \cdot \frac{4}{3} \cdot \frac{4}{5} \cdot \frac{6}{5} \cdot \frac{6}{7} \cdot \cdots$$

23.14 The discovery of the calculus in 1665 by Isaac Newton (1642–1727), and independently in 1675 by Gottfried Wilhelm Leibniz (1646–1716) highlighted this very productive century in which mathematical thinking developed again to about the standard the Greeks had reached. We shall not dwell here on the works of Newton and Leibniz. Their discovery of the calculus was a natural outgrowth of analytic geometry and the "method of indivisibles" which had been applied to many individual situations by earlier workers.

23.15 Now mathematicians, drunk with the power of the calculus, blindly applied its methods to all sorts of problems, completely ignoring the shaky foundations on which it had been constructed. Many contradictions arose, to be resolved in the nineteenth century with a theory of limits. But now with the application of the methods of algebra and analysis to geometry, modern geometry was born. And the productive seventeenth century came to an end.

Exercise Set 23

1. Prove synthetically the Desargues two-triangle theorem (Theorem 4.7) for the case where the two triangles lie in distinct intersecting planes.
2. Pascal's mystic hexagram theorem states that if a not-necessarily-convex hexagon is inscribed in a conic section, then the three points of intersection of pairs of opposite sides are collinear. Prove this theorem for the case where the conic section is a circle.
3. Given five points that lie on a conic section, use Pascal's mystic hexagram theorem (Exercise 23.2) to locate other points on the conic.
4. Use a desk calculator (or computer) to calculate the value of the product of the first ten (or 100) factors of Wallis' expression for $\pi/2$ (see 23.13). How rapidly does this product converge?
5. In 1806, while still a student, C. J. Brianchon (1785–1864) proved that if a hexagon is circumscribed about a conic section, then the three diagonals that join opposite vertices are concurrent. Show that Brianchon's theorem is the dual (see 4.9) of Pascal's theorem (Exercise 23.2).

SECTION 24 | INTRODUCTION TO SIMILARITIES

24.1 In this chapter we shall do for *similarities*—mappings that carry figures into similar figures—what we did for isometries in Chapter 2. We begin with a *homothety*, a simple stretch or magnification of the plane which multiplies all distances from a fixed point called the *center* by the same *ratio* $k \neq 0$. This mapping is a direct similarity and carries each line into a parallel line and has just its center for an invariant point. Conversely, if any two similar figures have corresponding sides parallel, then there is a homothety that maps one of them to the other and its center is the point of concurrence of the lines joining corresponding vertices (Fig. 24.1).

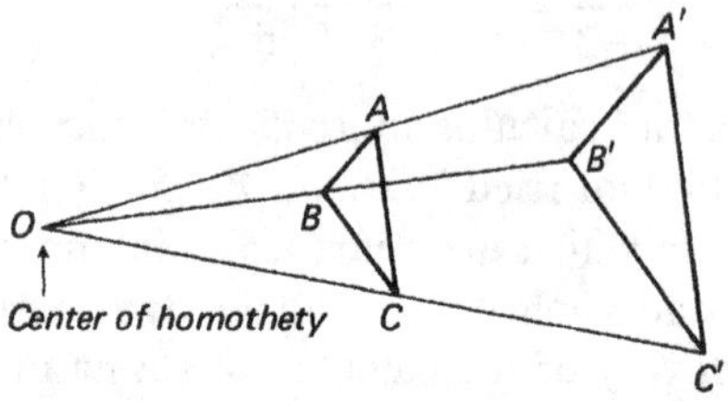

Figure 24.1

24.2 Two circles have two centers of homothety if the circles are not concentric. These centers are located on the line of centers of the circles and may be found by drawing parallel diameters AB and $A'B'$ in the two circles (Fig. 24.2). Then the lines

AA' and BB' meet at one center of homothety, the lines AB' and $A'B$ meet at the other. These centers of homothety, also called *centers of similitude*, divide harmonically the line of centers of the two circles.

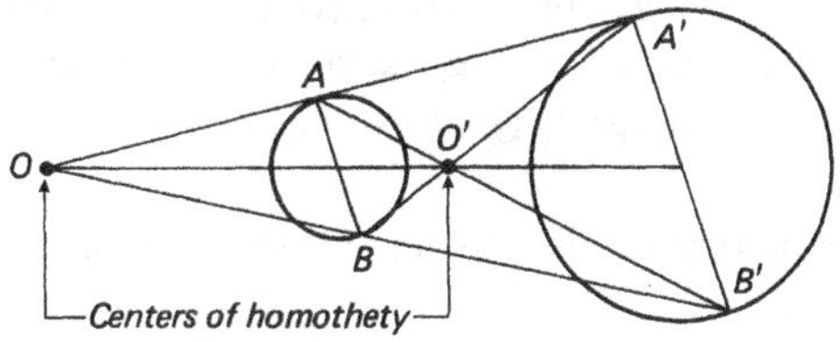

Figure 24.2

24.3 Two homotheties commute when and only when they have the same center or at least one of the ratios is $+1$. The product of two homotheties is either a homothety or a translation, the latter occurring when their ratios are reciprocals of one another (see Theorem 25.9).

24.4 A similarity is the composition of an isometry and a homothety. Any similarity that is not an isometry is either a rotation followed by a homothety with the same center or it is a reflection followed by a homothety whose center lies on the mirror, the first being direct, the second product opposite (see Theorems 26.9 and 26.10).

24.5 Any two given segments AB and $A'B'$ are related by just two similarities (with A' the image of A and B' the image of B), one direct and the other opposite. Any two similar triangles are related by just one similarity. In this sense, if similarity α maps triangle ABC to triangle $A'B'C'$, we agree that α^{-1} mapping triangle $A'B'C'$ to triangle ABC does not constitute a different similarity between these triangles.

24.6 The theory described above will be developed in the next two sections. With that foundation we shall then have a sufficient basis to proceed to three sections of applications to elementary geometry. A final section is devoted to the analytic representations of similarities, enabling us to use these transformations in the Cartesian plane. Thus this chapter parallels closely, but more briefly, Chapter 2.

Exercise Set 24

1. Show that a similarity preserves angles.
2. Where are the two centers of homothety for the base and summit of an isosceles trapezoid?
3. Show that the product of two homotheties with the same center is a homothety, and find its ratio and center.
4. Prove that if two similar triangles have their corresponding sides parallel, then the lines joining corresponding vertices concur, so they are related by a homothety.
5. Given two directly similar triangles, show how to find a rotation and a homothety (not necessarily with the same center) whose product maps one triangle to the other.

6. Given two oppositely similar triangles, show how to find a reflection and a homothety (whose center does not necessarily lie on the mirror) whose product maps one triangle to the other.
7. Find the centers of homothety of two nonintersecting circles with radii 3 and 5, and find the ratios in which these centers divide the line of centers of the circles.
8. Prove that the centers of homothety of two circles divide the line of centers of the circles harmonically.
9. Prove that the construction for the centers of homothety of two circles, given in 24.2, is correct.
10. Find a homothety other than the identity map (homothety with ratio $= +1$) that carries a given rectangle into itself (not necessarily point by point).
11. Denote the homothety with center A and ratio k by $H(A, k)$. Given that $O(0, 0)$, $P(1, 0)$, and $Q(0, 2)$ are points in the Cartesian plane, find the image of triangle OPQ under each homothety.
    a) $H(O, 1)$
    b) $H(O, 2)$
    c) $H(O, -1)$
    d) $H(O, \frac{1}{3})$
    e) $H(A, 2)$, where $A(2, 0)$
    f) $H(A, -\frac{1}{2})$
    g) $H(B, 2)$, where $B(4, 0)$
    h) $H(B, -\frac{1}{2})$
12. Find all homotheties that are involutoric.
13. Find the inverse of the homothety with a given center and ratio.
14. Show that, for a given homothety α, the smallest positive n for which $\alpha^n = \iota$
    a) is 1 iff $\alpha = \iota$,
    b) is 2 iff the ratio of the homothety is -1, and
    c) does not exist otherwise.

SECTION 25 HOMOTHETY

25.1 Definition A *homothety* $H(O, k)$, where O is a fixed point in the plane and k is a nonzero real number, is that transformation that maps point O to itself and maps any other point P to a point P', such that O, P, P' are collinear and $\mathbf{OP'} = k \cdot \mathbf{OP}$. Point O is called the *center* and k the *ratio* of the homothety.

The homothety is a building block for similarity mappings in about the same way as a reflection is a building block for isometries. One cannot obtain all similarity mappings from products of homotheties alone, but they are necessary and basic to similarities. In this section we shall study the elementary properties of homotheties and their products. In the next section we shall show the relationship between homotheties and similarities.

25.2 Theorem The homothety $H(O, k)$ maps a line segment AB into a parallel segment $A'B'$ with $A'B' = |k| \cdot AB$.

Let $H(O, k)$ map A to A', B to B', and any other point P on segment AB to P' (Fig. 25.2). Then $\measuredangle BOP = \measuredangle B'OP'$ and

$$\frac{OB'}{OB} = \frac{OP'}{OP} = |k|.$$

Hence triangles BOP and $B'OP'$ are similar, so $B'P'/BP = |k|$ and $B'P'$ is parallel to BP. Similarly $P'A'/PA = |k|$ and $P'A'$ is parallel to PA. Now, since BPA is a straight line, then so also is $B'P'A'$ a straight line, and $B'A'/BA = |k|$. □

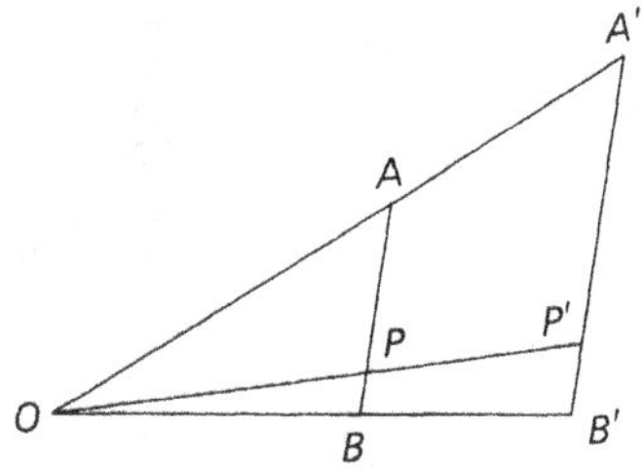

Figure 25.2

25.3 Corollary A homothety maps any given triangle into a similar triangle. In general, it maps each polygon into a similar polygon.

25.4 Corollary A homothety preserves angles between lines.

25.5 Theorem A homothety $H(O, k)$ maps a circle of radius r into another circle whose radius r' is $|k| \cdot r$, and whose center C' is the homothetic image of the center C of the given circle.

Let $H(O, k)$ map the center C of the given circle to C', and any point P on the circle to point P', as in Fig. 25.5. Since $CP = r$, the radius of the given circle, then $C'P' = |k| \cdot r$ by Theorem 25.2. That is, the locus of the images of point P on the given circle is another circle of radius $|k| \cdot r$ and center C', the homothetic image of C. □

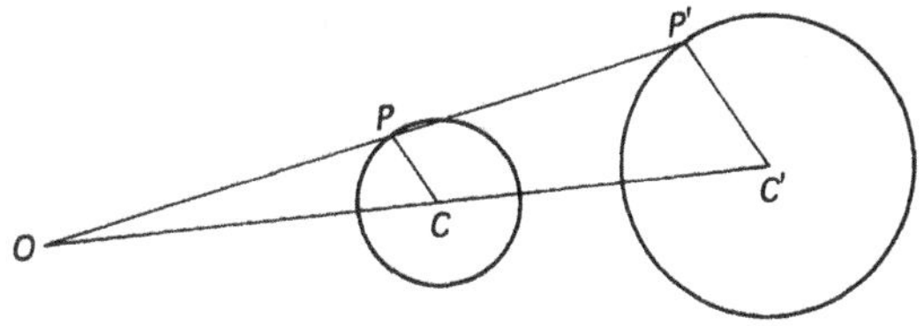

Figure 25.5

25.6 Theorem For any two given circles there is a homothety that maps one to the other. If the circles are unequal and not concentric, then there are two such homotheties.

If the unequal, nonconcentric circles of radii r and r' do not intersect, then the centers are at the meeting of their common external tangents (with ratio of homothety r'/r) and the meeting of their common internal tangents (with ratio $-r'/r$). In all cases we may locate these centers (Fig. 25.6) by taking the centers of the given circles as C and C', then erecting perpendiculars to the line of centers CC' at C and C'. Let these perpendiculars cut circle C at P and Q and circle C' at P' and Q'. The centers of homothety are at the intersections of PP' and QQ' and of PQ' and $P'Q$. We denote these centers by I and E, I being the *internal* center of similitude and lying between C and C' and E being the *external* center of similitude and lying outside segment CC'. ▯

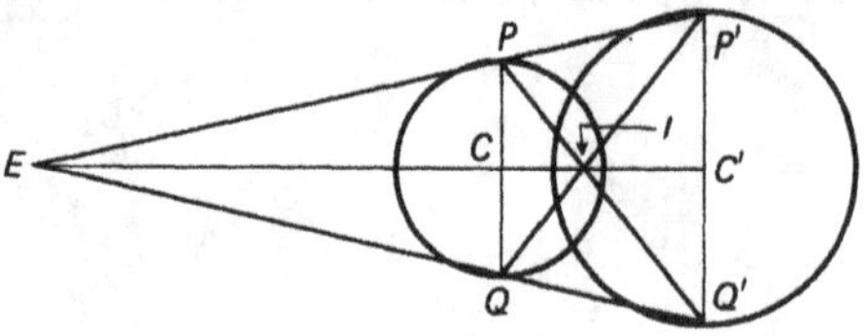

Figure 25.6

25.7 Theorem If two similar noncongruent triangles are so oriented that each side of one is parallel to the corresponding side of the other, then there is a homothety that maps one triangle to the other.

Let the triangles be ABC and $A'B'C'$, as in Fig. 25.7. Since pairs of corresponding sides are parallel, then the two lines of each pair meet on the line at infinity; that is, the two triangles are coaxial. By Desargues' two-triangle theorem (Theorem 4.7), these triangles are also copolar at a point O. It follows that point O serves as the center of homothety and that the ratio of homothety is $\mathbf{OA'/OA} = \pm A'B'/AB$. ▯

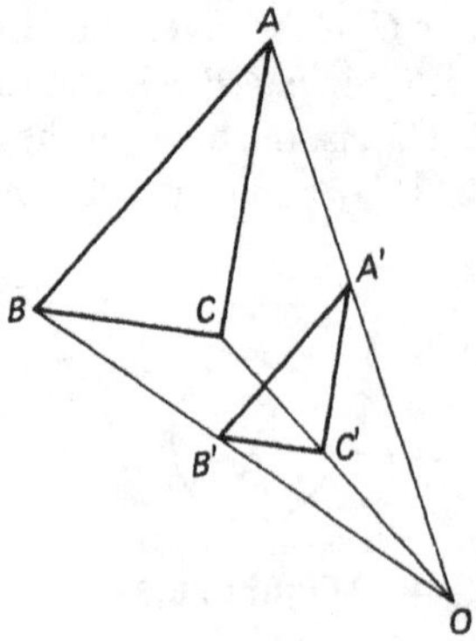

Figure 25.7

25.8 Theorem A homothety is determined by two distinct points and their images.

Let the homothety map points A and B to A' and B', respectively. Then the center of homothety is the point of intersection of the lines AA' and BB'. Its ratio is $\pm A'B'/AB$, the minus sign occurring if the center lies between A and A', the plus sign otherwise. ▯

▶**25.9 Theorem** If $jk \neq 1$, then the product of the two homotheties $H(O, k)$ and $H(Q, j)$ is the homothety $H(P, jk)$ where P is collinear with O and Q. If $jk = 1$, then P becomes ideal and the product of the two homotheties is a translation.

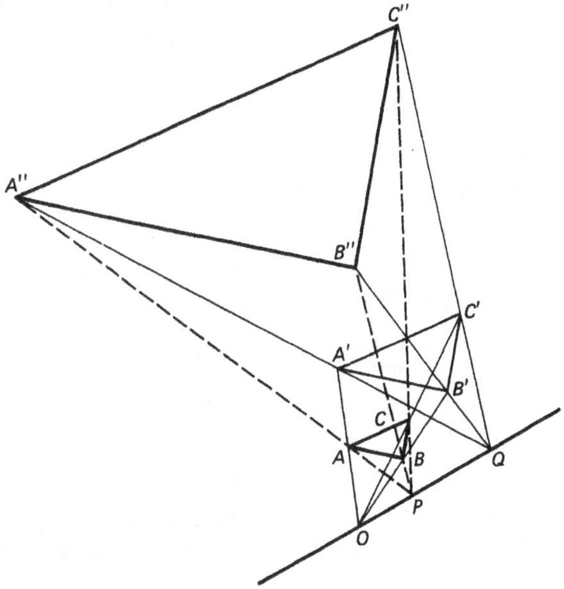

Figure 25.9

Since each homothety maps a triangle into a similar triangle with sides parallel to the given triangle, it follows that their product does the same. Thus either product is a homothety of ratio $jk \neq 1$, or a translation when $jk = 1$.

Let $H(O, k)$ map $\triangle ABC$ to $\triangle A'B'C'$ and let $H(Q, j)$ map $\triangle A'B'C'$ to $\triangle A''B''C''$ (see Fig. 25.9). Since AB, $A'B'$, and $A''B''$ are all parallel, they meet at an ideal point I. Thus triangles $AA'A''$ and $BB'B''$ are copolar at I. By the Desargues theorem, they are coaxial; that is, points O, P, Q are collinear. ▯

25.10 Theorem The product of two homotheties having the same center is commutative, and is a homothety with the same center and with ratio equal to the product of the ratios of the given homotheties.

25.11 Theorem The center of a homothety of ratio not equal to $+1$ is the only fixed point of the homothety. Lines through the center are the only fixed lines.

25.12 The concept of *direct* or *opposite* applies to similar figures as well as to congruent figures. Appropriate definitions are left for the exercises.

25.13 Theorem A homothety is a direct transformation.

25.14 Theorem A homothety of ratio $+1$ is the identity map.

25.15 Theorem The inverse of the homothety $H(O, k)$ is the homothety $H(O, 1/k)$.

▶**25.16 Theorem** All homotheties and translations form a transformation group.

▶**25.17 Theorem** All homotheties having the same center form a transformation group.

Exercise Set 25

1. Prove Corollary 25.3.
2. Prove Corollary 25.4.
3. Locate the center of homothety and find its ratio for two equal nonconcentric circles.
4. Prove that the centers of similitude of two circles lie on the line of centers of the circles.
5. Prove that, as stated in the proof of Theorem 25.6, the intersections of the common tangents of two nonintersecting circles are their centers of homothety.
6. Prove that the construction in the proof of Theorem 25.6 that applies "in all cases" does indeed give the centers of similitude.
7. In the proof of Theorem 25.8, show that the center lies between A and A' iff it lies between B and B'.
8. In Theorem 25.9, show that P divides OQ in the ratio $(j-1)/(j(k-1))$.
9. Use the result of Exercise 25.8 to show that when $j \neq 1$ and $k \neq 1$, then the homotheties of Theorem 25.9 do not commute. Illustrate this noncommutativity with a figure.
10. Prove Theorem 25.10.
11. Prove Theorem 25.11.
12. The product $H(B, 1/k) \cdot H(A, k)$ is a translation. What translation?
13. Prove Theorem 25.13.
14. Prove Theorem 25.14.
15. Prove Theorem 25.15.
16. Prove Theorem 25.16.
17. Prove Theorem 25.17.
18. Define *direct* and *opposite similar figures.*
19. Show that a product of three homotheties is a homothety or a translation.
20. Generalize Exercise 25.19 to a product of four or more homotheties.

SECTION 26 | SIMILARITY

26.1 Definition A *similarity* is a map of the plane that carries each point pair A, B into a point pair A', B', such that for some fixed positive real number k, $A'B' = k \cdot AB$. The number k is called the *ratio* of the similarity. The term *similitude* is also used for this mapping.

This section completes our study of similarities and their properties, so that we may proceed rapidly to the applications in Sections 27 to 29. Although the theory we develop here is not as extensive as that for isometries in Chapter 2, it is quite sufficient for our purposes.

26.2 Theorem A similarity of ratio 1 is an isometry.

26.3 Theorem A similarity maps segments into segments.

Let a similarity of ratio k map A to A' and B to B'. Take any point P between A and B and let the image of P be P'. Then

$$A'B' = k \cdot AB = k(AP + PB) = k \cdot AP + k \cdot PB = A'P' + P'B',$$

so P' lies on segment $A'B'$. ▯

26.4 Theorem A similarity is a transformation of the plane.

26.5 Theorem A similarity preserves angles.

Mark points B and C on the two sides of angle A to form a triangle ABC. Let the similarity map triangle ABC to triangle $A'B'C'$ by Theorem 26.3. Then

$$\frac{A'B'}{AB} = \frac{B'C'}{BC} = \frac{C'A'}{CA}.$$

Thus triangles ABC and $A'B'C'$ are similar by SSS. Hence $\measuredangle CAB \cong \measuredangle C'A'B'$ and the theorem follows. ▯

26.6 Corollary A similarity maps a triangle into a similar triangle. More generally, it maps a polygon into a similar polygon.

26.7 Theorem A similarity is determined by any three noncollinear points and their images.

26.8 Theorem There are exactly two similarities, one direct and one opposite, that map any point pair A, B into any other point pair A', B' (understanding that $A \neq B$, $A' \neq B'$, A' is the image of A, and B' is the image of B).

26.9 Theorem Each direct similarity of ratio k that is not an isometry is the product of a rotation and a homothety of ratio k having the same center. Furthermore, such a product is commutative.

Clearly such a product is a direct similarity. By Theorem 26.8, this similarity is determined by a segment AB and its image $A'B'$. Let AA' and BB' meet at Q, and draw the circles through A, B, Q, and through A', B', Q to meet again at O, as shown in Fig. 26.9. Then we have

$$\measuredangle AOB \cong \measuredangle AQB \cong \measuredangle A'QB' \cong \measuredangle A'OB'$$

and

$$\measuredangle BAO \cong \measuredangle BQO \cong \measuredangle B'QO \cong \measuredangle B'A'O$$

by the properties of angles inscribed in a circle. Hence triangles ABO and $A'B'O$ are similar. Now, if AB is rotated about O through angle AOA' to A_1B_1, then A_1B_1 is parallel to $A'B'$ and triangles OA_1B_1 and $OA'B'$ are similar, so B_1 lies on line OB'. Thus the homothety $H(O, k)$ maps A_1B_1 to $A'B'$. This rotation and homothety satisfy the theorem. Now, if the figure formed by O, A_1, B_1, A', B' is rotated about O through angle $A'OA$ (the inverse of $\measuredangle AOA'$), then A_1B_1 will coincide with AB. It follows that the homothety and rotation commute. ▯

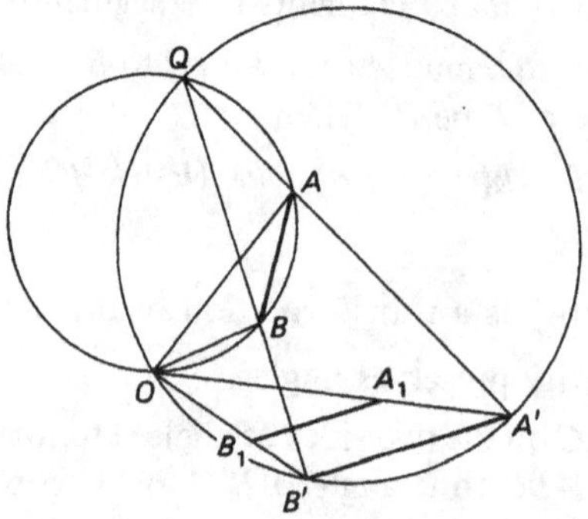

Figure 26.9

26.10 Theorem Any opposite similarity of ratio k that is not an isometry is the product of a reflection, and a homothety whose center lies on the mirror. Furthermore, such a product is commutative.

Let the similarity map segment AB to CD, and let the lines AB and CD meet at Q (Fig. 26.10). Of the two bisectors of the angles at Q, call that one n so that σ_n maps AB to EF where EF and CD have the same sense. Draw any other line n_1 parallel to (and

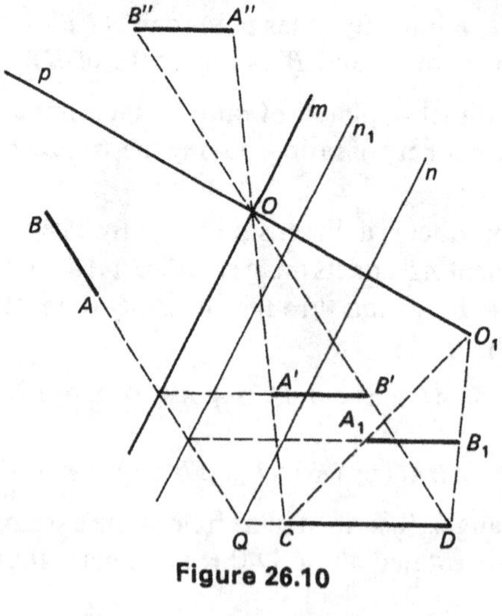

Figure 26.10

distinct from) n. Reflect AB in line n_1 to A_1B_1. Let O_1 be the center of the homothety that maps A_1B_1 to CD. Draw line p through O_1 perpendicular to n. Reflect AB in line p to $A''B''$. Now the center O of the homothety that carries $A''B''$ to CD lies on p. Draw line m through O and parallel to n, and let $\sigma_m(AB) = A'B'$. Then m is the mirror of reflection, and O is the center of homothety mapping AB to CD.

Proofs of the validity of the construction, and of the commutativity of the product of these transformations are left for the reader to supply.

To verify the construction, first show that all such points A_1 and A', reflections of A in mirrors parallel to n, lie on a line through A and perpendicular to n. Then show that all the centers O_1 lie on another line p perpendicular to n. Show that the reflection of AB to $A''B''$ in p makes $A''B''$ parallel to CD and opposite in sense, so that the center O of the homothety that carries $A''B''$ to CD lies on p. Show that $H(O, -k)\cdot\sigma_p$ maps AB (to $A''B''$ and then) to CD. Show that this map is equal to $H(O, k)\cdot\sigma_o\cdot\sigma_p = H(O, k)\cdot\sigma_m$.

To establish the commutativity, reflect the figure formed by O, $A'B'$, and CD in line m, and consider its image. □

▶**26.11 Theorem** All similarities form a transformation group.

▶**26.12 Theorem** All direct similarities form a transformation group.

Exercise Set 26

1. Prove Theorem 26.2.
2. Prove Theorem 26.4.
3. Prove that a similarity maps circles into circles.
4. Prove Corollary 26.6.
5. Prove Theorem 26.7 as a corollary to Theorem 13.6.
6. Prove Theorem 26.8.
7. a) Discuss the case when the circles in the proof of Theorem 26.9 are tangent.
   b) Discuss the case when one or both of the circles in the proof of Theorem 26.9 reduces to a straight line; that is, when A, B, Q are collinear, or when A', B', Q are collinear.
8. Prove the commutativity of Theorem 26.9 as suggested in the text.
9. On a clean sheet of paper trace the two segments of Fig. 26.9 labeled AB and $A'B'$. Relabel the first one BA; that is, interchange the labels on its endpoints. Perform the construction of Theorem 26.9 for these new segments.
10. Repeat Exercise 26.9 for the segments AB and CD of Fig. 26.10, and using the construction of Theorem 26.10.
11. Prove the construction of Theorem 26.10.
12. Prove the commutativity of Theorem 26.10.
13. Prove Theorem 26.11.
14. Prove Theorem 26.12.

15. In general there are two homotheties that map one circle to another, and a homothety is a direct similarity. Theorem 26.8 states there is just one direct similarity that maps one segment to another.
    a) Explain this apparent paradox.
    b) Find an opposite similarity mapping one circle to another.
    c) Are there more than two direct similarities that map one circle to another? Defend your answer.
    d) How many opposite similarities map one circle to another?
    e) In this same sense, how many direct and opposite similarities map one equilateral triangle to another? Here we are permitting vertex A of triangle ABC to map into any one of the three vertices of the image triangle.
    f) Repeat part (e) for a square.
    g) Repeat part (e) for a general rectangle.

SECTION 27 APPLICATIONS OF SIMILARITIES TO ELEMENTARY GEOMETRY

27.1 In this and the next section are presented applications of similarities to high school geometry. In Section 29, more advanced applications, applications in the domain of college geometry, are presented. Certainly many of these theorems, even into Section 29, are readily usable in high school geometry classrooms.

27.2 Theorem Three parallel lines cut off proportional segments from two transversals.

If the transversals are parallel, then the theorem follows immediately since opposite sides of the resulting parallelograms are congruent. So suppose transversals ABC and $A'B'C'$ meet at P (Fig. 27.2). Then $H(P, \mathbf{PB}/\mathbf{PA})$ maps triangle PAA' to PBB', and $H(P, \mathbf{PC}/\mathbf{PB})$ maps triangle PBB' to PCC'. Now

$$AB/BC = A'B'/B'C'$$

follows from similar triangles. ▯

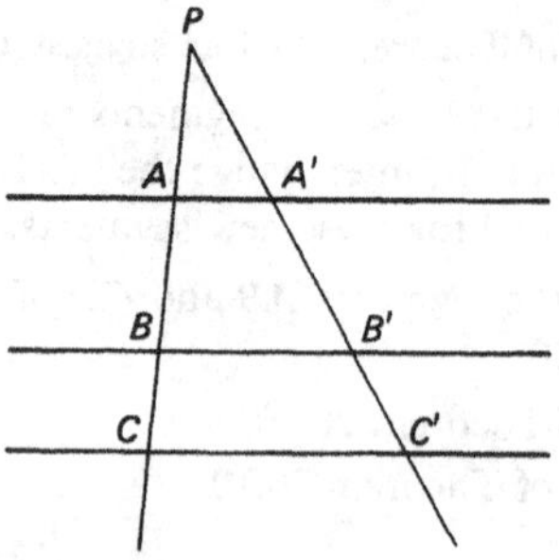

Figure 27.2

27.3 Corollary A line parallel to one side of a triangle cuts off proportional segments from the other two sides.

27.4 Theorem In trapezoid $ABCD$ with parallel sides AB and CD, let the diagonals meet at E. Then triangles ABE and CDE are similar.

Under the homothety $H(E, -AE/EC)$ point C maps to A, and D maps to a point on EB (Fig. 27.4). Since the line CD maps to the parallel through A, it follows that D maps to B. Hence the image of triangle CDE is triangle ABE, so $\triangle ABE \sim \triangle CDE$. ▯

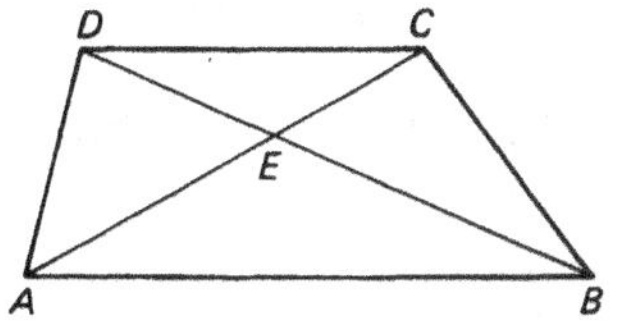

Figure 27.4

27.5 Theorem The common internal tangents of two nonintersecting circles meet on the circles' line of centers.

Let the tangents TT' and UU' meet at C as in Fig. 27.5. The homothety $H(C, -CT'/CT)$ maps one circle to the other. By Theorem 25.5, it maps one center to the other. But a point and its image are collinear with the center of the homothety. The theorem follows. ▯

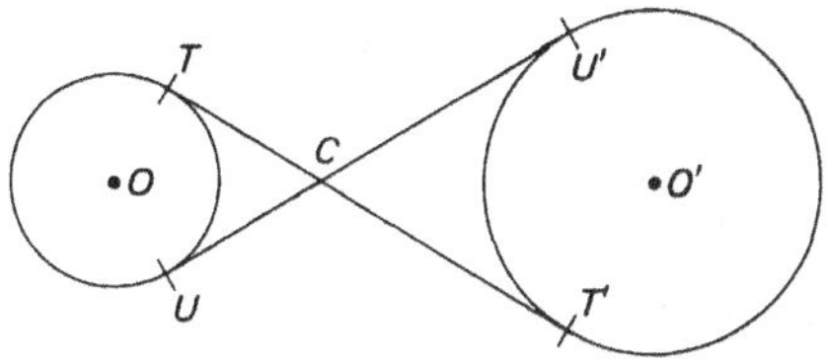

Figure 27.5

27.6 Theorem The figure formed by joining the midpoints of the sides of a square is a square having half the area of the given square.

Let O be the center of the square $ABCD$ (Fig. 27.6). Then O is the meeting of the medians $A'C'$ and $B'D'$. Since the diagonals and also the medians meet at right angles, and are bisected by the center, it follows that the similarity composed of a 45° rotation and a homothety of ratio OA/OA', both with center O, maps square $ABCD$ into quadrilateral $A'B'C'D'$. Since $OA/OA' = \sqrt{2}$, the theorem follows. ▯

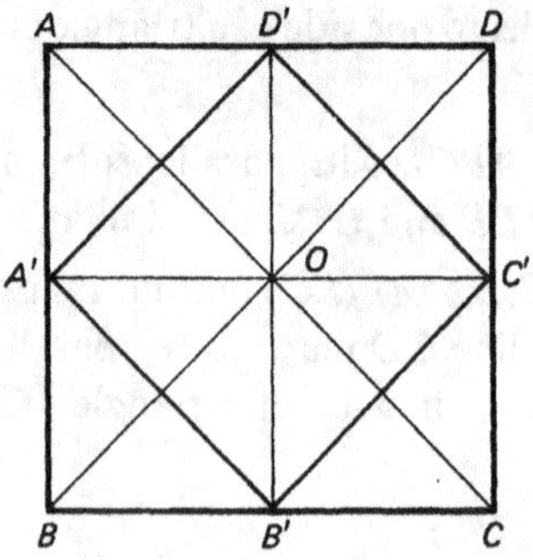

Figure 27.6

27.7 Problem Let P be a fixed point on a circle. Find the locus of the midpoints of all chords PA.

All such midpoints M lie on the homothetic image of the given circle under the homothety $H(P, 1/2)$ (Fig. 27.7). Hence their locus is a circle with radius half that of the given circle, and tangent to it internally at P. ▯

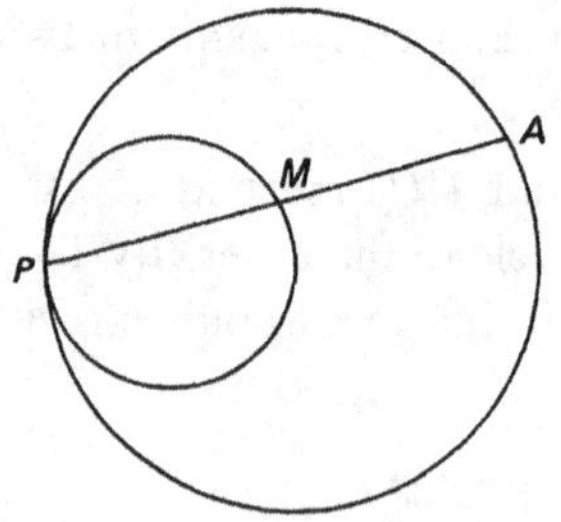

Figure 27.7

27.8 Theorem If PT is a tangent, and PAB a secant from an external point P to a circle, then $\mathbf{PA \cdot PB} = (PT)^2$.

Reflect triangle PAT in the internal bisector of angle P, and then apply the homothety $H(P, PB/PT)$ to the result. Then PT maps into PB, and A maps into a point on PT (Fig. 27.8). Since $\measuredangle PBT$ and $\measuredangle PTA$ are both measured by half of arc AT,

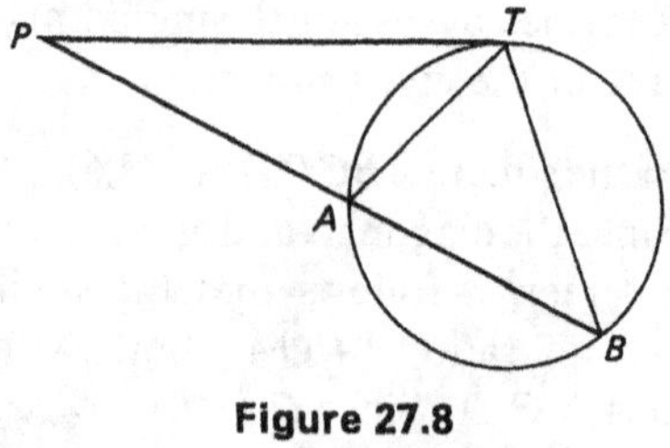

Figure 27.8

these angles are congruent. Thus the homothety and reflection map $\measuredangle PTA$ into $\measuredangle PBT$, and it follows that triangle PAT maps into triangle PTB. Now we have $PA/PT = PT/PB$, and the theorem follows. ▯

27.9 Theorem The altitude to the hypotenuse of a right triangle is the mean proportional between the segments into which it divides the hypotenuse.

Let CF be the altitude to the hypotenuse of right triangle ABC as shown in Fig. 27.9. In right triangle ACF, $\measuredangle ACF = 90° - \measuredangle A$. Hence $\measuredangle ACF = \measuredangle B$. Thus there is a similarity (a 90° rotation and homothety with center F) mapping triangle ACF to triangle CBF. Hence $AF/FC = FC/FB$, establishing the theorem. ▯

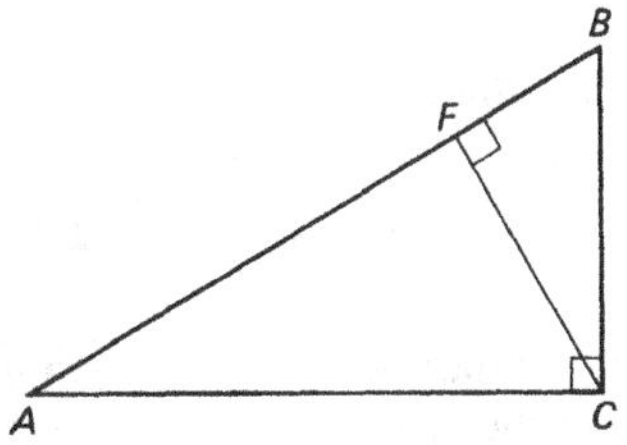

Figure 27.9

27.10 Theorem The median of a trapezoid has length equal to half the sum of the bases of the trapezoid.

Let M and N be the midpoints of the nonparallel sides of the trapezoid $ABCD$, so that MN is the median. Let the sides AD and BC meet at P (Fig. 27.10). Then $H(P, PM/PD)$ maps summit DC to median MN, and $H(P, PM/PA)$ maps base AB onto median MN. Hence

$$\frac{MN}{DC} = \frac{PM}{PD} \quad \text{and} \quad \frac{AB}{MN} = \frac{PA}{PM},$$

from which, by writing the first equation slightly differently, and then multiplying the two equations above side for side, we obtain

$$\frac{MN}{DC} = \frac{PD + DM}{PD} = 1 + \frac{DM}{PD} \quad \text{and} \quad \frac{AB}{DC} = \frac{PA}{PD} = \frac{PD + 2DM}{PD} = 1 + \frac{2DM}{PD}.$$

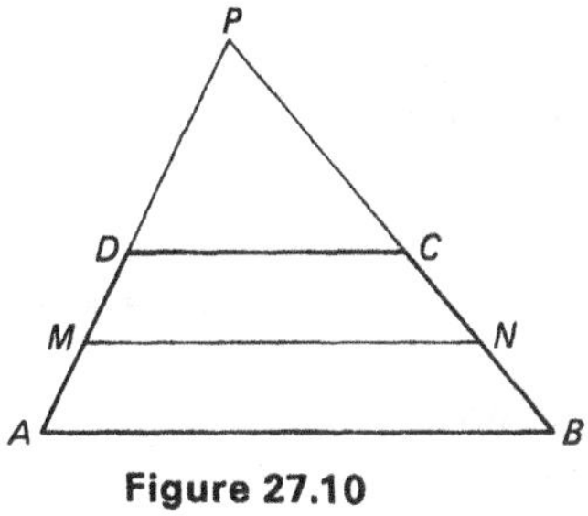

Figure 27.10

Solving each of these equations for DM/PD, obtain

$$\frac{DM}{PD}=\frac{MN}{DC}-1=\frac{1}{2}\left(\frac{AB}{DC}-1\right), \qquad \frac{MN}{DC}=\frac{1}{2}\left(\frac{AB}{DC}+1\right),$$

and finally,

$$MN=\tfrac{1}{2}(AB+DC). \ \square$$

Exercise Set 27

1. Prove Corollary 27.3.
2. Prove Theorem 27.6 by reflecting triangles $AA'D'$, $BB'A'$, $CC'B'$, and $DD'C'$ in their hypotenuses.
3. From a given point A, straight lines are drawn to points P on a given segment BC. Find the locus of all points M that divide these segments in a given ratio r.
4. Find the ratio of the areas of the two triangles ABE and CDE in the trapezoid of Fig. 27.4.
5. Prove that the diagonals of a trapezoid intersect at a point that divides them into proportional segments, and find the ratio of proportionality.
6. Prove that the segment joining the midpoints of the diagonals of a trapezoid has length $(b-s)/2$, when b and s are the lengths of the base and summit.
7. Chords AB and CD of a given circle meet at point E. Find the similarity that maps triangle ACE to triangle DBE.
8. Find the locus of the midpoints of all segments drawn to a circle from a fixed point P in the plane of the circle.
9. Prove that the common external tangents to two circles meet on their line of centers.
10. Two circles are tangent internally and have ratio 2:1 for their radii. Prove that a chord of the larger, from their point of tangency, is bisected by the smaller.
11. Two circles of radii 3 and 7 are tangent externally. Their common external tangents meet at P. Find the length of the tangent from P to the larger circle.
12. Given that PAB and PCD are any two secants from external point P to a circle, find the similarity that maps triangle PAC onto triangle PDB.
13. Extend side BA of rhombus $ABCD$ to a point E. Through E draw a parallel to BC, and through A a parallel to the diagonal BD, and let these parallels meet at F. Prove that triangle EAF is isosceles.
14. From point P on leg AC of right triangle ABC drop perpendicular PQ to the hypotenuse AB. Prove that triangles APQ and ABC are similar. State what similarity maps triangle APQ to ABC.
15. Drop perpendiculars from given points D and F on legs BC and CA of right triangle ABC to points E and G on the hypotenuse AB. Prove that triangles AFG and DBE are similar.
16. Medians AA' and BB' of triangle ABC meet at G. Prove that triangles ABG and $A'B'G$ are similar, and find their similarity map.

SECTION 28 | FURTHER ELEMENTARY APPLICATIONS

28.1 Theorem Let A be the midpoint of arc CD of a given circle. Let chord AB cut chord CD at M. Then $AM \cdot AB$ is independent of the location of B on the circle.

In Fig. 28.1 angles ACD and CBA are congruent, since they are measured by halves of the congruent arcs AD and CA. Thus triangles AMC and ACB are similar, so there is a similarity mapping one triangle onto the other. Then $AM/AC = AC/AB$, so $AM \cdot AB = (AC)^2$, a constant. ▯

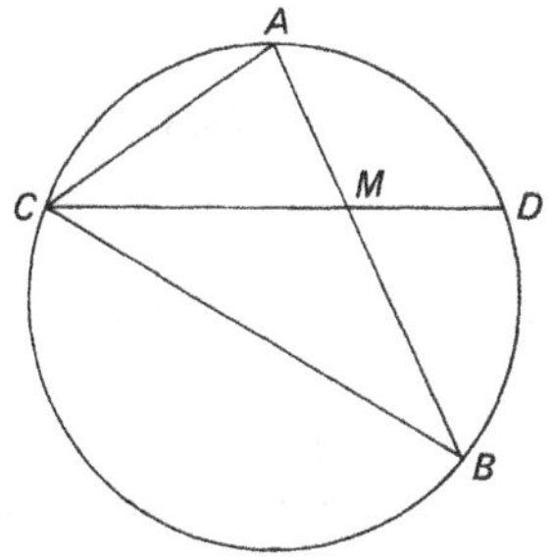

Figure 28.1

28.2 Theorem The triangles ACD and BCE formed from a triangle ABC by altitudes AD and BE are similar.

Since these right triangles share an acute angle at C (Fig. 28.2), it follows that there is a similarity (a reflection in the bisector of angle C, and a homothety with center C) that maps these triangles one to the other. ▯

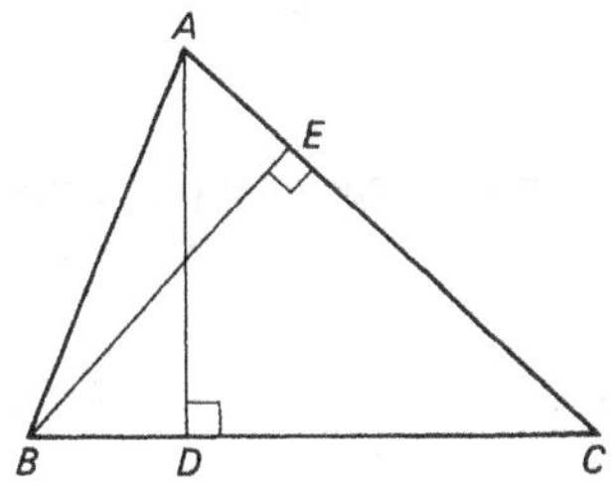

Figure 28.2

28.3 Problem Let P be a variable point on the semicircle of diameter AOB. Let the perpendicular from point B to the tangent at P meet line AP at point X. Find the locus of X.

Since OP is perpendicular to the tangent at P, then OP and BX are parallel as shown in Fig. 28.3. Hence $H(A, 2)$ maps OP to BX. Then the locus of X is the homothetic image of the semicircle on which P lies, a semicircle of center B and radius BA. ▯

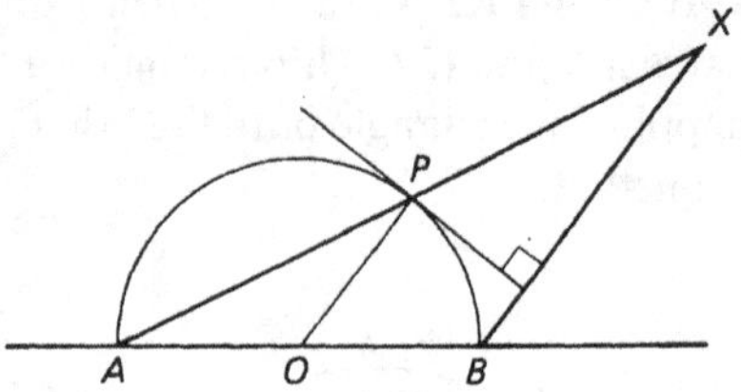

Figure 28.3

28.4 Theorem Let perpendiculars erected at arbitrary points on the sides of triangle ABC meet in pairs at points P, Q, R. Then triangle PQR is similar to the given triangle.

See Fig. 28.4. A 90° rotation of triangle ABC into triangle $A'B'C'$ leaves the sides of triangle $A'B'C'$ parallel to the corresponding sides of triangle PQR. By Theorem 25.7 an appropriate homothety will map triangle $A'B'C'$ into triangle PQR. The theorem follows. ▯

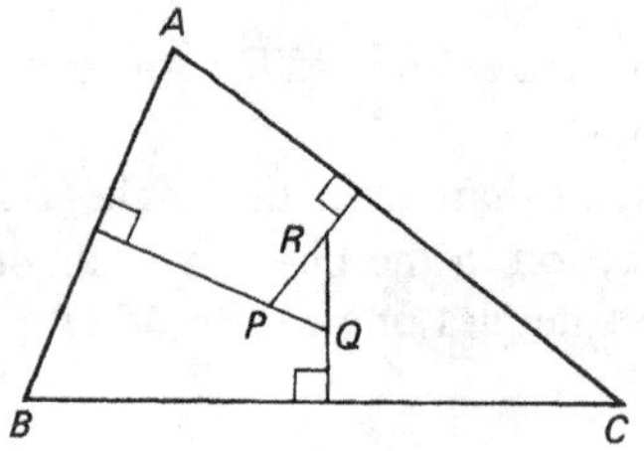

Figure 28.4

28.5 Theorem Let AB, CD, and EF each be perpendicular to BD so that AD and BC meet at E, and points A and C lie on the same side of BD. Then EF is half the harmonic mean of AB and CD.

Recall that the harmonic mean of a and b is $2ab/(a+b)$. Let $AB = a$, $CD = b$, $EF = x$, $BF = u$, and $FD = v$ (Fig. 28.5). Then $H(B, u/(u+v))$ maps triangle BCD to BEF, so $x/b = u/(u+v)$. Also $H(D, v/(u+v))$ maps triangle DAB to DEF, so $x/a = v/(u+v)$. Then we have

$$\frac{x}{a} + \frac{x}{b} = \frac{v}{u+v} + \frac{u}{u+v} = 1 \qquad \text{and} \qquad x = \frac{ab}{a+b},$$

so EF is half the harmonic mean of AB and CD. ▯

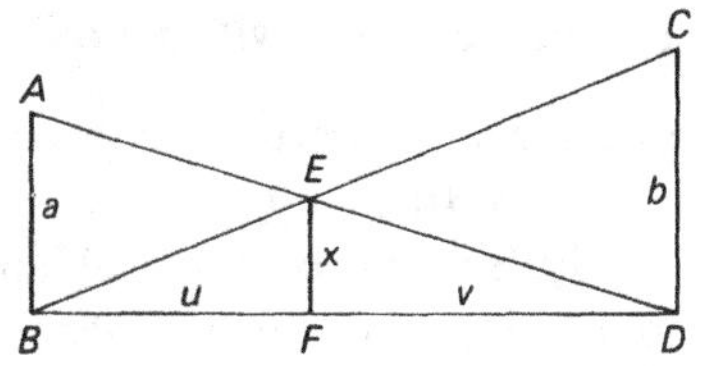

Figure 28.5

28.6 Theorem The bisector of an internal angle of a triangle divides the opposite side into segments proportional to the adjacent sides.

Let AU bisect angle A of triangle ABC (see Fig. 28.6). Let $H(B, BC/BU)$ map triangle BAU to triangle BPC. Then $\measuredangle BAU \cong \measuredangle APC$. It follows also that $BU/UC = BA/AP$. Furthermore, since AU is parallel to PC, then

$$\measuredangle PCA \cong \measuredangle CAU \cong \measuredangle BAU \cong \measuredangle APC,$$

so triangle APC is isosceles and $AP \cong AC$. Hence $BU/UC = BA/AC$. ▯

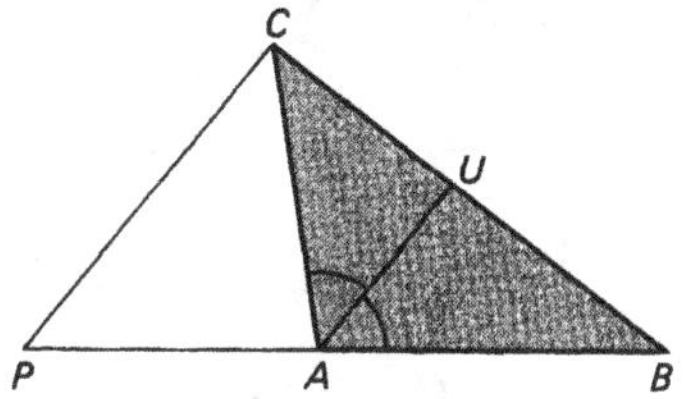

Figure 28.6

28.7 Theorem Given that the perpendicular bisectors of the sides concur, then the altitudes of a triangle concur.

Let $H(G, -2)$ map triangle ABC into triangle $A_1B_1C_1$ (Fig. 28.7). Then vertices A, B, C become the midpoints of the sides of triangle $A_1B_1C_1$, and since the corresponding sides are parallel, the altitudes AD, BE, CF become the perpendicular bisectors of the sides of triangle $A_1B_1C_1$. The theorem follows, since we assumed that the perpendicular bisectors of the sides of a triangle concur. ▯

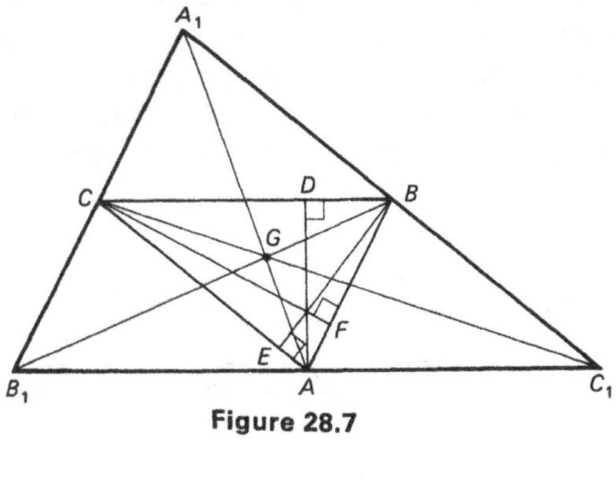

Figure 28.7

28.8 Compare the proof of Theorem 28.7 with Corollary 20.8.

28.9 The method of homothety is quite useful for constructions. The basic principle is illustrated in the next two items. In each case it is not possible to satisfy all the required conditions immediately. So we solve the problem partially by omitting one of the conditions. Then a homothety maps the partial solution into a complete solution. One is reminded of a gambit in chess, where a player gives up a piece to his opponent in order to gain a winning position.

28.10 Problem Locate points D and E on sides AB and AC of a given triangle ABC so that $BD \cong DE \cong EC$.

Arbitrarily choose points D' and E' on sides AB and AC (Fig. 28.10) so that $BD' \cong E'C$. [Here we give up the condition that $D'E'$ shall be congruent to the other two lengths.] Let the circle $D'(B)$ cut the parallel to BC through E' within the triangle

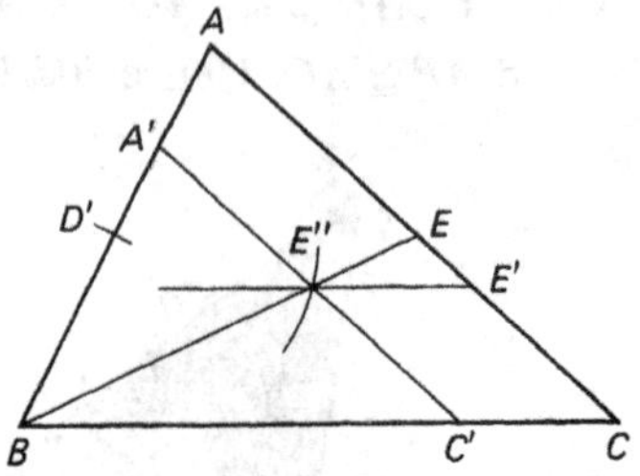

Figure 28.10

at E''. Draw through E'' a parallel $A'C'$ to AC, and observe that triangles ABC and $A'BC'$ are similar and that the problem is solved for triangle $A'BC'$. [We have regained now the condition $BD' \cong D'E'' \cong E''C'$ at the expense of the size of the triangle.] The homothety $H(B, BC/BC')$ maps this solution onto triangle ABC. This map may be accomplished by drawing BE'' to cut AC at E, one of the desired points. ▯

28.11 Problem Inscribe a square in a given semicircle.

In the solved problem, clearly the center O of the semicircle is the midpoint of one side of the square. Let us keep this condition and temporarily overlook the condition that two vertices must lie on the semicircle. On the diameter AOB of the semicircle, construct any convenient square $P'Q'R'S'$ with O the midpoint of the side $P'Q'$ lying on AB (Fig. 28.11). Now, with O as center, project vertices R' and S' onto

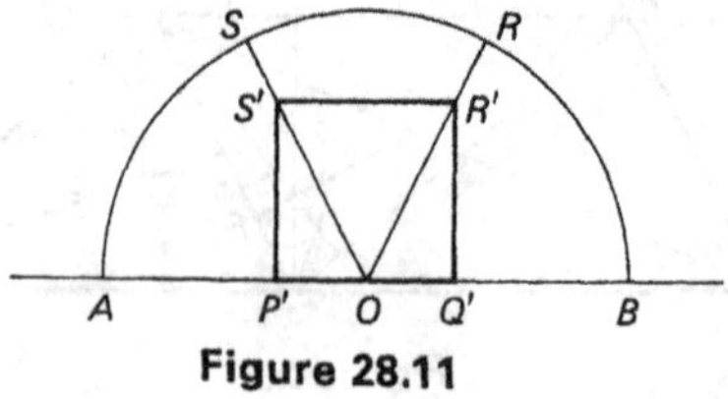

Figure 28.11

the semicircle to the desired points R and S. The other two vertices P and Q are merely the feet of the perpendiculars to AB from S and R. ▯

28.12 Compare the construction of Problem 28.11 with that of Problem 8.13.

Exercise Set 28

1. Chord AB is congruent to the radius of a given circle with center O, OM is perpendicular to AB at M, and D is the foot of the perpendicular dropped from M to OA. Find the area of triangle MDA.
2. In a circle chords AB and AC are congruent. Chord AD cuts segment BC at E. Prove triangles ABD and AEB are similar.
3. Three circles with centers A, B, C pass through O. Diameters OAA', OBB', OCC' are drawn. Prove that the sides of triangle $A'B'C'$ pass through the other points of intersection of the circles.
4. The base and summit of a trapezoid have lengths b and s. Find the ratio in which a line segment MN divides the altitude of the trapezoid when MN is parallel to the base, terminates on the sides, and has length m.
5. Take any point F on side AB of parallelogram $ABCD$, and let DF meet diagonal AC at E and side CB (extended) at G. Prove that $(DE)^2 = EF \cdot EG$.
6. Prove that the angle between the base of an isosceles triangle and the altitude to one of the congruent sides is congruent to half the vertex angle.
7. In isosceles triangle ABC with $AB \cong AC$, let the altitude from B meet the perpendicular to BC at C in point D. Take point E on BD so that $ED \cong EC$. Prove that triangles ABC and ECD are similar.
8. Take points X and Y on sides AB and AC of triangle ABC so that XY is parallel to BC. From the midpoint A' of side BC draw line $A'X$ to meet CA at M, and line $A'Y$ to meet BA at N. Prove that MN is parallel to BC.
9. Prove that the bisector of an external angle of a triangle divides the opposite side externally in segments proportional to the adjacent sides.
10. Take points D, E, F on sides BC, CA, AB of equilateral triangle ABC so that $BD \cong CE \cong AF$. Prove that triangle DEF is equilateral.
11. In triangle ABC, $\measuredangle A \cong \measuredangle B \cong 2\measuredangle C$ and AU is the bisector of angle A. Prove that $(AB)^2 = BU \cdot BC$.
12. Construct a triangle ABC given the measures of $\measuredangle A$, $a + c$, and $a + b$.
13. Construct a square so that one side lies along the base line BC of a given triangle ABC and its other two vertices lie on rays BA and CA outside the triangle.
14. In a given triangle inscribe a triangle homothetic to another given triangle.
15. In a given triangle inscribe a parallelogram homothetic to a given parallelogram.
16. Inscribe a square in a given circular sector.
17. Inscribe a square in a given triangle.
18. Inscribe a rectangle similar to a given rectangle:
    a) in a given circle
    b) in a given semicircle
    c) in a given triangle
    d) in a given square, so that one vertex of the rectangle lies on each side of the square.

SECTION 29 | ADVANCED APPLICATIONS

29.1 Theorem The medians of a triangle concur.

Since the medial triangle $A'B'C'$ has sides parallel to those of the given triangle ABC, there is a homothety mapping one triangle to the other. The center of the homothety is the point of concurrence of the lines joining corresponding vertices, that is, the medians of triangle ABC. □

29.2 Theorem The orthocenter H, the circumcenter O, and the centroid G of a triangle, all lie on a line called the *Euler line* of the triangle, and $\mathbf{HG} = 2\mathbf{GO}$.

Theorem 28.7 showed that $H(G, -\frac{1}{2})$ maps triangle ABC and its orthocenter H into triangle $A'B'C'$ and its orthocenter which is the circumcenter O for triangle ABC. Hence the theorem. □

29.3 Corollary The distance from the circumcenter to a side of a triangle is half the distance from the orthocenter to the opposite vertex.

29.4 Theorem *The ninepoint circle theorem.* The midpoints A', B', C' of the sides, the feet D, E, F of the altitudes, and the midpoints N_a, N_b, N_c of the segments joining the orthocenter H to the vertices of triangle ABC, all lie on a circle whose center N is the midpoint of the Euler segment HO.

We show that the homothety $H(H, 2)$ maps the nine listed points onto the circumcircle, from which the theorem all follows (Fig. 29.4).

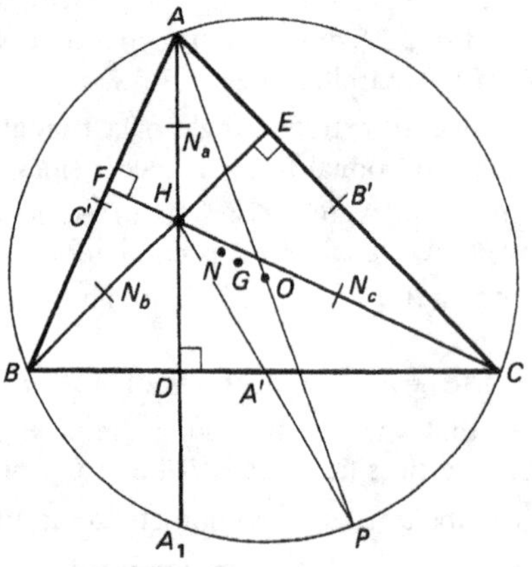

Figure 29.4

Let diameter AO cut the circle again at P. Then $\measuredangle ABP = \measuredangle ACP = 90°$, since each angle is inscribed in a semicircle. Now PC and BE are both perpendicular to AC, and BP and FC are both perpendicular to AB. Thus $BPCH$ is a parallelogram, so its diagonals bisect one another; that is, HP passes through A' and $HP = 2HA'$. Hence the homothety $H(H, 2)$ maps A' to P on the semicircle. Similarly B' and C' are also mapped onto the circumcircle by $H(H, 2)$.

Let altitude AD cut the circumcircle again at A_1. Then

$$\sphericalangle A_1CB = \sphericalangle A_1AB = 90° - \sphericalangle ABC = \sphericalangle BCF,$$

since the first two angles are each inscribed in arc BA_1, and the last equalities are seen from right triangles BAD and BCF. Since CD is perpendicular to HA_1 and $\sphericalangle A_1CD \cong \sphericalangle DCH$, triangle CHA_1 is isosceles, so its altitude CD bisects its base HA_1. Hence $H(H, 2)$ maps D to A_1. Similarly the other feet E and F of the altitudes to triangle ABC are mapped to the circumcircle by $H(H, 2)$.

Clearly $H(H, 2)$ maps N_a, N_b, N_c to A, B, C on the circumcircle. Therefore the nine points A', B', C', D, E, F, N_a, N_b, N_c all lie on a circle half the size of the circumcircle, and with center of homothety H. Thus $H(H, 2)$ also maps the center N of this ninepoint circle to the center O of the circumcircle. That is, N lies midway between H and O. ▯

29.5 Theorem The altitude to the hypotenuse of a right triangle divides the hypotenuse into segments proportional to the squares of the adjacent sides.

Let CF be the altitude to the hypotenuse of right triangle ABC. Since triangles ACF and CBF are similar (shown in Fig. 27.9), then

$$\frac{AC}{AF} = \frac{CB}{CF} \quad \text{and} \quad \frac{AC}{CF} = \frac{CB}{BF}.$$

By multiplying these equations side for side, obtain

$$\frac{AC^2}{AF} = \frac{CB^2}{BF}. \ ▯$$

29.6 Problem Construct a right triangle given its perimeter and the ratio of the squares of its legs.

Let p denote the perimeter and m/n the given ratio as shown in Fig. 29.6a. On a base line mark segments $A'F' = m$ and $F'B' = n$, and draw a semicircle whose center shall be O on $A'B'$ as diameter (Fig. 29.6b). Let the perpendicular to $A'B'$ at F' cut

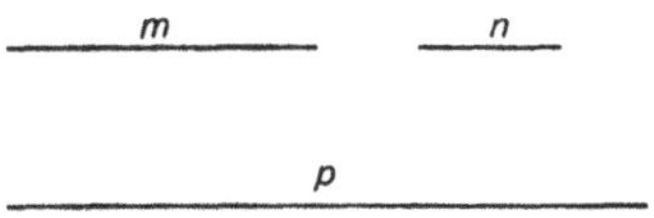

Figure 29.6a

the semicircle at C'. Then triangle $A'B'C'$ is similar to the desired triangle, by Theorem 29.5. On the tangent at B' to the semicircle, mark $B'P'$ equal to the perimeter of triangle $A'B'C'$, and also $B'Q = p$. Let the perpendicular to $B'Q$ at Q meet OP' at P. The parallel to $B'P'$ through P cuts line $A'B'$ at B. Points A and C are located homothetic to A' and C' in center O and lying on the circle $O(B)$. The proof of the construction is left as an exercise. ▯

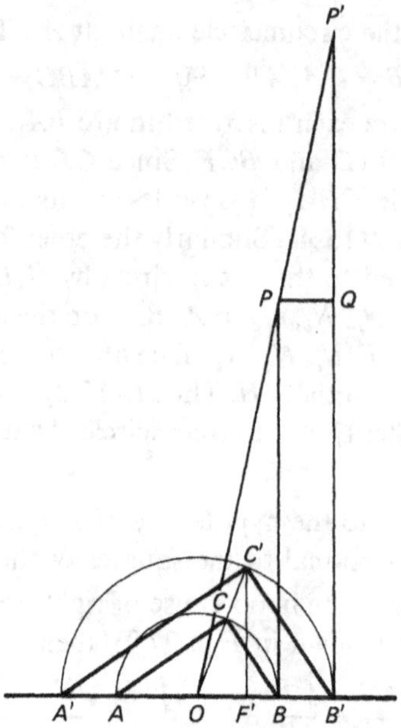

Figure 29.6b

29.7 Problem Construct a triangle given its angles and its area m^2.

Given its angles, one can construct a similar triangle $A'B'C'$, and its altitude $A'D'$, as shown in Fig. 29.7a, along with the given length m. Now $\frac{1}{2}BC \cdot AD = m^2$,

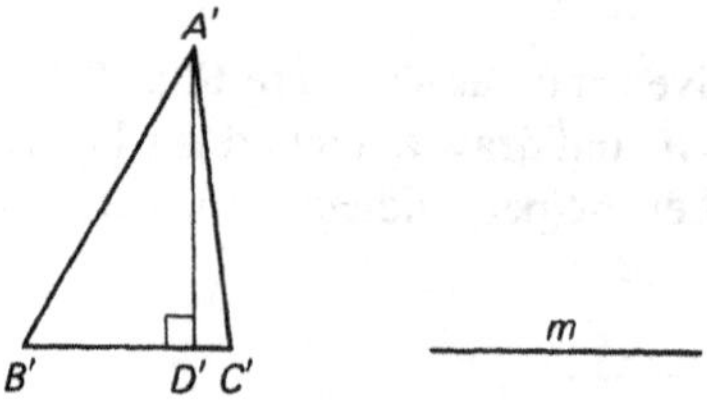

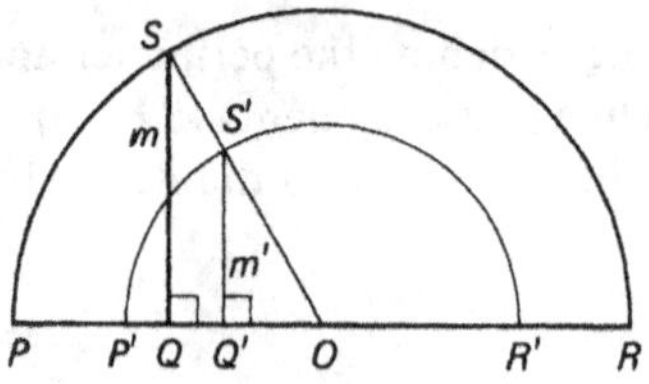

Figure 29.7a

Figure 29.7b

so m is the mean proportional between $\frac{1}{2}BC$ and AD. Construct the mean proportional m' between $\frac{1}{2}B'C'$ and $A'D'$ on a line segment $P'Q'R'$, where $P'Q' \cong \frac{1}{2}B'C'$ and $Q'R' \cong A'D'$ (Fig. 29.7b). Draw the semicircle having center O on $P'R'$ as diameter. Let the semicircle cut the perpendicular to $P'R'$ at Q' in point S'. Then $Q'S'$ is the mean proportional m' between $\frac{1}{2}B'C'$ and $A'D'$. Locate S on OS' so that $QS = m$ and QS is perpendicular to $P'R'$ at Q. Then QS and $Q'S'$ are homothetic in center O. The circle with center O and radius OS cuts the line $P'R'$ at points P and R so that $PQ \cong \frac{1}{2}BC$ and $QR \cong AD$. The desired triangle is easily constructed by marking A on $A'D'$ so that $AD' \cong QR$. Then AB and AC are drawn parallel to $A'B'$ and $A'C'$. ▯

29.8 Problem Through a given point P draw a line to pass through the inaccessible point of intersection of two given lines m and n.

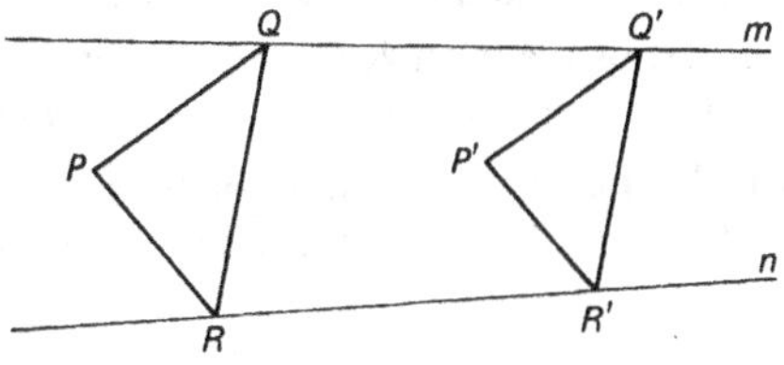

Figure 29.8

Draw two lines PQ and PR through P so that Q lies on m, and R on n, as in Fig. 29.8. Now draw $Q'R'$ from m to n and parallel to QR. Draw lines parallel to PQ and PR through Q' and R' to meet at P'. Then triangles PQR and $P'Q'R'$ are homothetic, so line PP' passes through the point of intersection of QQ' and RR'. ▯

Exercise Set 29

1. In triangle ABC, show that G and H divide NO in the same ratios internally and externally, ratios $\frac{1}{3}$ and $-\frac{1}{3}$.
2. In triangle ABC, show that $H(G, -2)$ also maps the ninepoint circle to the circumcircle, hence H and G are the two centers of homothety for these circles.
3. Prove that the four Euler lines of the four triangles of an orthocentric quadrangle $ABCH$ are concurrent. Find their point of concurrence.
4. Prove Corollary 29.3.
5. Prove the construction of Problem 29.6.
6. Prove the construction of Problem 29.7.
7. Through one of the two points of intersection of two circles, draw a line on which the two circles intercept congruent chords (but not the common chord of the two circles).
8. Through one of the points of intersection of two circles, draw a line on which the two circles intercept chords having a given ratio r.
9. Given three concentric circles, draw a secant so that the segment between the first and second circles is congruent to that between the second and third circles.
10. Construct a triangle, given an angle, the length of the bisector of this angle, and the ratio of the two segments into which this bisector divides the opposite side.
11. Construct a trapezoid given the nonparallel sides, the angle between them, and the ratio of the parallel sides.
12. Construct a square given the sum of its side and diagonal.
13. Construct a right triangle, given the perimeter and the ratio of its legs.
14. Prove that the four centroids of an orthocentric quadrangle form an orthocentric quadrangle homothetic to the given one.
15. Prove that the ninepoint circle of an orthocentric quadrangle is concentric with that of the orthocentric quadrangle formed from the four centroids.
16. Prove that the four points of an orthocentric quadrangle are the centroids of another homothetic orthocentric quadrangle. [This is the converse of Exercise 29.14.]

17. The circumcenters and the centroids of an orthocentric quadrangle form two orthocentric quadrangles having the same ninepoint center, the center of similitude of these quadrangles. Prove this theorem and find the ratio of similitude.
18. Prove that the orthocenter of the triangle formed by the centroids of triangles HBC, HCA, HAB (H is the orthocenter of triangle ABC) is the centroid of triangle ABC.
19. Prove that the locus of the third vertex of all triangles directly similar to a given triangle and having the first vertex at a fixed point and the second vertex along a straight line, is a straight line.
20. If the second vertex of the triangle of Exercise 29.19 is to lie on a circle instead of a straight line, prove that the locus of the third vertex is a homothetic circle.
21. Construct a triangle similar to a given triangle, having one vertex at a fixed point and the other two vertices lying on two fixed lines.
22. Construct a triangle similar to a given triangle and having its three vertices lying on three given lines.
23. Prove that the median AA' and the segment $B'C'$ joining the midpoints of sides CA and AB of a triangle ABC bisect each other.
24. Find the locus of the centroid of a triangle having one side and the circumcircle fixed.
25. Prove that the product of two sides of a triangle is equal to the product of the circumdiameter and the altitude to the third side. [This is Theorem 6.20.]

▶SECTION 30 ANALYTIC REPRESENTATIONS OF SIMILARITIES

30.1 Theorem The homothety $H(O, k)$ centered at the origin $O(0, 0)$ is determined by the equations

$$\begin{cases} x' = kx \\ y' = ky. \end{cases}$$

Clearly the indicated equations multiply every distance from the origin by the factor $|k|$. □

30.2 By translating from point $P(a, b)$ to the origin, applying the homothety $H(O, k)$, then translating back to P again, we obtain the equations for a homothety with ratio k and center P:

$$\begin{cases} x' = k(x - a) + a \\ y' = k(y - b) + b, \end{cases}$$

which reduce to those indicated in Theorem 30.3.

30.3 Theorem A homothety $H(P, k)$ centered at point $P(a, b)$ has the analytic representation

$$\begin{cases} x' = kx + (1 - k)a \\ y' = ky + (1 - k)b, \end{cases}$$

which is a homothety centered at the origin followed by a translation.

30.4 We see that each homothety can be written as a homothety centered at the origin followed by a translation. Observe also that, when k is factored from each right-hand side, the homothety can be written as a translation followed by a homothety centered at the origin:

30.5 Theorem A homothety $H(P, k)$ centered at point $P(a, b)$ has the analytic representation

$$\begin{cases} x' = k(x + a/k - a) \\ y' = k(y + b/k - b), \end{cases}$$

which is a translation followed by a homothety centered at the origin.

By Exercise 22.14, every isometry has an equation of the form

$$\begin{cases} x' = ax - by + c \\ y' = \pm(bx + ay) + d \qquad \text{where } a^2 + b^2 = 1, \end{cases}$$

direct if the plus sign holds, and opposite if the minus sign holds. Each similarity, then, being the product of a homothety and an isometry, can be written in the following form.

30.6 Theorem Each similarity can be written analytically in the form

$$\begin{cases} x' = ax - by + c \\ y' = \pm(bx + ay) + d, \end{cases}$$

where $(a^2 + b^2)^{1/2} = k$, the ratio of the homothety. It is direct if the plus sign holds, and opposite if the minus sign holds.

30.7 Theorem Every set of equations of the form given in Theorem 30.6 represents a similarity, provided $a^2 + b^2 \neq 0$.

30.8 Example Find equations for the two similarities that map the segment OA to the segment BC where $O(0, 0)$, $A(1, 0)$, $B(2, 3)$, and $C(5, 7)$. For the direct isometry as written in Theorem 30.6, obtain the equations

$$\begin{matrix} 2 = c \\ 3 = d \end{matrix} \quad \text{and} \quad \begin{matrix} 5 = a + c \\ 7 = b + d, \end{matrix}$$

which yield

$$\begin{cases} x' = 3x - 4y + 2 \\ y' = 4x + 3y + 3. \end{cases}$$

For the opposite isometry of Theorem 30.6, we have

$$\begin{matrix} 2 = c \\ 3 = d \end{matrix} \quad \text{and} \quad \begin{matrix} 5 = a + c \\ 7 = -b + d, \end{matrix}$$

which yield

$$\begin{cases} x' = 3x + 4y + 2 \\ y' = 4x - 3y + 3. \end{cases}$$

Exercise Set 30

1. Prove Theorem 30.1.
2. Show algebraically the equivalence of the forms of Theorems 30.3 and 30.5.
3. Show geometrically the equivalence of the forms of Theorems 30.3 and 30.5.
4. Prove Theorem 30.6.
5. Prove Theorem 30.7.
6. Find the value of the ratio k in the similarities of Example 30.8.
7. Find the angle of rotation for each of the similarities of Example 30.8.
8. Write the similarities of Theorem 30.6 analytically as products of (1) a homothety centered at the origin, (2) a rotation about the origin, (3) a reflection if necessary, and (4) a translation.
9. Show that the product of two similarities of the forms given in Theorem 30.6 is another similarity of that same form.
10. Find the direct similarity that maps segment AB to segment $A'B'$ where these four points have the coordinates:
    a) $A(1, 0)$, $B(2, 0)$, $A'(1, 0)$, $B'(2, 0)$
    b) $A(1, 0)$, $B(2, 0)$, $A'(2, 0)$, $B'(1, 0)$
    c) $A(1, 0)$, $B(2, 3)$, $A'(-1, 2)$, $B'(-3, -3)$
    d) $A(1, 0)$, $B(2, 3)$, $A'(-3, -3)$, $B'(-1, 2)$
11. Find the opposite similarity that maps each segment AB to $A'B'$ for the sets of points given in Exercise 30.10.
12. Find equations for the similarity that carries triangle ABC, where $A(0, 0)$, $B(1, 0)$, and $C(0, 2)$, into triangle $A'B'C'$, where
    a) $A'(3, 0)$, $B'(3, 2)$, $C'(7, 0)$
    b) $A'(\frac{9}{2}, 4)$, $B'(4, 4)$, $C'(\frac{9}{2}, 5)$
    c) $A'(0, 0)$, $B'(3, 0)$, $C'(0, 6)$
    d) $A'(-3, -2)$, $B'(-4, -2)$, $C'(-3, -4)$
    e) $A'(-5, 5)$, $B'(-6, \frac{7}{2})$, $C'(-2, 3)$

4 VECTORS AND COMPLEX NUMBERS IN GEOMETRY

SECTION 31 THE SEARCH FOR THE MEANING OF COMPLEX NUMBERS

31.1 Throughout the history of mathematics it seems that the rule has been to use each new entity for a few hundred years before deciding just what it is you are using. The early thinking by mathematicians about complex numbers was not unlike the muddy ideas expressed in other areas of mathematics, but occasionally it was dotted with a bit of insight.

31.2 In an attempt to justify using negative, irrational, and imaginary numbers, the English mathematician George Peacock (1791–1858) stated his absurd "principle of permanence of forms," that "equal general expressions of arithmetic are to remain equal when the letters no longer denote simple quantities, and hence also when the operation is changed"! He was trying to state that the same rules applied to negatives, etc., as apply to positive rational numbers. For example, since $a+b=b+a$ whenever a and b are positive rational numbers, then $a+b=b+a$ should hold for all numbers.

Such fuzzy thinking never really contributed to the advancement of mathematics. And especially not when done as recently as the nineteenth century.

31.3 The earliest glimmerings of the existence of imaginary numbers appeared some 1100 years ago, when the Hindu Mahavira made the very intelligent statement that "a negative number has no square root." Furthermore, he wisely made no attempt to work with these nonexistent square roots of negative numbers.

31.4 Luca Pacioli (1445–1509) stated the rule that the product of two negative numbers is a positive number, but used that fact only rarely. And to the student with little understanding of negative numbers, this rule is far from obvious. He is likely to feel that a product such as $(-2) \times (-3)$ means (not 2) $\times$ (not 3), and if you multiply something that is *not* 2 by something *not* 3, then the result certainly is *not* 6. Thus he concludes that $(-2) \times (-3) = -6$.

31.5 Rafael Bombelli (1526–1573) improved algebraic notation, wrote roots of cubic equations as sums of imaginary numbers, and formulated rules for handling imaginaries.

31.6 René Descartes (1596–1650) stated that a polynomial equation of degree n has no more than n roots, and it was he who applied those unfortunate words "real" and "imaginary" to numbers. At this time imaginaries were considered to be "uninterpretable and even self-contradictory." That they were used with ever-increasing faith is one of the great mysteries of human nature.

31.7 Roger Cotes (1682–1716), by simple formalism, gave the forerunner to the famous theorem of De Moivre. The first real bit of mathematics of complex numbers thus was born. After Cotes' death Newton lamented, "If Cotes had lived we should have learnt something."

31.8 This brings us to Abraham De Moivre (1667–1754), whose famous theorem states

$$(\cos\theta + i\sin\theta)^n = \cos n\theta + i\sin n\theta$$

for n a positive integer, introducing complex numbers into trigonometry. It is said that he died of an arithmetic progression. Suddenly, at the ripe old age of 87, he found that he was requiring some 15 minutes more sleep each night than the night before. When the terms of this progression reached 24 hours, he expired in his sleep.

31.9 A firm believer in the manipulation of formulas, Leonhard Euler (1707–1783) was the first to state De Moivre's theorem in its present form, and to extend it to all values of n. It was in 1777 that he suggested i for $\sqrt{-1}$ and called $a^2 + b^2$ the *norm* of $a + bi$. Euler was a rather ordinary person. For a mathematician. When in Berlin, he answered the Queen's questioning only in monosyllables. When she asked why, he replied that it was because he had "just come from a country where every person who speaks is hanged."

31.10 In order to give some sort of meaning to complex numbers, Caspar Wessel (1745–1818), a Norwegian who spent much of his life as a surveyor for the Danish Academy of Sciences, formulated their geometric interpretation in the plane. He interpreted addition and multiplication as translation and rotation (see Sections 33 and 34). His very complete paper appeared in 1799, two years after his discovery. Unfortunately it lay unnoticed by the mathematical world for nearly 100 years.

31.11 As so often occurs in the history of science, another person, the bookkeeper Jean Robert Argand (1768–1822) of Geneva, arrived at the same interpretation in 1806. His paper, although not as clear as Wessel's, was widely read. Hence the plane of complex numbers is called an *Argand diagram*.

31.12 Now, simply replacing coordinates (a, b) of points in the plane by complex numbers $a + bi$ provides a nice picture, aids the imagination, and has worthwhile applications in geometry, trigonometry, and electrical engineering, but in no way does it prove that complex numbers exist in algebra. There was still no justification for using an imaginary number as a root of an equation. But the nineteenth century marked the beginning of efforts to solve the problems in the foundations of mathematics. So the time was ripe for someone to find the true meaning of complex numbers in mathematics, and again two men, independently of one another, provided identical answers.

31.13 In 1825 Carl Friedrich Gauss (1777–1855), the greatest mathematician of modern times, stated that "the true metaphysics of $\sqrt{-1}$ is illusive." Thus he recognized the problem at hand, a giant step forward! Since his doctoral dissertation was a proof of the fundamental theorem of algebra—that each polynomial with complex coefficients has at least one complex zero—he was quite well versed in imaginary numbers.

31.14 In 1831 he found the "true metaphysics of $\sqrt{-1}$." From a basis in the real number system he defined an algebra of ordered pairs (a, b) wherein equality, addition, and multiplication are given by

$$(a, b) = (c, d) \quad \text{iff} \quad a = c \text{ and } b = d,$$

$$(a, b) + (c, d) = (a + c, b + d),$$

and

$$(a, b)(c, d) = (ac - bd, ad + bc).$$

In the resulting algebra one proves quite readily that these ordered pairs (a, b) behave like complex numbers $a + bi$. Thus these ordered pairs of real numbers *are* complex numbers. The mathematician is now satisfied of their existence. This is the "true metaphysics of $\sqrt{-1}$."

31.15 Gauss did not publish his work, not unusual for him, and quite independently the Irishman William Rowan Hamilton (1805–1865) performed the same feat in 1835 and published his work. Later Gauss claimed the earlier discovery, so the complex plane is sometimes called the *Gauss plane*.

31.16 It would seem that the ordered pair notation (a, b) is nothing more than a thin disguise for $a + bi$, but the algebraic theory of the former is rigorous, eliminating that mysterious entity i, which stands for the (nonexistent?) $\sqrt{-1}$. How many high school graduates even today believe that "i stands for the square root of -1, a number you cannot take the square root of"? Even as late as 1873, the Larousse dictionary stated that imaginary numbers are "impossible" and that algebra up to then had found "only two impossible entities: the negative and the imaginary." It is a curiosity that "the imaginary" came to be understood before "the negative."

Exercise Set 31

1. a) From the two equations

$$\sqrt[3]{a + \sqrt{-b}} = p + \sqrt{-q} \quad \text{and} \quad \sqrt[3]{a - \sqrt{-b}} = p - \sqrt{-q},$$

Bombelli obtained

$$\sqrt[3]{a^2 + b} = p^2 + q.$$

Show this.

b) Use part (a) to find $p^2 + q$ when $a + \sqrt{-b} = 2 + \sqrt{-121}$. Then solve for q and substitute that value into the equation $(p + \sqrt{-q})^3 = 2 + \sqrt{-121}$, obtaining $4p^3 - 15p = 2$, which has a root $p = 2$. Now show that

$$\sqrt[3]{2 + \sqrt{-121}} + \sqrt[3]{2 - \sqrt{-121}} = 4.$$

2. Prove De Moivre's Theorem (see 31.8).
3. Assuming De Moivre normally required 8 hours of sleep, how many days was it before he expired after his progression started (see 31.8)?
4. Using Gauss' definitions for equality, addition, and multiplication as given in 31.14, show that
   a) $(a, b) + (0, 0) = (a, b)$
   b) $(a, b)(1, 0) = (a, b)$
   c) $(a, b) + (-a, -b) = (0, 0)$
   d) $(a, b)(a/(a^2 + b^2), -b/(a^2 + b^2)) = (1, 0)$ when $(a, b) \neq (0, 0)$
   e) $(0, 1)(0, 1) = (-1, 0)$
5. Interpret the ordered pairs (a, b) of Exercise 31.4 as complex numbers $a + bi$, and show that each equation in that exercise is true for complex numbers. Especially, what is the interpretation for part (e)?

SECTION 32 | INTRODUCTION TO COMPLEX NUMBERS

32.1 Many educated people retain a subtle distrust of complex numbers. They have not overcome the feeling that "i is a symbol for the square root of a number (-1) that you cannot take the square root of." Thus complex numbers are thought to be mysterious quantities (or *non*quantities) that are not understood at all. They are thought to be entities that are purely *imaginary* (in the nontechnical sense of that word), and that have no conceivable use.

32.2 Nothing could be further from the truth! The analysis of electric and electronic circuitry relies most heavily on the complex number system. Similar analysis can be made in mechanics, too. It is with great reluctance that we do not take the space here to show such applications. Instead, we shall investigate the very elegant applications of complex numbers to plane geometry and trigonometry.

32.3 It is hoped that, after reading the material in this chapter, you will have a much finer appreciation of the *realness* of complex numbers, that you will feel comfortable in their presence, and that you will be able to apply them to appropriate problems in geometry and, perhaps to a lesser extent, in trigonometry. In particular, you should understand just what is meant by that mysterious equation $i^2 = -1$, and by the plus sign in the complex number $2 + 3i$. Does it mean addition? If so, then what is the sum of 2 and $3i$? Do you get 5 *somethings*? Again, since $2 + 3i \neq 3 + 2i$, which is the greater? Is $2 + 3i > 3 + 2i$, or is $3 + 2i > 2 + 3i$? Just what is meant by saying that the complex numbers cannot be ordered?

32.4 In short, this chapter proposes to show the reader that the complex numbers are just as *real* to work with as the real numbers. To this end, we shall approach the geometry of complex numbers through a discussion of vectors in the plane.

32.5 A complex number may be thought of as a polynomial $a + bi$ in the *symbol* i, with real coefficients a and b, subject to the condition that

$$i^2 = -1.$$

Equality, addition, and multiplication for complex numbers are exactly the same as for polynomials, remembering that i^2 is always replaced by -1:

$$a + bi = c + di \quad \text{iff} \quad a = c \text{ and } b = d,$$
$$(a + bi) + (c + di) = (a + c) + (b + d)i,$$

and

$$\begin{aligned}(a + bi)(c + di) &= ac + adi + bci + bdi^2 \\ &= ac + (ad + bc)i + bd(-1) \\ &= (ac - bd) + (ad + bc)i.\end{aligned}$$

This interpretation of complex numbers as polynomials is quite satisfactory to the mathematician, but it does not provide the student with any sort of concrete picture of what complex numbers mean. He certainly does not expect ever to meet one on the street corner. The polynomial idea is convenient mathematically, easy to talk about, but completely sterile for young students.

32.6 Real numbers are easily pictured as points on a number line or *real axis*. Furthermore, this interpretation is easily modified to allow real numbers as vectors along the real axis. Thus $5 - 3 = 2$ means "travel 5 units in the positive direction and then 3 units in the negative direction. The result is the same as traveling 2 units in the positive direction." (See Fig. 32.6.) Thus 5, -3, and 2 can be thought of as vectors (directed lengths) of 5, 3, and 2 units along the real axis in the appropriate directions.

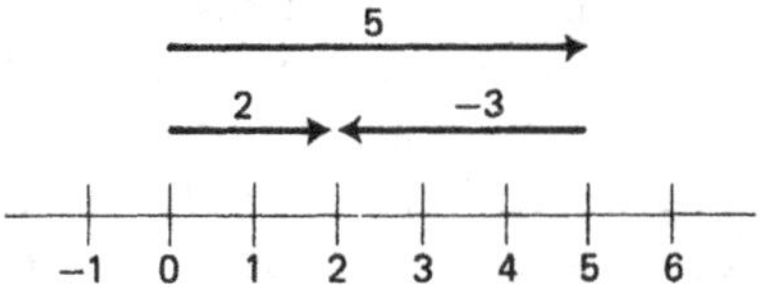

Figure 32.6

32.7 In just the same way, we associate points in the plane with complex numbers. The point $P(a, b)$ in the Cartesian plane is associated with the complex number $a + bi$, and hence with the vector $\vec{OP}$ from the origin O to the point P. Then the x-axis is called the *real axis* (Re) and the y-axis the *imaginary axis* (Im) (see Fig. 32.7a). When we consider the two points P and Q as representing the complex numbers $a + bi$ and $c + di$, the sum $(a + c) + (b + d)i$ may be shown vectorially quite nicely. When we let S be the point representing that sum, then $POQS$ is a parallelogram; that is, $\vec{OS}$ is the vector sum of $\vec{OP}$ and $\vec{OQ}$ (see 33.6 through 33.9). Figure 32.7b shows this addition.

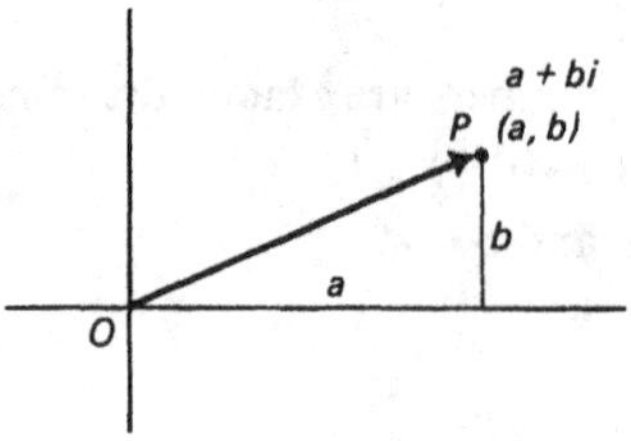

Figure 32.7a

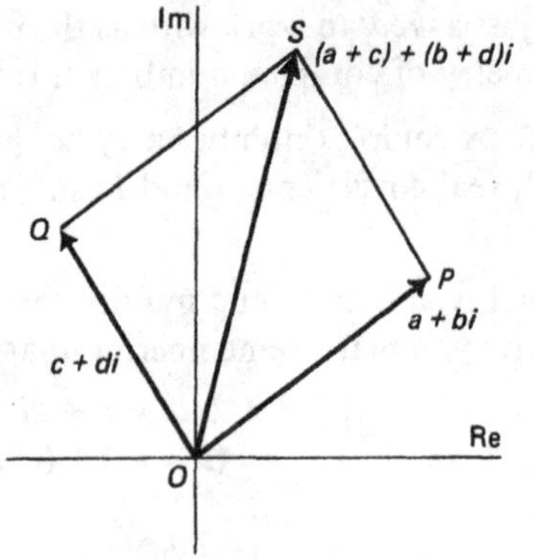

Figure 32.7b

32.8 Vectors in the plane certainly are real entities; a distance in a given direction is not simply a figment of the imagination. Hence we shall begin our study of complex numbers by the slightly round-about route of vectors and vector addition, which will lead naturally into complex numbers and their addition.

32.9 To obtain an added bonus, we define multiplication of vectors from a natural geometric desire to represent rotations and homotheties on vectors. This multiplication turns out to be the usual multiplication of complex numbers (see 33.10 through 33.12 and 34.8 through 34.13). Thus from the geometric vector interpretation, we obtain the complex numbers. This geometric picture introduces a "reality" to the complex number system that few high school students are ever given the opportunity to appreciate.

32.10 Although our primary purpose in studying vectors here is to lead into the complex number system, the great importance and usefulness of vectors as such should not be overlooked. Many theorems can be proved beautifully by means of vectors. Some examples are given in 33.15 to 33.21, 34.17, 34.24, and the exercises in Sections 33 and 34. Again, vector methods do not provide the only tool in mathematics, but they do provide a tool that is both useful and important.

32.11 Section 35 leads you from vectors into complex numbers, and the remainder of this chapter develops and illustrates the geometry of complex numbers. The teacher who understands this material is much better prepared to impart to his students a genuine feeling for complex numbers.

Exercise Set 32

1. Order (>) among the real numbers has the following three properties:
   1) *Trichotomy.* Every real number x satisfies exactly one of the three statements, $x > 0$, $x = 0$, or $-x > 0$.
   2) *Closure of the positive numbers under addition.* If $a > 0$ and $b > 0$, then $a + b > 0$.
   3) *Closure of the positive numbers under multiplication.* If $a > 0$ and $b > 0$, then $ab > 0$.

The following chain of theorems demonstrates why the complex numbers cannot be ordered in the same way as are the real numbers. Prove each theorem to complete this demonstration.

a) It cannot be true that $-1 > 0$.
b) $1 > 0$.
c) Assume that the complex numbers do satisfy the three properties listed above. Show that it is not true that $i > 0$.
d) It is not true that $-i > 0$.
e) $i \neq 0$.
f) The three order properties are not satisfied by i.
g) The set of all complex numbers cannot satisfy the three order properties.
h) It makes no sense to discuss whether $3 + 2i > 2 + 3i$ or $2 + 3i > 3 + 2i$ is true (see 32.3).

2. Calculate the following.
a) $(2 + 3i) + (3 - 2i)$
b) $(2 + 5i) + (2 - 5i)$
c) $(2 + 5i) - (7 + 6i)$
d) $(2 + 3i)(3 + 2i)$
e) $(2 + 3i)(3 - 2i)$
f) $(2 + 3i)(2 - 3i)$
g) $(2 + i)/(1 - i)$
h) $(3 + 2i)/(4 + 3i)$
i) $(1 + i)^{12}$
j) $(5 + 10i)/(2 + i)$
k) Is the answer to part (b) a complex number? Explain.

3. Solve for real numbers x and y.
a) $2x + 3i = 5 - yi$
b) $7x + 8yi = 1 + i$
c) $5 + x + y = 2y + (2x - y)i$
d) $x^2 - y^2 + 2xyi = -1$
e) $x^2 - y^2 + 2xyi = i$
f) $(x + yi)^2 = i$
g) $(x + yi)^2 = -i$
h) $(x + yi)^2 = 7 - 24i$
i) $(x + yi)^2 = -7 + 24i$

4. Show that:
a) $i^2 = -1$
b) $i^3 = -i$
c) $i^4 = 1$
d) $i^{4k+m} = i^m$ for integral k
e) $1/i = i^{-1} = -i$
f) $i^{-m} = i^{4k-m}$ for integral k

5. Evaluate:
a) i^5
b) i^6
c) i^7
d) i^8
e) i^{42}
f) i^{7183}
g) i^{-2}
h) i^{-3}
i) i^{-4}
j) i^{-5}
k) i^{-6}
l) i^{-7}
m) i^{-8}
n) i^{-35}
o) i^{-2161}
p) $5i^9 - 8i^5 + 7i^4 - 3i^{-3} + 7$

6. What is the result of traveling 5 yd east and then 10 yd north? What is the relation between such travels and the complex number $5 + 10i$?

7. Interpret $2 - 3i$ in terms of traveling as in Exercise 32.6.

8. Write out your own explanation of just what "$i^2 = -1$" means to you. Save your essay and read it after completing this chapter. Similarly, give an interpretation to the plus sign in a complex number such as $2 + 3i$.

9. Is i a number? Explain.
10. One is often told "you cannot do thus-and-so" by mathematics teachers when the statement is correct only in an appropriate context. Some examples of such things that "cannot be done" are given below. Explain the circumstances under which you cannot, and under which you can, do each listed task, and then perform the task.
    a) Take the square root of -5.
    b) Factor $x^2 - 7$.
    c) Factor $x^2 + 2$.
    d) Factor $x^2 + 4x + 5$.
    e) Cancel the b's in the quotient ab/bc of two two-digit numbers to get a/c. Example: $1\not{6}/\not{6}4 = 1/4$.
    f) Find a statement that cannot be true and cannot be false.
    g) Find two nonparallel lines that have no points in common.
    h) Find numbers a and b so that $a^2 + ab + b^2$ is a square number.
    i) Find three distinct lines m, n, p so that n and p are each perpendicular to m, but n and p are not parallel.

SECTION 33 | VECTORS

33.1 Definition A *vector* in the plane is an ordered pair (a, b) of real numbers a and b. Vectors are denoted by boldface lower-case Roman letters **u**, **v**, **w**, . . . or by point pairs with an arrow overbar $\overrightarrow{AB}$.

33.2 Definition Vectors $\mathbf{u} = (a, b)$ and $\mathbf{v} = (c, d)$ are called *equal* iff $a = c$ and $b = d$.

33.3 More informally, a vector is a directed distance; the vector $(5, -3)$ means "5 units in the x-direction and -3 units in the y-direction." We do distinguish between the directed segment $\overline{AB}$ going specifically from the given point A to the given point B and the vector $\overrightarrow{AB}$ determined by that directed segment. Thus $\overline{AB}$ refers to the specific directed segment from A to B, whereas $\overrightarrow{AB}$ simply is that direction and distance with no reference to specific initial and terminal points. We write $\overline{AB} = \overline{CD}$ only when $A = C$ and $B = D$. We write $\overrightarrow{AB} = \overrightarrow{CD}$ whenever $ABDC$ is a (perhaps degenerate) parallelogram. We never write $\overline{AB} = \overrightarrow{CD}$.

33.4 Definition Given the points $P_1(x_1, y_1)$ and $P_2(x_2, y_2)$, we write

$$\overrightarrow{P_1P_2} = (x_2 - x_1, y_2 - y_1).$$

33.5 That is, $\overrightarrow{P_1P_2}$ is the vector representing the directed distance necessary to travel from P_1 to P_2 (Fig. 33.5). It follows immediately that the vector $\overrightarrow{OP} = (x, y)$ when O and P have coordinates $O(0, 0)$ and $P(x, y)$.

33.6 Definition Let A, B, C be points such that $\overrightarrow{AB} = \mathbf{v}$ and $\overrightarrow{BC} = \mathbf{w}$ for given vectors **v** and **w**. We define *vector addition* by

$$\mathbf{v} + \mathbf{w} = \overrightarrow{AC}.$$

33.7 Informally, we think of adding the vectors **v** and **w** by placing the head of vector **v** on the tail of vector **w**, then taking as their sum the vector from the tail of **v**

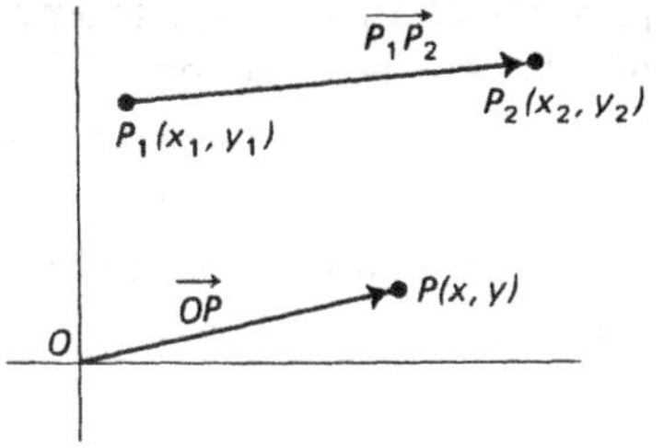

Figure 33.5

to the head of **w** as in Fig. 33.7. That vector addition is associative and commutative is stated next, the commutativity illustrating that the sum **v** + **w** is given by the *parallelogram law*; that is, in the parallelogram having **v** and **w** as sides, **v** + **w** is that diagonal that emanates from the vertex where the two vector tails meet.

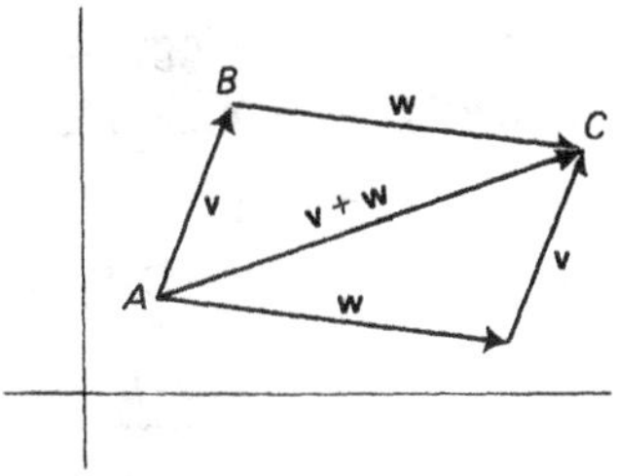

Figure 33.7

33.8 Theorem Vector addition is commutative and associative.

The proof of this theorem is indicated in Figs. 33.7 and 33.8, the details being left for the reader to supply. ▯

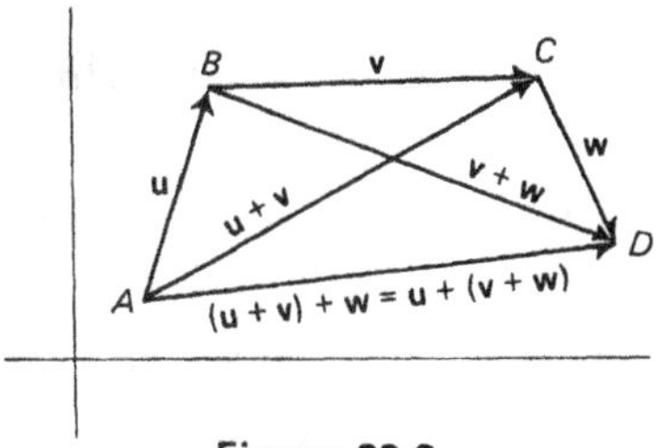

Figure 33.8

33.9 Theorem When $\mathbf{u} = (a, b)$ and $\mathbf{v} = (c, d)$, then

$$\mathbf{u} + \mathbf{v} = (a + c, b + d).$$

33.10 Definition When k is a real number and $\mathbf{v} = (a, b)$ is a vector, we define *scalar multiplication* as follows. We say that the *scalar product* of k and $\mathbf{v}$ is

$$k\mathbf{v} = k(a, b) = (ka, kb).$$

33.11 Thus scalar multiplication of a vector by a real number (or *scalar*) k is simply a stretch or homothety of ratio k. The direction of the vector is unchanged when $k > 0$ and is reversed when $k < 0$. The magnitude or length of the vector is multiplied by the factor $|k|$.

33.12 Theorem When c and k are real numbers and $\mathbf{u}$ and $\mathbf{v}$ are vectors, then
1) $c(k\mathbf{v}) = (ck)\mathbf{v} = (kc)\mathbf{v} = k(c\mathbf{v})$,
2) $k(\mathbf{u} + \mathbf{v}) = k\mathbf{u} + k\mathbf{v}$,
3) $(c + k)\mathbf{v} = c\mathbf{v} + k\mathbf{v}$.

33.13 Definition When $\mathbf{u}$ and $\mathbf{v}$ are vectors, then we write

$$\mathbf{u} - \mathbf{v} = \mathbf{u} + (-1)\mathbf{v}.$$

This operation on vectors is called *vector subtraction.*

33.14 The properties developed so far are quite adequate to prove many theorems concerning points and lengths in polygonal figures. Some examples follow.

▶**33.15 Theorem** The diagonals of a parallelogram bisect each other.

Let the parallelogram be $ABCD$ with $\vec{AB} = \vec{DC} = \mathbf{u}$ and $\vec{BC} = \vec{AD} = \mathbf{v}$.

33.16 *First proof.* Let the diagonals meet at E as in Fig. 33.16. Now

$$\vec{AC} = \vec{AB} + \vec{BC} = \mathbf{u} + \mathbf{v} \quad \text{and} \quad \vec{DB} = \vec{DA} + \vec{AB} = -\mathbf{v} + \mathbf{u}.$$

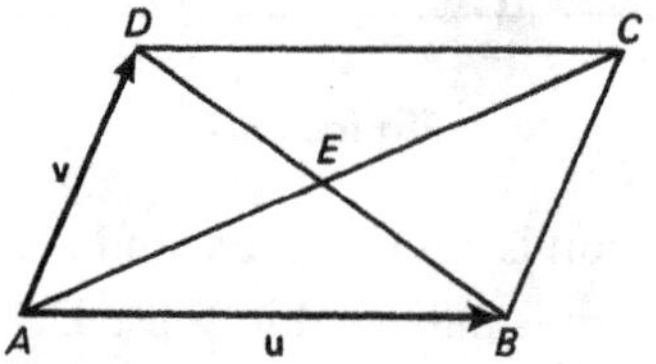

Figure 33.16

Since E is part way along AC and part way along DB, there are positive constants m and n, each less than 1, such that

$$\vec{AE} = m\vec{AC} = m(\mathbf{u} + \mathbf{v}) \quad \text{and} \quad \vec{DE} = n\vec{DB} = n(\mathbf{u} - \mathbf{v}).$$

Now we consider $\vec{AE}$ from another point of view:

$$\vec{AE} = \vec{AD} + \vec{DE} = \mathbf{v} + n(u - \mathbf{v}).$$

Equating these two forms for $\vec{AE}$, we have

$$m(\mathbf{u} + \mathbf{v}) = \mathbf{v} + n(\mathbf{u} - \mathbf{v}),$$

so

$$(m - n)\mathbf{u} = (1 - n - m)\mathbf{v}.$$

This last equation can be true only if the coefficients of both vectors are zero, since neither vector is a multiple of the other. Thus

$$m - n = 0 \quad \text{and} \quad 1 - n - m = 0,$$

from which we obtain $m = n = \frac{1}{2}$. Hence

$$\overrightarrow{AE} = \tfrac{1}{2}\overrightarrow{AC} \quad \text{and} \quad \overrightarrow{DE} = \tfrac{1}{2}\overrightarrow{DB},$$

establishing the theorem. ▯

33.17 *Second proof.* Let E be the midpoint of AC, and F the midpoint of BD. Then $\overrightarrow{AE} = \frac{1}{2}\overrightarrow{AC} = \frac{1}{2}(\mathbf{u}+\mathbf{v})$ and $\overrightarrow{AF} = \overrightarrow{AD} + \frac{1}{2}\overrightarrow{DB} = \mathbf{v} + (\mathbf{u}-\mathbf{v})/2 = (\mathbf{u}+\mathbf{v})/2 = \overrightarrow{AE}$. Since the vectors $\overrightarrow{AE}$ and $\overrightarrow{AF}$ are equal, it follows that points E and F coincide; that is, the two diagonals meet at their midpoints. ▯

▶**33.18 Theorem** The medians of a triangle concur at a trisection point of each median.

Let us show that the points 2/3 of the way along each of the three medians coincide. To that end (see Fig. 33.18), let $\mathbf{u} = \overrightarrow{AB}$ and $\mathbf{v} = \overrightarrow{AC}$. Then $\overrightarrow{BC} = \mathbf{v} - \mathbf{u}$. Now, for medians AA', BB', CC', in turn, we have

$$\begin{aligned}
\tfrac{2}{3}\overrightarrow{AA'} &= \tfrac{2}{3}(\overrightarrow{AB} + \overrightarrow{BA'}) = \tfrac{2}{3}(\overrightarrow{AB} + \tfrac{1}{2}\overrightarrow{BC})\\
&= \tfrac{2}{3}\overrightarrow{AB} + \tfrac{1}{3}\overrightarrow{BC} = \tfrac{2}{3}\mathbf{u} + \tfrac{1}{3}(\mathbf{v}-\mathbf{u})\\
&= (\mathbf{u}+\mathbf{v})/3,\\
\overrightarrow{AB} + \tfrac{2}{3}\overrightarrow{BB'} &= \overrightarrow{AB} + \tfrac{2}{3}(\overrightarrow{BA} + \overrightarrow{AB'})\\
&= \overrightarrow{AB} + \tfrac{2}{3}(\overrightarrow{BA} + \tfrac{1}{2}\overrightarrow{AC})\\
&= \mathbf{u} + \tfrac{2}{3}(-\mathbf{u} + \tfrac{1}{2}\mathbf{v}) = (\mathbf{u}+\mathbf{v})/3,\\
\overrightarrow{AC} + \tfrac{2}{3}\overrightarrow{CC'} &= \overrightarrow{AC} + \tfrac{2}{3}(\overrightarrow{CA} + \overrightarrow{AC'}) = \overrightarrow{AC} + \tfrac{2}{3}(\overrightarrow{CA} + \tfrac{1}{2}\overrightarrow{AB}) = \mathbf{v} + \tfrac{2}{3}(-\mathbf{v} + \tfrac{1}{2}\mathbf{u})\\
&= (\mathbf{u}+\mathbf{v})/3.
\end{aligned}$$

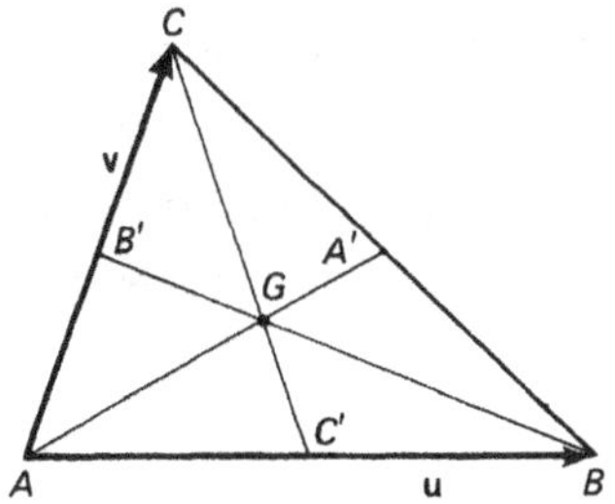

Figure 33.18

Since these last three vectors are all equal and have the same initial points, their terminal points coincide; that is, the medians concur at a trisection point of each median. ▯

▶**33.19 Theorem** The altitudes of a triangle concur.

Let us this time define the triangle ABC in terms of the vectors $\overrightarrow{OA} = \mathbf{u}$, $\overrightarrow{OB} = \mathbf{v}$, and $\overrightarrow{OC} = \mathbf{w}$ from its circumcenter to its vertices as in Fig. 33.19. Then the vectors $\mathbf{u}$, $\mathbf{v}$, $\mathbf{w}$ all have the same length, the circumradius. Letting A_1 be the fourth vertex of the parallelogram $BOCA_1$, then $BOCA_1$ is a rhombus whose diagonals are perpendicular and bisect each other at point A', the midpoint of side BC. Furthermore, vector $OA_1 = \mathbf{v} + \mathbf{w}$. Let point H be defined by making AOA_1H a parallelogram. Hence

$$\vec{OH} = \vec{OA} + \vec{AH} = \vec{OA} + \vec{OA_1} = \mathbf{u} + \mathbf{v} + \mathbf{w}.$$

Since OA_1 is perpendicular to BC, then so also is AH perpendicular to BC. Furthermore, by the symmetry of the representation for $\vec{OH}$, it follows that BH is perpendicular to AC, and CH is perpendicular to AB. Hence H is the meeting of the three altitudes. ▯

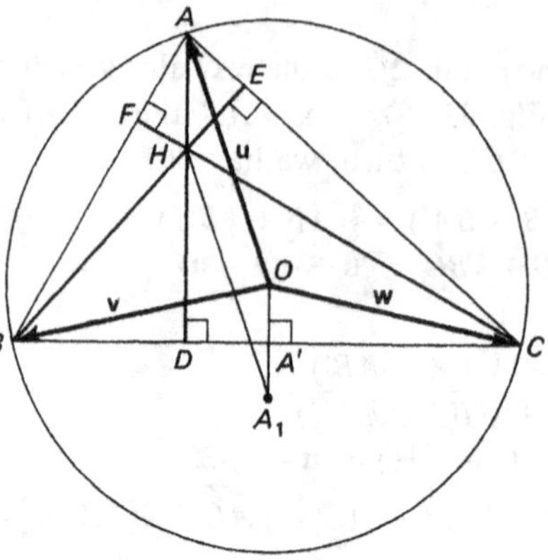

Figure 33.19

33.20 We call special attention to the equation

$$\vec{OH} = \mathbf{u} + \mathbf{v} + \mathbf{w},$$

proved in Theorem 33.19. It states that the vector from the circumcenter to the orthocenter of a triangle is equal to the sum of the vectors from the circumcenter to the three vertices.

▶**33.21 Theorem** The orthocenter, circumcenter, and centroid of a triangle are collinear and $\vec{HG} = 2\vec{GO}$.

Referring to the notation in the proof of Theorem 33.19 and Fig. 33.19, we have

$$\vec{OH} = \mathbf{u} + \mathbf{v} + \mathbf{w}.$$

Now $\vec{OG}$ may be calculated from the equations of Theorems 33.18 and 33.19, since $\vec{AG} + (\vec{AB} + \vec{AC})/3$. Thus

$$\begin{aligned}\vec{OG} &= \vec{OA} + \vec{AG}\\ &= \vec{OA} + (\vec{AB} + \vec{AC})/3\\ &= \mathbf{u} + ((\mathbf{v} - \mathbf{u}) + (\mathbf{w} - \mathbf{u}))/3\\ &= (\mathbf{u} + \mathbf{v} + \mathbf{w})/3\\ &= \tfrac{1}{3}\vec{OH},\end{aligned}$$

from which the theorem follows. ▯

Exercise Set 33

1. Prove Theorem 33.8.
2. Show that the set of all vectors is closed under vector addition; that is, show that the sum of any two vectors is always another vector.
3. Prove Theorem 33.9.
4. Prove Theorem 33.12.
5. Show that the set of all vectors in the plane is closed under vector subtraction, which is neither commutative nor associative.
6. Show that if $\mathbf{u} + \mathbf{v} = \mathbf{w}$, then $\mathbf{u} = \mathbf{w} - \mathbf{v}$.
7. Show that vector addition is *well defined*, that is, if A, B, C, D, E, F are six points such that $\overrightarrow{AB} = \overrightarrow{DE}$ and $\overrightarrow{BC} = \overrightarrow{EF}$, then $\overrightarrow{AB} + \overrightarrow{BC} = \overrightarrow{DE} + \overrightarrow{EF}$.
8. Show that the vector $\mathbf{0} = (0, 0)$ is the additive identity, and that each vector $\mathbf{v} = (a, b)$ has the additive inverse $-\mathbf{v} = (-1)\mathbf{v} = (-a, -b)$.
9. Prove that the midpoints of the sides of a quadrilateral form a parallelogram.
10. Prove that the segment joining the midpoints of two sides of a triangle is parallel to and equal to half the third side.
11. Assuming that O is any point in the plane of line AB, and that M is the midpoint of segment AB, show that $\overrightarrow{OM} = (\overrightarrow{OA} + \overrightarrow{OB})/2$. (Compare with Theorem 2.12.)
12. Given that O is any point in the plane of line AB, and that P is a point on line AB that divides segment AB in the ratio r/s, show that $\overrightarrow{OP} = (r\overrightarrow{OB} + s\overrightarrow{OA})/(r + s)$.
13. When O is any point in the plane of a triangle ABC, show that
$$\overrightarrow{OA} + \overrightarrow{OB} + \overrightarrow{OC} = \overrightarrow{OA'} + \overrightarrow{OB'} + \overrightarrow{OC'},$$
where A', B', C' are the midpoints of the sides of the triangle.
14. Prove that the lines joining a given vertex of a parallelogram to the midpoints of the two opposite sides trisect that diagonal not through the given vertex.
15. Prove that in any triangle the three medians, as vectors, form a triangle.
16. In a triangle, do the angle bisectors, as vectors, form a triangle?
17. Show that the segment joining the midpoints of the nonparallel sides of a trapezoid is parallel to the bases and congruent to half their sum.

SECTION 34 | VECTOR MULTIPLICATION

34.1 By the Pythagorean theorem, the length (or *magnitude*) of the vector $\mathbf{v} = (a, b)$ is given by $(a^2 + b^2)^{1/2}$. We use the familiar absolute value bars to designate this length.

34.2 Definition If $\mathbf{v} = (a, b)$, then we define the *absolute value* or *magnitude* $|\mathbf{v}|$ of the vector $\mathbf{v}$ by
$$|\mathbf{v}| = (a^2 + b^2)^{1/2}.$$

34.3 Theorem 34.4 states that absolute value as applied to vectors has the usual properties of absolute value.

34.4 Theorem If $\mathbf{u}$ and $\mathbf{v}$ are vectors and c is a scalar, then
1) $|\mathbf{v}| \geqslant 0$,
2) $|\mathbf{v}| = 0$ iff $\mathbf{v} = (0, 0)$, denoted by $\mathbf{0}$,
3) $|c\mathbf{v}| = |c||\mathbf{v}|$,
4) $|\mathbf{u} + \mathbf{v}| \leqslant |\mathbf{u}| + |\mathbf{v}|$,
5) $|\mathbf{u} - \mathbf{v}| \geqslant |\mathbf{u}| - |\mathbf{v}|$.

34.5 Definition If $|\mathbf{u}| = 1$, then $\mathbf{u}$ is called a *unit vector.*

34.6 Theorem Each unit vector $\mathbf{u}$ has the form $(\cos\theta, \sin\theta)$ where θ is the trigonometric angle in standard position made by the vector $\mathbf{u}$ when it is considered to emanate from the origin (see Fig. 34.6).

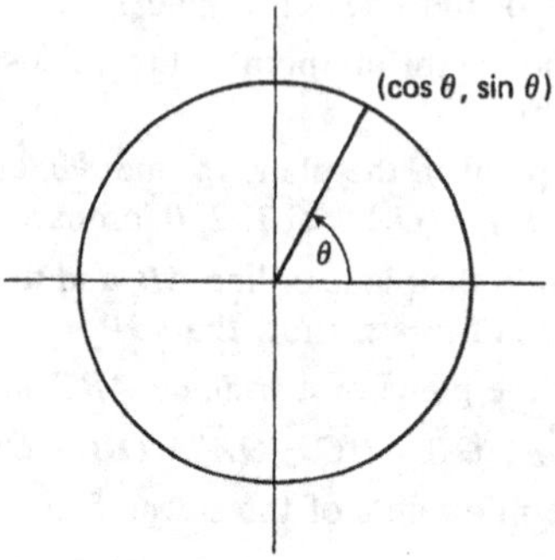

Figure 34.6

34.7 Theorem Each vector $\mathbf{v}$ can be written in the form $c\mathbf{u}$ where $\mathbf{u}$ is a unit vector and c is a nonnegative scalar. If $\mathbf{v} \neq \mathbf{0}$, this representation is unique.

Simply take $c = |\mathbf{v}|$ and $\mathbf{u} = (1/c)\mathbf{v}$. □

34.8 It is convenient to define vector multiplication in two stages. First, we consider a unit vector as a rotation about the origin through angle θ where the unit vector $\mathbf{u} = (\cos\theta, \sin\theta)$. The product of the two unit vectors $\mathbf{u} = (\cos\theta, \sin\theta)$ and $\mathbf{v} = (\cos\phi, \sin\phi)$ becomes that unit vector representing the product of the two rotations. We have

$$\begin{aligned}\mathbf{uv} &= (\cos(\theta+\phi), \sin(\theta+\phi))\\ &= (\cos\theta\cos\phi - \sin\theta\sin\phi, \sin\theta\cos\phi + \cos\theta\sin\phi)\end{aligned}$$

by applying the formulas from trigonometry for $\cos(\theta+\phi)$ and $\sin(\theta+\phi)$. (See also Exercise 21.6.) The formal definition follows. See Fig. 34.8, where, letting $\overrightarrow{OQ} = \mathbf{u}$, $\overrightarrow{OP'} = \mathbf{v}$, $\overrightarrow{OQ'} = \mathbf{uv}$, and $P(1, 0)$, we have triangles OPQ and $OP'Q'$ congruent by SAS.

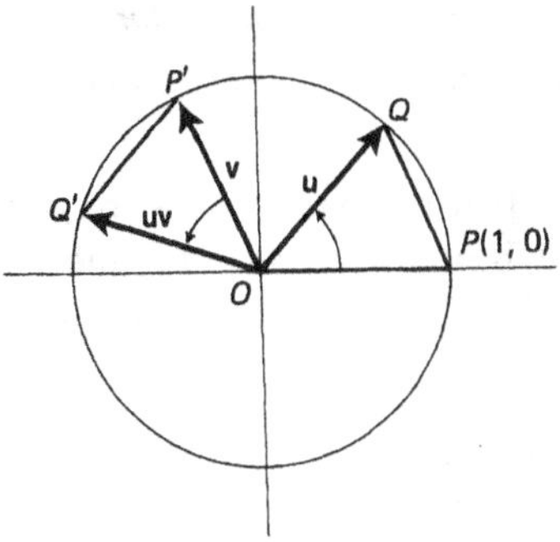

Figure 34.8

34.9 Definition If $\mathbf{u} = (a, b)$ and $\mathbf{v} = (c, d)$ are unit vectors, then we define their *product** $\mathbf{uv}$ by

$$\mathbf{uv} = (ac - bd, ad + bc).$$

34.10 Theorem Unit vector multiplication is commutative, associative, and distributive over vector addition, has the identity (1, 0), denoted by **1**, and each unit vector $\mathbf{u} = (a, b)$ has a multiplicative inverse given by $\mathbf{u}^{-1} = (a, -b)$.

Clearly the vector $\mathbf{1} = (1, 0)$ is equal to $(\cos 0, \sin 0)$, so this vector is the multiplicative identity. This result is important enough for us to restate the notation in the next definition. Since $(a, b)(a, -b) = (a^2 + b^2, -ab + ab) = (1, 0)$ whenever (a, b) is a unit vector, it follows that $\mathbf{u}^{-1} = (a, -b)$ is the multiplicative inverse of $\mathbf{u} = (a, b)$. (Note that $\mathbf{u}^{-1}$ is the reflection of $\mathbf{u}$ in the x-axis as mirror (see Fig. 34.10). The rest of

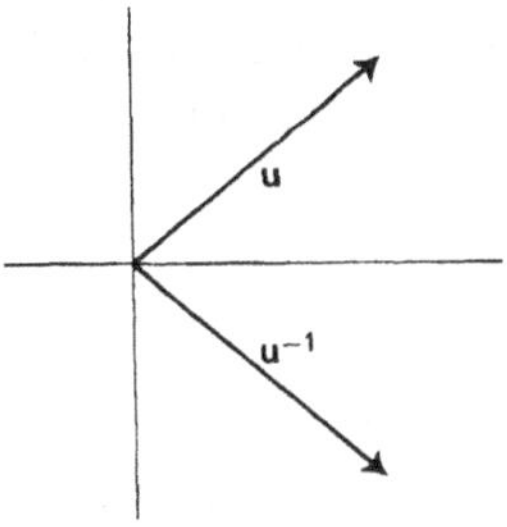

Figure 34.10

this proof, the commutativity, associativity, and distributivity, follow either from the algebraic equations or from the geometric interpretation of unit vector multiplication as the product of rotations. □

* Vector multiplication, as defined here, is never denoted by a raised dot as in $\mathbf{u} \cdot \mathbf{v}$ or by a cross as in $\mathbf{u} \times \mathbf{v}$. These notations are reserved for two other vector products, the so-called *scalar* or *dot product* $\mathbf{u} \cdot \mathbf{v} = ac + bd$, and the *cross product*, which is defined in space of three dimensions.

34.11 Definition Let $\mathbf{1} = (1, 0)$, called the *vector one.*

Recalling Definition 34.5, we see that $\mathbf{1}$ is a unit vector, but that not every unit vector is $\mathbf{1}$.

34.12 Now we turn our attention to multiplication of any two vectors $\mathbf{v}_1$ and $\mathbf{v}_2$. By Theorem 34.7, let $\mathbf{u}_1$ and $\mathbf{u}_2$ be unit vectors and c_1 and c_2 be scalars, both nonnegative, such that $\mathbf{v}_1 = c_1\mathbf{u}_1$ and $\mathbf{v}_2 = c_2\mathbf{u}_2$. It seems natural to define their product by

$$\mathbf{v}_1\mathbf{v}_2 = (c_1 c_2)(\mathbf{u}_1\mathbf{u}_2).$$

Geometrically, $\mathbf{v}_1\mathbf{v}_2$ is a rotation and a homothety, which can be displayed as in Fig. 34.12. Let $\overrightarrow{OP} = \mathbf{1}$, $\overrightarrow{OQ} = \mathbf{v}_1$, $\overrightarrow{OP'} = \mathbf{v}_2$, and $\overrightarrow{OQ'} = \mathbf{v}_1\mathbf{v}_2$. Then triangles OPQ and $OP'Q'$ are similar. That is, $\mathbf{v}_1\mathbf{v}_2$ is the result of rotating $\mathbf{v}_1$ through the angle θ_2 of

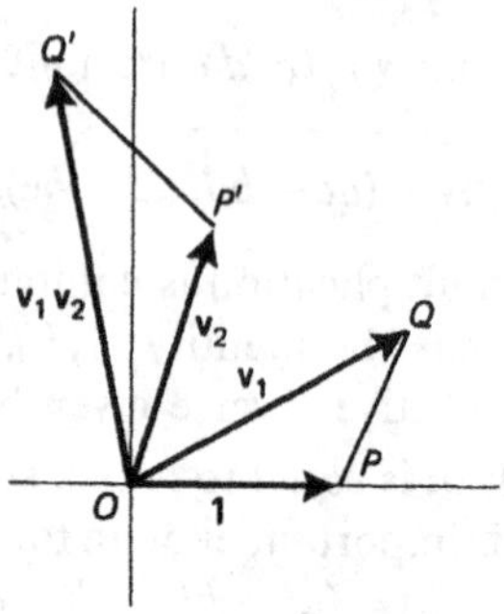

Figure 34.12

$\mathbf{v}_2$ and magnifying it by the ratio $|\mathbf{v}_2|$. Curiously, this definition of vector multiplication gives rise to the same algebraic formulation as for unit vectors. This fact we now state as our formal definition of vector multiplication.

34.13 Definition If $\mathbf{u} = (a, b)$ and $\mathbf{v} = (c, d)$ are any two vectors, then we define their *product* by

$$\mathbf{uv} = (ac - bd, ad + bc).$$

34.14 Theorem Vector multiplication is commutative and associative, distributive over vector addition, and has identity $\mathbf{1} = (1, 0)$. Each vector $\mathbf{v} = (a, b) \neq (0, 0)$ has the multiplicative inverse

$$\mathbf{v}^{-1} = \left(\frac{a}{a^2 + b^2}, \frac{-b}{a^2 + b^2}\right).$$

The expression for the inverse $\mathbf{v}^{-1}$ of vector $\mathbf{v}$ follows from the equation

$$(a, b)(a, -b) = (a^2 + b^2, 0) = (|\mathbf{v}|^2)\mathbf{1}.$$

Thus

$$\mathbf{v}^{-1} = (1/|\mathbf{v}|^2)(a, -b). \ \square$$

34.15 Theorem $|\mathbf{uv}| = |\mathbf{u}||\mathbf{v}|$.

34.16 We close this section with a pair of applications.

▶**34.17 Example** Find a vector condition for two vectors $\mathbf{u} = (a, b) \neq \mathbf{0}$ and $\mathbf{v} = (c, d) \neq \mathbf{0}$ to be *orthogonal* (perpendicular) (Fig. 34.17).

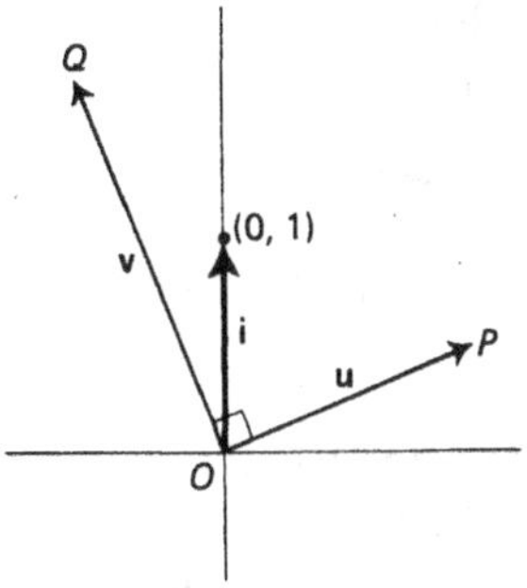

Figure 34.17

Let us denote the vector $(0, 1) = (\cos 90°, \sin 90°)$ by $\mathbf{i}$. Then $\mathbf{i}$ is the vector corresponding to a 90° rotation. Since $\mathbf{u}$ and $\mathbf{v}$ are orthogonal, then a rotation of either $\pm 90°$ carries $\mathbf{u}$ into the direction of $\mathbf{v}$. Thus we may write

$$\mathbf{v} = k\mathbf{iu} \qquad \text{for some scalar } k \neq 0,$$

as the condition for perpendicularity of the vectors $\mathbf{u}$ and $\mathbf{v}$. Then

$$\mathbf{v} = k(0, 1)(a, b) = (0, k)(a, b) = (-kb, ka).$$

That is,

$$c = -kb \qquad \text{and} \qquad d = ka.$$

Eliminating k between these two equations, we obtain

$$ac + bd = 0$$

as a necessary and sufficient condition for the two nonzero vectors $\mathbf{u} = (a, b)$ and $\mathbf{v} = (c, d)$ to be orthogonal. In terms of the dot product (see the footnote to Definition 34.9), this equation states $\mathbf{u} \cdot \mathbf{v} = 0$. In terms of vector multiplication, if we write $\bar{\mathbf{u}} = (a, -b)$, called the *conjugate* of $\mathbf{u}$ (its mirror image in the x-axis), this condition for orthogonality may be written as

$$\mathbf{u}\bar{\mathbf{v}} + \bar{\mathbf{u}}\mathbf{v} = \mathbf{0},$$

for the expression on the left becomes

$$\begin{aligned}\mathbf{u}\bar{\mathbf{v}} + \bar{\mathbf{u}}\mathbf{v} &= (a, b)(c, -d) + (a, -b)(c, d) \\ &= (ac + bd, -ad + bc) + (ac + bd, ad - bc) \\ &= (2ac + 2bd, 0) \\ &= 2(ac + bd, 0).\end{aligned}$$

This last vector is $\mathbf{0}$ iff $ac + bd = 0$.

34.18 It should be stated formally that two notations were introduced in this last example. We do so.

34.19 Definition Let $\bar{\mathbf{v}} = (a, -b)$ denote the *vector conjugate* of the vector $\mathbf{v} = (a, b)$.

34.20 Theorem $\overline{\mathbf{u}+\mathbf{v}} = \bar{\mathbf{u}} + \bar{\mathbf{v}}$ and $\overline{\mathbf{uv}} = \bar{\mathbf{u}}\bar{\mathbf{v}}$.

34.21 Theorem $|\mathbf{v}\bar{\mathbf{v}}| = |\mathbf{v}|^2$.

34.22 Theorem $|\mathbf{u}\bar{\mathbf{u}} + \mathbf{v}\bar{\mathbf{v}}| = |\mathbf{u}\bar{\mathbf{u}}| + |\mathbf{v}\bar{\mathbf{v}}|$.

34.23 Definition Let $\mathbf{i} = (0, 1)$.

▶**34.24 Example** *Ceva's theorem.* Three cevians AD, BE, CF for triangle ABC are concurrent iff

$$\frac{\mathbf{BD}}{\mathbf{DC}} \cdot \frac{\mathbf{CE}}{\mathbf{EA}} \cdot \frac{\mathbf{AF}}{\mathbf{FB}} = +1.$$

We shall prove only that the equation is true when the cevians concur. The converse can be established by vector methods as suggested in the exercises.

Let $\overrightarrow{AB} = \mathbf{u}$ and $\overrightarrow{AC} = \mathbf{v}$. Let m and n be scalars such that $\overrightarrow{AF} = m\mathbf{u}$ and $\overrightarrow{AE} = n\mathbf{v}$ (see Fig. 34.24). Then

$$\overrightarrow{FC} = \overrightarrow{FA} + \overrightarrow{AC} = -m\mathbf{u} + \mathbf{v} \quad \text{and} \quad \overrightarrow{EB} = \overrightarrow{EA} + \overrightarrow{AB} = -n\mathbf{v} + \mathbf{u}.$$

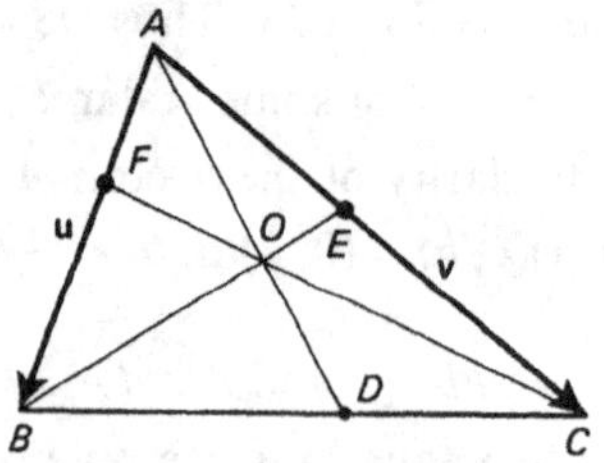

Figure 34.24

Then there are scalars r and s such that

$$\begin{aligned}\overrightarrow{AO} &= \overrightarrow{AF} + \overrightarrow{FO} = \overrightarrow{AF} + r\overrightarrow{FC} = m\mathbf{u} + r(-m\mathbf{u} + \mathbf{v}) \\ &= \overrightarrow{AE} + \overrightarrow{EO} = \overrightarrow{AE} + s\overrightarrow{EB} = n\mathbf{v} + s(-n\mathbf{v} + \mathbf{u}).\end{aligned}$$

Hence

$$m\mathbf{u} - rm\mathbf{u} + r\mathbf{v} = n\mathbf{v} - sn\mathbf{v} + s\mathbf{u}$$

and

$$(m(1-r) - s)\mathbf{u} = (n(1-s) - r)\mathbf{v}.$$

Since $\mathbf{u}$ and $\mathbf{v}$ have distinct directions, it follows that each coefficient must be zero; that is,

$$m(1-r) - s = 0 \quad \text{and} \quad n(1-s) - r = 0.$$

Thus

$$r = \frac{n(1-m)}{1-mn} \quad \text{and} \quad s = \frac{m(1-n)}{1-mn},$$

so we have

$$\overrightarrow{AO} = \frac{m(1-n)}{1-mn}\mathbf{u} + \frac{n(1-m)}{1-mn}\mathbf{v}.$$

Now there are also real numbers p and q, such that

$$\overrightarrow{BD} = p\overrightarrow{BC} = p(\mathbf{v}-\mathbf{u}) \qquad \text{and} \qquad \overrightarrow{BD} = \overrightarrow{BA} + \overrightarrow{AD} = -\mathbf{u} + q\overrightarrow{AO}.$$

Then we have

$$p(\mathbf{v}-\mathbf{u}) = -\mathbf{u} + q\frac{m(1-n)}{1-mn}\mathbf{u} + q\frac{n(1-m)}{1-mn}\mathbf{v}.$$

By collecting terms, obtain

$$[(1-p)(1-mn) - qm(1-n)]\mathbf{u} = [-p(1-mn) + qn(1-m)]\mathbf{v},$$

and again the coefficients of **u** and **v** must both be zero, yielding

$$p = \frac{n(1-m)}{m+n-2mn} \qquad \text{and} \qquad q = \frac{1-mn}{m+n-2mn}$$

and hence

$$\overrightarrow{BD} = \frac{n(1-m)}{m+n-2mn}(\mathbf{v}-\mathbf{u}).$$

Finally

$$\frac{\mathbf{BD}}{\mathbf{DC}}\cdot\frac{\mathbf{CE}}{\mathbf{EA}}\cdot\frac{\mathbf{AF}}{\mathbf{FB}} = \frac{\dfrac{n(1-m)}{m+n-2mn}}{1-\dfrac{n(1-m)}{m+n-2mn}}\cdot\frac{-1+n}{-n}\cdot\frac{m}{1-m}$$

$$= \frac{n(1-m)}{m+n-2mn-n(1-m)}\cdot\frac{n-1}{-n}\cdot\frac{m}{1-m}$$

$$= \frac{n(1-m)}{m-mn}\cdot\frac{n-1}{-n}\cdot\frac{m}{1-m}$$

$$= +1. \ \square$$

34.25 The examples given in these last two sections indicate that vector methods can be quite useful for certain problems, but there are cases when other methods are easier. Vectors are a convenient and helpful tool in geometry, but they are not *the* universal tool. In the next few sections we shall modify our approach slightly in order to increase this usefulness.

Exercise Set 34

1. Prove Theorem 34.4.
2. Prove Theorem 34.6.
3. Prove that $|\mathbf{v}| = |-\mathbf{v}|$.

4. Prove that the formula given in Definition 34.9 agrees with the concept of multiplication given in the discussion of 34.8.
5. Complete the proof of Theorem 34.10.
6. Complete the proof of Theorem 34.14.
7. Prove Theorem 34.15.
8. Prove Theorem 34.20.
9. Prove Theorem 34.21.
10. Prove Theorem 34.22.
11. Prove the converse part of Ceva's theorem (Example 34.24) by vector methods.
12. Let vectors **u** and **v** form the legs CA and CB of a right triangle ABC. Show that the Pythagorean relation is equivalent to the condition given in Example 34.17.
13. Show that the midpoint of the hypotenuse of a right triangle is equidistant from the three vertices.
14. Let vectors **u** and **v** form the sides CA and CB of triangle ABC. Show that the bisector of angle C satisfies the equation

$$\mathbf{t}_c = \frac{|\mathbf{u}|\mathbf{v} + |\mathbf{v}|\mathbf{u}}{|\mathbf{u}| + |\mathbf{v}|}.$$

15. Prove that the three angle bisectors of a triangle concur.
16. As in Exercise 34.14, find an expression for each angle bisector, as a vector, in terms of the vectors $\overrightarrow{PA}, \overrightarrow{PB}, \overrightarrow{PC}$ from any point P in the plane to the vertices of the triangle.
17. In parallelogram $ABCD$, show that

$$(AB)^2 + (BC)^2 + (CD)^2 + (DA)^2 = (AC)^2 + (BD)^2.$$

18. Prove that the diagonals of a rectangle have equal lengths.
19. Prove that the diagonals of a rhombus are perpendicular.
20. Prove that every point on the perpendicular bisector of a segment is equidistant from the ends of the segment.
21. Prove that when two medians of a triangle are congruent, then the triangle is isosceles.

SECTION 35 | VECTORS AND COMPLEX NUMBERS

35.1 Continuing our development of the algebra of plane vectors, let us examine more closely the two unit vectors $\mathbf{1} = (1, 0)$ and $\mathbf{i} = (0, 1)$. These vectors form a *basis* for plane vectors; that is, each vector **v** can be represented uniquely as a *linear combination* of **1** and **i**, as a sum of the form

$$\mathbf{v} = a\mathbf{1} + b\mathbf{i},$$

where a and b are scalars.

35.2 Theorem Let $\mathbf{v}^2$ denote $\mathbf{v}\mathbf{v}$. Then $\mathbf{1}^2 = \mathbf{1}$.

For $\mathbf{1}^2 = (1, 0)(1, 0) = (1 \cdot 1 - 0 \cdot 0, 1 \cdot 0 + 0 \cdot 1) = (1, 0) = \mathbf{1}$. □

35.3 Geometrically, **1** is a rotation through 0°; that is, **1** is the identity rotation. This idea will lead us quite naturally into complex numbers.

35.4 Theorem $\mathbf{i}^2 = -\mathbf{1}$.

For $\mathbf{i}^2 = (0, 1)(0, 1) = (0 \cdot 0 - 1 \cdot 1, 0 \cdot 1 + 1 \cdot 0) = (-1, 0) = -\mathbf{1}$. □

35.5 Another important geometric result is given in Theorem 35.4. It states that, since **i** is a 90° rotation, then $\mathbf{i}^2$ is the product of two 90° rotations, hence a 180° rotation. Since a 180° rotation is represented vectorially by $(\cos 180°, \sin 180°) = (-1, 0) = -\mathbf{1}$, we have $\mathbf{i}^2 = -\mathbf{1}$. Again, this equation simply states that the product of two 90° rotations is a 180° rotation.

35.6 Theorem The real numbers x and the vectors $x\mathbf{1}$ are *ring isomorphic*; that is, letting the real number x correspond to the vector $x\mathbf{1}$, and denoting this correspondence by $x \sim x\mathbf{1}$, we have

1) $x = y$ iff $x\mathbf{1} = y\mathbf{1}$,
2) $(x + y) \sim (x\mathbf{1} + y\mathbf{1})$,
3) $(xy) \sim (x\mathbf{1})(y\mathbf{1})$.

The statement $x\mathbf{1} = y\mathbf{1}$ is equivalent to

$$x(1, 0) = y(1, 0) \quad \text{and} \quad (x, 0) = (y, 0),$$

which is true iff $x = y$. Similarly part (2) is established, since $x\mathbf{1} + y\mathbf{1} = (x + y)\mathbf{1}$ by Theorem 33.12 and $(x + y) \sim (x + y)\mathbf{1}$. Also

$$(x\mathbf{1})(y\mathbf{1}) = (xy)\mathbf{1}^2 = (xy)\mathbf{1} \quad \text{and} \quad xy \sim (xy)\mathbf{1}. \ \square$$

35.7 Theorem 35.6 states that the real numbers x and the vectors $x\mathbf{1}$ along the x-axis behave the same algebraically, so we may treat them the same. Whether we perform arithmetic operations on real numbers or on the corresponding vectors, the results are the same, so it makes no difference in which system we work. This is clear when we realize that the geometry of vectors along the real line is an interpretation or a model of the algebra of real numbers. It means that, algebraically, we could simply omit writing the vector **1** in the form $x\mathbf{1}$ without causing confusion. The convenience of the ideas presented in these last three theorems is stated in the next theorem.

35.8 Theorem When vectors $\mathbf{u} = (a, b)$ are written as linear combinations of the unit vectors **1** and **i**, as $\mathbf{u} = a\mathbf{1} + b\mathbf{i}$, and when the vector **1** is omitted, as $\mathbf{u} = a + b\mathbf{i}$, then these vectors can be added and multiplied just as real polynomials in **i** are added and multiplied, provided $\mathbf{i}^2$, whenever it appears, is replaced by -1.

Letting $\mathbf{u} = a + b\mathbf{i}$ and $\mathbf{v} = c + d\mathbf{i}$, we have

$$\begin{aligned}(a + b\mathbf{i}) + (c + d\mathbf{i}) &= (a + c) + (b + d)\mathbf{i} \\ &= (a + c, b + d) \\ &= \mathbf{u} + \mathbf{v}\end{aligned}$$

and

$$\begin{aligned}(a + b\mathbf{i})(c + d\mathbf{i}) &= ac + ad\mathbf{i} + bc\mathbf{i} + bd\mathbf{i}^2 \\ &= ac + bd(-1) + (ad + bc)\mathbf{i} \\ &= (ac - bd) + (ad + bc)\mathbf{i} \\ &= (ac - bd, ad + bc) \\ &= (a, b)(c, d) \\ &= \mathbf{uv}.\end{aligned}$$

Thus the results are the same whichever technique is used to add or to multiply. □

35.9 Example To multiply the two vectors $\mathbf{u} = (2, -3)$ and $\mathbf{v} = (1, 5)$, we need no longer remember the complicated formula given in the definition of vector multiplication (Definition 34.13). Rather, we use the *method* of Theorem 35.8 to obtain

$$\mathbf{uv} = (2 - 3\mathbf{i})(1 + 5\mathbf{i}) = 2 + 15 + (-3 + 10)\mathbf{i} = 17 + 7\mathbf{i} = (17, 7).$$

35.10 This multiplication and addition remind one of the corresponding operations on complex numbers. Indeed, considering **i** as the imaginary unit instead of as a vector, the algebraic operations are identical. It follows that one interpretation for the complex number system is the system of vectors in the plane, just as real numbers can be interpreted as vectors along a line.

35.11 We now turn our attention to the geometry of complex numbers, first interpreting in the complex number plane the vector concepts we have developed.

35.12 Definition To each point Z of the plane we associate the complex number $z = a + bi$ where Z has coordinates (a, b). The plane thus interpreted is called the *Gauss plane*, point Z is the *image* of the complex number z, and z is the *affix* of Z (see Fig. 35.12).

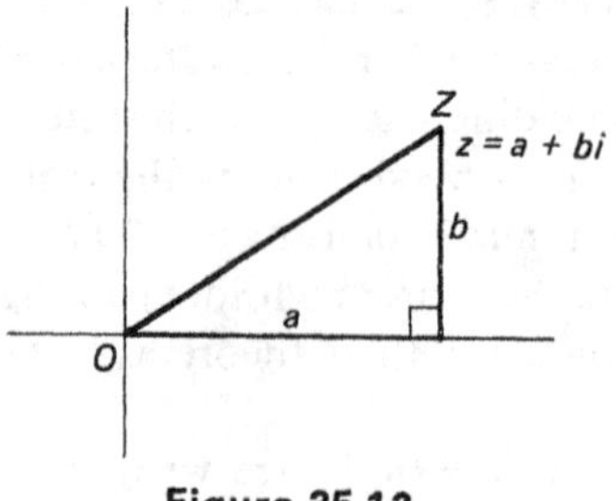

Figure 35.12

35.13 It follows that

$$\overrightarrow{OZ} = a\mathbf{1} + b\mathbf{i} = (a, b) \quad \text{iff} \quad \text{the affix of } Z \text{ is } a + bi.$$

Also, of course, the sum of two complex numbers corresponds to the vector sum of their associated vectors, and the product of two complex numbers corresponds to the vector product of their vectors.

35.14 Now is an appropriate time for you to reinvestigate the questions concerning complex numbers posed in 32.3. You should be in a much better position now to find logical answers.

35.15 Let us reexamine the vector properties established in Sections 33 and 34 as they apply to complex numbers. Two complex numbers $a + bi$ and $c + di$ are equal, $a + bi = c + di$, iff $a = c$ and $b = d$ (Definition 33.2). Their sum is given by (Theorem 33.9)

$$(a + bi) + (c + di) = (a + c) + (b + d)i.$$

Addition of complex numbers is commutative and associative (Theorem 33.8) and has identity $0 = 0 + 0i$ (Exercise 33.8). The additive inverse of $z = a + bi$ is $-z = -a - bi$ (Exercise 33.8).

The product of $a + bi$ and $c + di$ is given by (Definition 34.13),

$$(a + bi)(c + di) = (ac - bd) + (ad + bc)i.$$

This multiplication is commutative, associative, and distributive over addition (Theorem 34.14). It has identity $1 = 1 + 0i$, and each nonzero complex number $z = a + bi$ has the multiplicative inverse (Theorem 34.14),

$$z^{-1} = \frac{1}{a + bi} = \frac{a}{a^2 + b^2} - \frac{b}{a^2 + b^2}\,\mathrm{i}.$$

Theorem 35.4 states that $i^2 = -1$.

35.16 In Theorems 34.6 and 34.7 it was seen that each complex number $z = a + bi$ can be written in the form (see Fig. 35.16)

$$z = r(\cos\theta + i \sin\theta)$$

where

$$a = r\cos\theta \qquad \text{and} \qquad b = r\sin\theta,$$

and where

$$r = (a^2 + b^2)^{1/2}, \qquad \sin\theta = b/r, \qquad \cos\theta = a/r, \qquad \tan\theta = b/a.$$

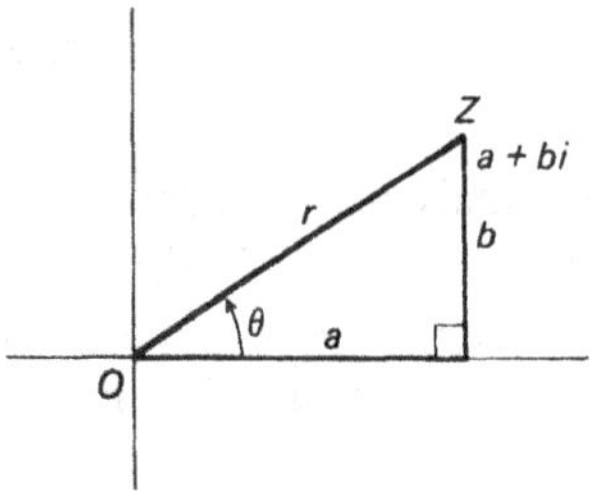

Figure 35.16

From the discussion in 34.8, it is clear that if

$$z = r(\cos\theta + i\sin\theta) \qquad \text{and} \qquad w = s(\cos\phi + i\sin\phi),$$

then

$$zw = rs(\cos(\theta + \phi) = i\sin(\theta + \phi)).$$

The absolute value of $z = a + bi$ is given by

$$|z| = |a + bi| = (a^2 + b^2)^{1/2} = (z\bar{z})^{1/2},$$

where $\bar{z} = a - bi$, the *complex conjugate* of z (Definitions 34.2 and 34.19 and Theorem 34.21). And for complex numbers z and w, we have (Theorems 34.4 and 34.14)

$$|z| \geqslant 0,$$

$$|z| = 0 \quad \text{iff} \quad z = 0,$$

$$|z| - |w| \leqslant |z + w| \leqslant |z| + |w|,$$

and

$$|zw| = |z||w|.$$

Also $\overline{z + w} = \bar{z} + \bar{w}$ and $|z|^2 = |\bar{z}|^2 = z\bar{z}$, by Theorems 34.20 and 34.21.

35.17 Definition If $z = r(\cos\theta + i\sin\theta) = a + bi$, then $a + bi$ is called the *rectangular form* of z, a is its *real part* and b is its *imaginary coefficient*. We write

$$\text{Re}(z) = a \qquad \text{and} \qquad \text{Im}(z) = b.$$

The symbol i is called the *imaginary unit*. Also $r(\cos\theta + i\sin\theta)$ is called the *polar* or *trigonometric form* for z, r is its *absolute value* (see Definition 34.2) or its *amplitude*, and θ is its *angle* or *argument*. It is convenient to define cis θ (read "sis theta") or $e^{i\theta}$ by

$$\text{cis}\,\theta = e^{i\theta} = \cos\theta + i\sin\theta.$$

The expression $re^{i\theta}$ is called the *exponential form* of z.

35.18 Definition In the complex number $z = x + yi$, if $x = 0$, then z is called *pure imaginary*; if $y = 0$, then z is called *real*.

35.19 The rectangular form $a + bi$ is especially useful for purposes of addition and subtraction, not as convenient for multiplication and division, and almost useless for powers and roots. On the other hand, with the trigonometric and exponential forms of complex numbers addition is next to impossible, but multiplication, division, raising to a power, and taking roots are easily performed. Hence it is important to be familiar with all three forms presented in Definition 35.17. The convenience of the trigonometric and exponential forms is illustrated in Exercises 35.6 through 35.11 and 35.18 through 35.23.

35.20 In a course in complex analysis, it is customary to use only radian measure for angles θ, especially when using the form $e^{i\theta}$. Since we are concerned here only with the convenience of such forms to geometry, we shall make no such restriction. In fact, we shall generally use degree measure for our angles.

Exercise Set 35

1. Multiply the vectors (2, 3) and (4, 5), then multiply the corresponding complex numbers and show that the products correspond.
2. Repeat Exercise 35.1 using addition instead of multiplication.
3. Answer the questions posed in 32.3. Compare your answers with those given earlier. Again, save these answers until this chapter has been completed.

4. Show that when $z = a + bi \neq 0$, then $z^{-1} = (a - bi)/(a^2 + b^2)$.
5. Show that when $z \neq 0$, then $z^{-1} = \bar{z}/|z|^2$.
6. Show that $e^{i\theta}e^{i\phi} = e^{i(\theta+\phi)}$.
7. Prove De Moivre's theorem: If n is a natural number, then $(\text{cis}\,\theta)^n = \text{cis}\,n\theta$.
8. Show that $(e^{i\theta})^n = e^{in\theta}$ for all natural numbers n.
9. Prove that $(\text{cis}\,\theta)/(\text{cis}\,\phi) = \text{cis}(\theta - \phi)$.
10. Show that both De Moivre's theorem (Exercise 35.7) and the statement of Exercise 35.8 hold true for n any integer.
11. Rewrite the statement of Exercise 35.9 in exponential form.
12. Evaluate $(5\,\text{cis}\,30°)^4$.
13. Evaluate $(\frac{1}{2} + i\sqrt{3}/2)^{27}$.
14. Show that $e^{i\pi} + 1 = 0$, an interesting identity relating five of the most important constants in mathematics!
15. Show that $z = \bar{z}$ iff z is real.
16. Show that $z = -\bar{z}$ iff z is pure imaginary.
17. Show that $z + \bar{z}$, $z\bar{z}$, and $(z - \bar{z})/i$ are all real.
18. Show that if $w = r^{1/n}\text{cis}((\theta + 360k°)/n)$ for any integers $n > 0$ and k, then $w^n = r\,\text{cis}\,\theta$. Hence w is an nth root of $r\,\text{cis}\,\theta$.
19. Find the three cube roots of 8.
20. Find the five fifth roots of -1.
21. Find the two square roots of i.
22. Find the two square roots of $-i$.
23. Find the four fourth roots of -1.
24. Show that:
    a) $|z| = |\bar{z}|$
    b) $\overline{z + w} = \bar{z} + \bar{w}$
    c) $\overline{(zw)} = (\bar{z})(\bar{w})$
    d) $\overline{(z/w)} = \bar{z}/\bar{w}$
    e) $\bar{\bar{z}} = z$
25. Show that $|z + w|^2 + |z - w|^2 = 2(|z|^2 + |w|^2)$. Compare with Exercise 34.17.
26. Solve $2z + \bar{z} = 5 + i$ for z, making use of the fact that whenever a given equation $w = z$ is true, then the equation $\bar{w} = \bar{z}$ is also true.
27. Solve $(5 - 2i)z + \bar{z} = 6 - 16i$ for z.
28. Solve $|z| - z = 1 + 2i$ for z.
29. Solve $|z| + z = 1 + 5i$ for z.
30. Show that if $a\bar{z} + b = z$ has no solution z, then $|a| = 1$.

SECTION 36 | TRIANGLES IN THE GAUSS PLANE

36.1 In this and the remaining sections of this chapter, all lower-case italic letters shall denote complex numbers; i, of course, is still reserved for the imaginary unit.

Unless stated otherwise, upper-case italic letters denote the points whose affixes are the corresponding lower-case italic letters. Thus we now begin to show the association between plane geometry and the algebra of complex numbers.

36.2 Theorem $\mathrm{Re}(z) = (z + \bar{z})/2$ and $\mathrm{Im}(z) = (z - \bar{z})/2i$.

36.3 Theorem The image of the point whose affix is z under a:
1) halfturn about the origin is $-z$,
2) reflection in the x-axis is $\bar{z}$,
3) reflection in the y-axis is $-\bar{z}$,
4) rotation through angle θ about the origin is $ze^{i\theta} = z \operatorname{cis} \theta$,
5) translation through vector $\overrightarrow{OW}$ is $z + w$,
6) rotation through 90° about the origin is iz.

The proofs of these statements are left for the exercises. □

36.4 Theorem 33.6 tells how the complex numbers z and w should be added. Geometrically, $z + w$ is the affix of the fourth vertex S of the parallelogram having affixes z, 0, w for three consecutive vertices (see Fig. 36.4a). Figure 36.4b shows the corresponding construction for the product zw: Let U be the image of 1. Construct triangle OWP similar to triangle OUZ. Then P is the image of zw.

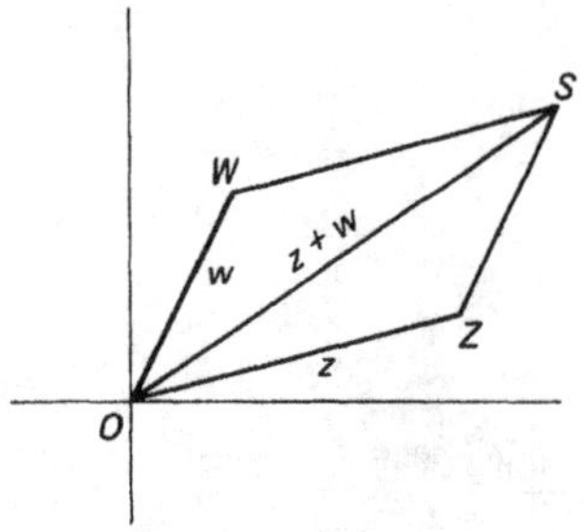

Figure 36.4a

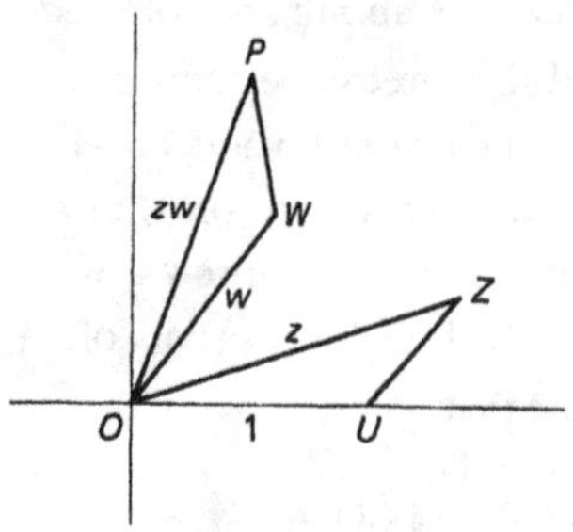

Figure 36.4b

▶**36.5 Problem** To construct $w - z$, take $w - z$ as the affix of the fourth vertex D of the parallelogram $OZWD$ (see Fig. 36.5).

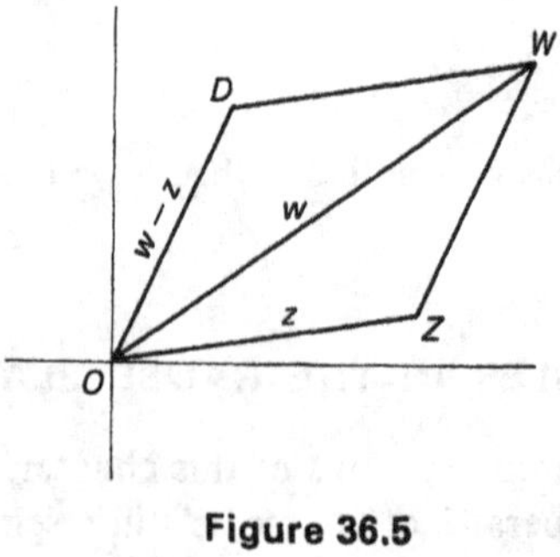

Figure 36.5

▶**36.6 Problem** To construct z^{-1} when $z \neq 0$, let 1 be the affix of U and construct R so that triangles ORU and OUZ are similar. Then R is the image of z^{-1} (see Fig. 36.6).

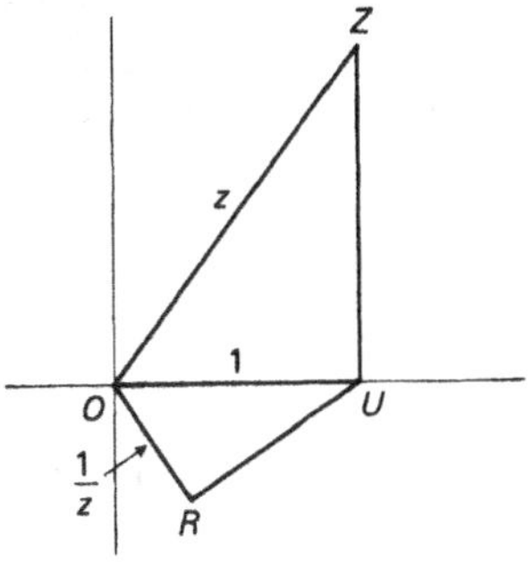

Figure 36.6

Proofs of these last two constructions are left as exercises. ▯

36.7 Theorem A given triangle ABC is equilateral iff

$$(b-a)\lambda^2 = (c-b)\lambda = a-c,$$

where λ equals either

$$(-1+i\sqrt{3})/2 \qquad \text{or} \qquad (-1-i\sqrt{3})/2.$$

Since λ equals either $e^{120°i}$ or $e^{240°i}$, then $(b-a)\lambda$ rotates side AB through either 120° or 240° into side BC, and $(c-b)\lambda$ rotates side BC through the same angle into side CA (see Fig. 36.7). In either case the triangle ABC is equilateral, counterclockwise in the first case and clockwise in the second. Conversely, each equilateral triangle is formed by such rotations. ▯

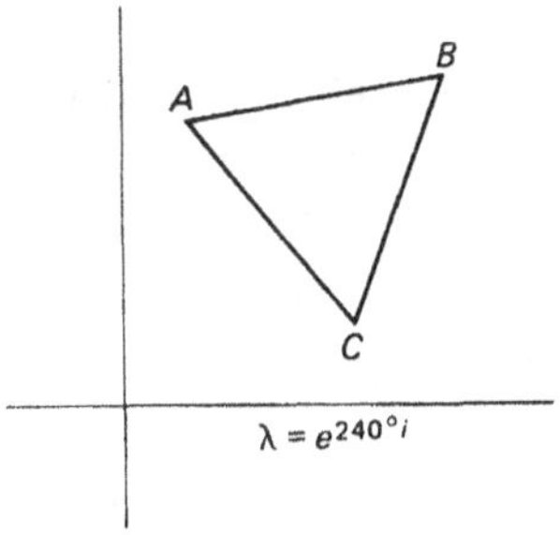

Figure 36.7

36.8 Before continuing our development, let us digress briefly to recall some of the basic properties of determinants. Persons or classes who have not yet had the opportunity to study this topic may wish to spend some extra time here learning the basic concepts of determinant algebra. Although the development given below is sufficient for our modest needs, most texts on analytic geometry or on college algebra contain a fuller treatment of the subject. Put simply, a determinant is a number associated with a square array or *matrix* of numbers in the manner prescribed by the definitions that follow.

36.9 Definition A *second-order determinant* is defined by

$$\begin{vmatrix} a & b \\ c & d \end{vmatrix} = ad - bc.$$

36.10 Definition A *third-order determinant* is defined by

$$\begin{vmatrix} a & b & c \\ d & e & f \\ g & h & j \end{vmatrix} = a\begin{vmatrix} e & f \\ h & j \end{vmatrix} - b\begin{vmatrix} d & f \\ g & j \end{vmatrix} + c\begin{vmatrix} d & e \\ g & h \end{vmatrix}$$
$$= aej - ahf - bdj + bfg + cdh - ceg.$$

36.11 Example

$$\begin{vmatrix} 1 & 2 \\ 3 & 4 \end{vmatrix} = 1 \cdot 4 - 3 \cdot 2 = -2.$$

Also

$$\begin{vmatrix} 1 & 2 & 3 \\ 0 & -2 & 1 \\ 3 & -1 & 0 \end{vmatrix} = 1\begin{vmatrix} -2 & 1 \\ -1 & 0 \end{vmatrix} - 2\begin{vmatrix} 0 & 1 \\ 3 & 0 \end{vmatrix} + 3\begin{vmatrix} 0 & -2 \\ 3 & -1 \end{vmatrix} = 1 \cdot 1 - 2(-3) + 3 \cdot 6 = 25.$$

Although we do not need it as yet, it is convenient to define a fourth-order determinant at this time.

▶**36.12 Definition** A *fourth-order determinant* is defined by

$$\begin{vmatrix} a & b & c & d \\ e & f & g & h \\ j & k & l & m \\ n & o & p & q \end{vmatrix} = a\begin{vmatrix} f & g & h \\ k & l & m \\ o & p & q \end{vmatrix} - b\begin{vmatrix} e & g & h \\ j & l & m \\ n & p & q \end{vmatrix} + c\begin{vmatrix} e & f & h \\ j & k & m \\ n & o & q \end{vmatrix} - d\begin{vmatrix} e & f & g \\ j & k & l \\ n & o & p \end{vmatrix}.$$

36.13 Fifth- and higher-order determinants are defined in just this same manner in terms of determinants of the next-lower order. Of the many properties of determinants, we select only a few basic ones to present here.

36.14 Theorem The value of a determinant is unchanged if a multiple of one row is added to a different row, or if a multiple of one column is added to a different column.

36.15 Theorem If all the elements of a row or of a column of a determinant are multiplied by a constant, then the value of the determinant is multiplied by that constant.

36.16 Corollary If all the elements of a row or of a column of a determinant are zeros, then the value of the determinant is zero.

36.17 Theorem If two rows or two columns of a determinant are proportional, then the value of the determinant is zero.

36.18 Having presented these few basic properties of determinants, we are now able to return to the geometry of complex numbers. Some geometric conditions lend themselves to most elegant determinant forms.

36.19 Theorem Two (perhaps degenerate) triangles ABC and DEF are directly similar iff

$$\begin{vmatrix} a & d & 1 \\ b & e & 1 \\ c & f & 1 \end{vmatrix} = 0.$$

Let the two triangles be similar and suppose that the similarity that maps triangle ABC to DEF is composed of a rotation through angle θ and a homothety of ratio r, and let $z = r\operatorname{cis}\theta$. Then $e - d = z(b - a)$ and $f - d = z(c - a)$. (Note that the centers of the rotation and the homothety need not be located.) Now

$$\begin{vmatrix} a & d & 1 \\ b & e & 1 \\ c & f & 1 \end{vmatrix} = \begin{vmatrix} a-a & d-d & 1 \\ b-a & e-d & 1 \\ c-a & f-d & 1 \end{vmatrix} = \begin{vmatrix} 0 & 0 & 1 \\ b-a & z(b-a) & 1 \\ c-a & z(c-a) & 1 \end{vmatrix} = 0$$

by subtracting multiples of the last column from the other two columns (Theorem 36.14) and then applying Theorem 36.17.

Conversely, if the determinant is zero, then

$$0 = \begin{vmatrix} 0 & 0 & 1 \\ b-a & e-d & 1 \\ c-a & f-d & 1 \end{vmatrix} = \begin{vmatrix} b-a & e-d \\ c-a & f-d \end{vmatrix} = (b-a)(f-d) - (c-a)(e-d),$$

from which it follows that there is a complex number z such that

$$e - d = z(b - a) \qquad \text{and} \qquad f - d = z(c - a).$$

But then triangles ABC and DEF are similar. □

36.20 Corollary A triangle ABC is equilateral iff

$$\begin{vmatrix} a & b & 1 \\ b & c & 1 \\ c & a & 1 \end{vmatrix} = 0.$$

36.21 Corollary Two triangles ABC and DEF are oppositely similar iff

$$\begin{vmatrix} a & \bar{d} & 1 \\ b & \bar{e} & 1 \\ c & \bar{f} & 1 \end{vmatrix} = 0.$$

36.22 Corollary A triangle ABC is isosceles with apex A iff

$$\begin{vmatrix} a & \bar{a} & 1 \\ b & \bar{c} & 1 \\ c & \bar{b} & 1 \end{vmatrix} = 0.$$

Exercise Set 36

1. Prove Theorem 36.2.
2. Prove Theorem 36.3.
3. Prove the construction of Problem 36.5.
4. Prove the construction of Problem 36.6.
5. Show that when $z = re^{i\theta}$, then
   a) $\cos\theta = (\text{Re}(z))/|z| = (z + \bar{z})/2(z\bar{z})^{1/2}$
   b) $\sin\theta = (\text{Im}(z))/|z| = (z - \bar{z})/2i(z\bar{z})^{1/2}$
   c) $\tan\theta = (\text{Im}(z))/(\text{Re}(z)) = (z - \bar{z})/i(z + \bar{z})$
6. Prove Theorem 36.14.
7. Prove Theorem 36.15.
8. Prove Corollary 36.16.
9. Prove Theorem 36.17.
10. a) Prove that a second-order determinant is zero iff its rows are proportional; that is, iff there is a constant k such that each element of one of the rows is k times the corresponding element of the other row.
    b) Show that "rows" can be replaced by "columns" in part (a).
    c) Show that parts (a) and (b) are not true for a third-order determinant.
11. Prove Corollary 36.20.
12. Prove Corollary 36.21.
13. Prove Corollary 36.22.
14. Show that the triangle whose vertices have affixes a, b, and $zb + (1 - z)a$ is directly similar to the triangle with vertices whose affixes are 0, 1, and z.
15. Show that

$$\cos \measuredangle BAC = \frac{(b-a)(\bar{c}-\bar{a}) + (\bar{b}-\bar{a})(c-a)}{2\sqrt{(b-a)(\bar{b}-\bar{a})}\sqrt{(c-a)(\bar{c}-\bar{a})}}.$$

16. Find an expression similar to that in Exercise 36.15 for
    a) $\sin \measuredangle BAC$ b) $\tan \measuredangle BAC$
17. Given that $\lambda = \text{cis}\,60°$ in Theorem 36.7, then show that the figure cannot be a triangle.
18. Show that triangle ABC is equilateral iff $(c - b)^2 = (b - a)(a - c)$.
19. Show that triangle ABC is equilateral iff

$$1/(c - b) + 1/(b - a) + 1/(a - c) = 0.$$

20. What is true of triangle ABC if

$$\begin{vmatrix} a & \bar{a} & 1 \\ b & \bar{b} & 1 \\ c & \bar{c} & 1 \end{vmatrix} = 0?$$

21. What is true of triangle ABC if

$$\begin{vmatrix} a & \bar{b} & 1 \\ b & \bar{c} & 1 \\ c & \bar{a} & 1 \end{vmatrix} = 0?$$

22. Show that when two rows or two columns of a determinant are interchanged, then the value of the determinant is multiplied by −1.

23. Show that

$$\begin{vmatrix} a & b & c+d \\ e & f & g+h \\ j & k & l+m \end{vmatrix} = \begin{vmatrix} a & b & c \\ e & f & g \\ j & k & l \end{vmatrix} + \begin{vmatrix} a & b & d \\ e & f & h \\ j & k & m \end{vmatrix} .$$

24. Generalize Exercise 36.23.

25. Investigate how to expand a third- or higher-order determinant in terms of rows other than the first row (as defined in Definitions 36.10 and 36.12) and also in terms of columns.

26. Define a fifth-order determinant. How many terms appear in the complete expansion of a second-, third-, fourth-, and fifth-order determinant?

27. For any two points A and B in the plane, let us define $A*B = C$, where point C is taken such that triangle ABC is directly similar to a given triangle whose affixes are 0, 1, z.
   a) Show that $*$ is a binary operation on the points of the plane.
   b) Show that $*$ is not commutative.
   c) Show that $*$ is not associative.
   d) Show that $A*A = A$; that is, if $A*A$ is to be defined meaningfully, then we must have $A*A = A$.
   e) Show that, given any two of the three points A, B, and C, we can solve the equation $A*B = C$ for the third point.
   f) Prove the *medial* property: $(A*B)*(C*D) = (A*C)*(B*D)$.
   g) Prove the "distributive law": $A*(B*C) = (A*B)*(A*C)$.
   h) Show that $(A*B)*A = A*(B*A)$. When the given triangle is equilateral, find this common value.

28. Prove that $|z+w|^2 = |z|^2 + |w|^2 + 2\text{Re}(z\bar{w})$.

29. Prove that $|z-w|^2 = |z|^2 + |w|^2 - 2\text{Re}(z\bar{w})$.

30. Prove that $|z+w|^2 + |z-w|^2 = 2(|z|^2 + |w|^2)$.

SECTION 37 | LINES IN THE GAUSS PLANE

37.1 Theorem Points A, B, C are collinear iff

$$\begin{vmatrix} a & \bar{a} & 1 \\ b & \bar{b} & 1 \\ c & \bar{c} & 1 \end{vmatrix} = 0.$$

37.2 Theorem An equation for the line passing through points A and B is

$$\begin{vmatrix} z & \bar{z} & 1 \\ a & \bar{a} & 1 \\ b & \bar{b} & 1 \end{vmatrix} = 0.$$

Clearly the equation is satisfied when $z = a$ and when $z = b$. The theorem then follows by Theorem 37.1. ▯

As an alternative method of proof, expand the determinant to obtain the equation

$$(\bar{a}-\bar{b})z-(a-b)\bar{z}+(a\bar{b}-\bar{a}b)=0.$$

Writing $x+iy$ for z, so $\bar{z}=x-iy$, we find that the equation reduces to

$$(\bar{a}-\bar{b}-a+b)x+i(\bar{a}-\bar{b}+a-b)y+(a\bar{b}-\bar{a}b)=0.$$

Since each of these coefficients is pure imaginary, when we multiply through by i, we obtain an equation of the form $\alpha x+\beta y+\gamma=0$ for real α, β, γ. This is an equation for a straight line. ▯

Theorem 37.2 gives us a form for every straight line, and we observe the next result as a direct corollary to the proof given for that theorem, yielding a simple equation for a line.

37.3 Corollary Each line in the Gauss plane has an equation of the form

$$az+\bar{a}\bar{z}+b=0 \qquad \text{where } b \text{ is real.}$$

37.4 Theorem The line $az+\bar{a}\bar{z}+b=0$, with b real, is satisfied by $-b/2a$ and by $i\bar{a}-b/2a$, and so is perpendicular to the line $O\bar{A}$.

That the two points satisfy the equation is straightforward algebra, since b is real. Then the line joins these two points, so it is parallel to the line from the origin to the image of $i\bar{a}$, a 90° rotation of the line $O\bar{A}$. ▯

37.5 Theorem Two lines $a_1z+\bar{a}_1\bar{z}+b_1=0$, and $a_2z+\bar{a}_2\bar{z}+b_2=0$, with b_1 and b_2 real, are parallel iff a_1/a_2 is real.

37.6 Theorem The two lines of Theorem 37.5 are perpendicular iff a_1/a_2 is pure imaginary.

Testing for parallelism or perpendicularity two lines whose equations are of the form given in Theorem 37.3 is quite simple, according to Theorems 37.5 and 37.6. We may test similarly any two lines whose equations are given in parametric form, as indicated in Theorem 37.7 and Corollaries 37.8 and 37.9.

37.7 Theorem The line through A and parallel to OB has the parametric equation, with real parameter t,

$$z=a+tb.$$

37.8 Corollary The lines $z=a+tb$ and $z=c+ud$, with real parameters t and u, are parallel iff b/d is real.

37.9 Corollary The lines of Corollary 37.8 are perpendicular iff b/d is pure imaginary.

▶**37.10 Theorem** The area of triangle ABC is equal to

$$\pm\frac{i}{4}\begin{vmatrix} a & \bar{a} & 1 \\ b & \bar{b} & 1 \\ c & \bar{c} & 1 \end{vmatrix},$$

whichever is positive.

Let $a-c=\alpha$ and $b-c=\beta$. Then the area of triangle ABC is given by $\frac{1}{2}|\alpha||\beta|\sin C$, so let us find an expression for $\sin C$ (see Fig. 37.10). Now $\beta=(|\beta|/|\alpha|)e^{iC}\alpha$, so we have $e^{iC}=\beta|\alpha|/|\beta|\alpha$. It follows that

$$\sin C=\operatorname{Im}\left(\frac{\beta|\alpha|}{\alpha|\beta|}\right)=\frac{|\alpha|}{|\beta|}\operatorname{Im}\left(\frac{\beta}{\alpha}\right)=\frac{|\alpha|}{|\beta|}\cdot\frac{1}{2i}\left(\frac{\beta}{\alpha}-\frac{\bar{\beta}}{\bar{\alpha}}\right)=\frac{|\alpha|^2}{|\alpha\beta|}\cdot\frac{1}{2i}\left(\frac{\beta}{\alpha}-\frac{\bar{\beta}}{\bar{\alpha}}\right)$$

$$=\frac{\alpha\bar{\alpha}}{|\alpha\beta|}\cdot\frac{1}{2i}\left(\frac{\beta}{\alpha}-\frac{\bar{\beta}}{\bar{\alpha}}\right)=\frac{1}{2i|\alpha\beta|}(\beta\bar{\alpha}-\alpha\bar{\beta}).$$

Now the area K is given by

$$K=\tfrac{1}{2}|\alpha||\beta|\sin C=\frac{|\alpha\beta|}{4i|\alpha\beta|}(\beta\bar{\alpha}-\alpha\bar{\beta})=\frac{1}{4i}(\beta\bar{\alpha}-\alpha\bar{\beta})$$

$$=\frac{1}{4i}[(b-c)(\bar{a}-\bar{c})-(a-c)(\bar{b}-\bar{c})]$$

$$=\frac{1}{4i}(b\bar{a}-a\bar{b}+c\bar{b}-b\bar{c}+a\bar{c}-c\bar{a})=\frac{i}{4}\begin{vmatrix} a & \bar{a} & 1\\ b & \bar{b} & 1\\ c & \bar{c} & 1\end{vmatrix}.$$

We obtained the plus sign in this case because triangle ABC was oriented counterclockwise, as shown in Fig. 37.10. [Where did we use this information?] The minus sign holds when the triangle is clockwise oriented. □

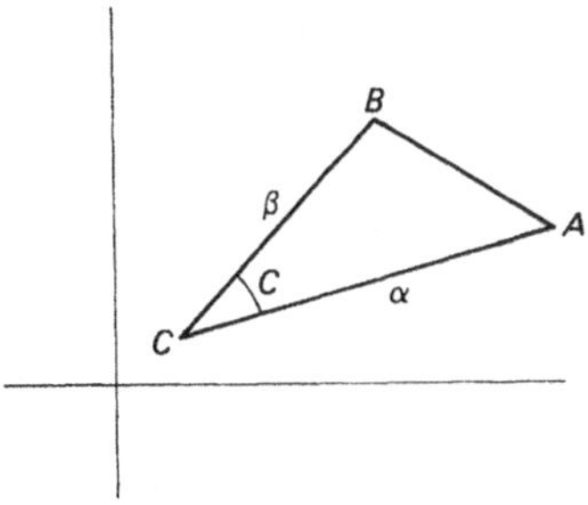

Figure 37.10

37.11 The formula of Theorem 37.10 is convenient for finding areas of triangles when three specific points are given. Its theoretical uses also abound; for example, the area of such a triangle is zero iff the three vertices are collinear. Hence a test for the collinearity of three points A, B, and C is that the determinant of Theorem 37.10 is equal to zero. (See Theorem 37.1.)

Exercise Set 37

1. Prove Theorem 37.1.
2. Show that the line through A and parallel to OB has the equation
$$\begin{vmatrix} z & \bar{z} & 1 \\ a & \bar{a} & 1 \\ b & \bar{b} & 0 \end{vmatrix} = 0.$$
3. through 9. Prove Theorems and Corollaries 37.3 through 37.9.
10. Let $A(a_1, a_2)$, $B(b_1, b_2)$, $C(c_1, c_2)$ be three points in the Cartesian plane. Show that the area of triangle ABC is given by the familiar formula from analytic geometry
$$K = \pm\tfrac{1}{2}\begin{vmatrix} a_1 & a_2 & 1 \\ b_1 & b_2 & 1 \\ c_1 & c_2 & 1 \end{vmatrix}.$$
11. Show that when the triangle ABC is clockwise oriented in Theorem 37.10, then the minus sign holds.
12. Show that the three lines $a_k z + \bar{a}_k \bar{z} + b_k = 0$, $k = 1, 2, 3$ and b_k real, are concurrent iff
$$\begin{vmatrix} a_1 & \bar{a}_1 & b_1 \\ a_2 & \bar{a}_2 & b_2 \\ a_3 & \bar{a}_3 & b_3 \end{vmatrix} = 0.$$
13. Find the point of intersection of the two lines
$$a_1 z + \bar{a}_1 \bar{z} + b_1 = 0 \quad \text{and} \quad a_2 z + \bar{a}_2 \bar{z} + b_2 = 0.$$
14. Show that the lines OA and OB are perpendicular iff $a\bar{b} + \bar{a}b = 0$.
15. Show that the two lines of Theorem 37.5 (and of Exercise 37.13) are perpendicular iff $a_1\bar{a}_2 + \bar{a}_1 a_2 = 0$.
16. Show that the line through point A, and perpendicular to line OB, has the equation
$$\begin{vmatrix} z & \bar{z} & 1 \\ a & \bar{a} & 1 \\ b & -\bar{b} & 0 \end{vmatrix} = 0.$$
17. Find the area of the triangle whose vertices have the affixes:
    a) $0, 1, 3 - 2i$ b) $0, 1, z$
    c) $3, 2i, 3 + 2i$ d) $5 - 3i, 2 + i, -1 + 2i$
18. a) Show that the centroid of triangle ABC has affix $(a + b + c)/3$.
    b) On the sides of a triangle ABC and external to it, construct triangles ABC', BCA', CAB' all similar to a given triangle PQR. Prove the centroids of triangles ABC and $A'B'C'$ coincide.
    c) If the triangle ABC of part (b) is equilateral, show that triangle $A'B'C'$ is also equilateral.
19. Show that if r is real, then $c = (a + rb)/(1 + r)$ is the affix of the point that divides segment AB in the ratio r.
20. Let $A_1B_1C_1$ and $A_2B_2C_2$ be any two triangles. Let $A_3B_3C_3$ be a triangle such that A_3, B_3, C_3 divide A_1A_2, B_1B_2, C_1C_2, respectively, in the same ratio r. Show that the centroids G_i of the three triangles $A_iB_iC_i$ are collinear, and find the ratio in which G_3 divides G_1G_2.

SECTION 38 | THE CIRCLE

In this section we shall treat both circles and lines simultaneously, for their equations arise together. The treatment will follow along the same path as that for lines in Section 37.

38.1 Theorem An equation for the circle with center C and radius r is

$$z\bar{z} - \bar{c}z - c\bar{z} = r^2 - c\bar{c}.$$

This equation is easily derived by squaring both sides of the equation $|z - c| = r$. ▯

38.2 Corollary Any circle has an equation of the form

$$z\bar{z} + az + \bar{a}\bar{z} + b = 0,$$

where b is real. The affix of its center is $-\bar{a}$, and its radius is equal to $(a\bar{a} - b)^{1/2}$.

38.3 Theorem An equation of the form

$$z = \frac{at + b}{ct + d}, \qquad \text{with } ad - bc \neq 0$$

and where t is a real parameter, represents a straight line when $c\bar{d}$ is real, or a circle in all other cases. Furthermore, each line or circle can be written in this parametric form.

The proof of this theorem is straightforward, but quite involved. Hence we shall break up the proof into three steps: (1) eliminate the parameter t and write the resulting equation in a convenient form; then we shall examine this equation when (2) $c\bar{d}$ is real, and (3) in all other cases.

Proof. 1) Taking the conjugate of each number in the given equation and recalling that t is real (so $\bar{t} = t$), we obtain

$$z = \frac{at + b}{ct + d} \qquad \text{and} \qquad \bar{z} = \frac{\bar{a}t + \bar{b}}{\bar{a}t + \bar{d}}.$$

Now solve both equations for t to obtain

$$t = \frac{b - zd}{zc - a} = \frac{\bar{b} - \bar{z}d}{\bar{z}\bar{c} - \bar{a}}.$$

Thus we have

$$z\bar{z}(\bar{c}d - c\bar{d}) + z(c\bar{b} - \bar{a}d) + \bar{z}(a\bar{d} - b\bar{c}) + (b\bar{a} - a\bar{b}) = 0,$$

and finally,

$$(*) \quad z\bar{z}i(\bar{c}d - \overline{\bar{c}d}) + zi(c\bar{b} - \bar{a}d) + \bar{z}i\overline{(c\bar{b} - \bar{a}d)} + i(b\bar{a} - \overline{\bar{a}b}) = 0.$$

In this last equation the coefficient of $z\bar{z}$, and also the constant term, are real.
2) If $c\bar{d}$ is real, then its conjugate $\bar{c}d$ is also real, so $\bar{c}d = \overline{\bar{c}d}$. Then equation (*) becomes

$$zi(c\bar{b} - \bar{a}d) + \bar{z}i\overline{(c\bar{b} - \bar{a}d)} + i(\bar{a}b - \overline{\bar{a}b}) = 0,$$

a straight line by Theorem 37.3.

3) In every other case, $\bar{c}d - \overline{\bar{c}d} \neq 0$, so we may divide equation (*) by that quantity, obtaining

$$z\bar{z} + z\frac{i(c\bar{b} - \bar{a}d)}{i(\bar{c}d - \overline{\bar{c}d})} + \bar{z}\frac{\overline{i(c\bar{b} - \bar{a}d)}}{\overline{i(\bar{c}d - \overline{\bar{c}d})}} + \frac{b\bar{a} - \overline{\bar{a}b}}{\bar{c}d - \overline{\bar{c}d}} = 0,$$

an equation for a circle, since $i(\bar{c}d - \overline{\bar{c}d})$ is real. ▯

38.4 Definition In the parametric equation of Theorem 38.3, no finite value of t yields $z = a/c$. Yet this point lies on the line or circle whenever $c \neq 0$. Let us rectify this omission by defining $t = \infty$ as follows. Replacing t by $1/u$ and clearing fractions, we obtain

$$z = \frac{at + b}{ct + d} = \frac{a/u + b}{c/u + d} = \frac{a + bu}{c + du}.$$

In this last form, set $u = 0$ to get $z = a/c$. Hence we agree that $t = \infty$ means $u = 0$ where $u = 1/t$, and we permit ∞ as a value for a real parameter. To that end, it becomes apparent that when $t = -d/c$, then z does not exist. So we add just one *ideal point* ∞ to the Gauss plane, and we write $a/0 = \infty$ when $a \neq 0$. Now the equation of Theorem 38.3 maps each real number including ∞ to a complex number including ∞.

38.5 Theorem The parametric equation $z = (at + b)/(ct + d)$ is determined to within a constant factor in numerator and denominator by any three values of the parameter t and the corresponding z values.

Geometrically, three points determine a line or a circle. The algebra is left as an exercise. ▯

38.6 It is customary and convenient to consider the z values corresponding to $t = 0$, 1, and ∞, calling them z_0, z_1, and z_∞. In the equation of Theorem 38.5 in particular, note that

$$z_0 = \frac{b}{d}, \qquad z_1 = \frac{a + b}{c + d}, \qquad \text{and} \qquad z_\infty = \frac{a}{c}.$$

These equations yield

$$(c + d)z_1 = cz_\infty + dz_0,$$

in which either c or d may be chosen arbitrarily. Then the other three parameters are uniquely determined.

38.7 The parametric equation of Theorem 38.3 represents either no locus, a single point, or the entire plane when $ad - bc = 0$ and when written in the form

$$(ct + d)z = at + b,$$

depending on the nature of the constants a, b, c, d. The details are not especially important to us.

38.8 Theorem The locus of all points Z whose distances from two fixed points A and B are in the ratio $k \neq 1$, k real, is a circle.

Setting $|z-a|/|z-b| = k$ and squaring, we obtain

$$\frac{(z-a)(\bar{z}-\bar{a})}{(z-b)(\bar{z}-\bar{b})} = k^2,$$

which may be rewritten as

$$(1-k^2)z\bar{z} - (\bar{a}-k^2\bar{b})z - (a-k^2 b)\bar{z} + (a\bar{a} - k^2 b\bar{b}) = 0,$$

an equation for a circle. □

38.9 Theorem The circle of Theorem 38.8 can be written in the parametric form, with real parameter t,

$$\frac{z-a}{z-b} = ke^{it}.$$

For this equation holds iff $|z-a|/|z-b| = k$. □

▶**38.10 Theorem** If $A \neq B$, then the equation of Theorem 38.9 represents a circle when $k \neq 1$, k real, and a line, the perpendicular bisector of AB, when $k = 1$.

▶**38.11 Theorem** For fixed t and with real parameter k, the equation of Theorem 38.9 represents a circle or line through points A and B.

Exercise Set 38

1. Prove Theorem 38.1.
2. Prove Corollary 38.2.
3. Prove Theorem 38.5.
4. Prove Theorem 38.9.
5. Prove Theorem 38.10.
6. Prove Theorem 38.11.
7. Find the center and radius of the circle of Theorem 38.3, part (3).
8. Show that each line can be written in the parametric form of Theorem 38.3.
9. Show that each circle can be written in the parametric form of Theorem 38.3.
10. Find parametric equations for the lines or circles given that z_0, z_1, and z_∞ are:
    a) $0, 1, \infty$ b) $0, 1, i$
    c) $0, 1+i, 3-i$ d) $2, 1+i, 3-i$
    e) $1+i, 1+3i, 2-i$ f) a, b, c
11. Investigate the various cases discussed in 38.7.
12. Find the values of k for which $z = a$ and for which $z = b$ in Theorem 38.11.
13. Show that when $|a| = 1$ or $|b| = 1$, then $|(a-b)/(1-\bar{a}b)| = 1$.
14. For given a and b, find the smallest value of $|(z-a)(z-b)|$.
15. If a and b are the affixes of two vertices of a square, locate the other two vertices for each possible position of the square.
16. The polynomial equation $z^6 - z^5 + z^3 - 7z^2 + 6z - 6 = 0$ has two roots whose sum is zero. Find them.
17. Locate the circumcenter and find the circumradius of triangle ABC.
18. Show that if $az + b\bar{z} = c$ has no solution, then $|a| = |b|$. Is the converse statement true?
19. Solve the equation of Exercise 38.18.
20. Reread Section 31, especially 31.14 and Exercises 31.4 and 31.5.

▶SECTION 39 ISOMETRIES AND SIMILARITIES IN THE GAUSS PLANE

As in the closing sections of Chapters 2 and 3, this last section of Chapter 4 is devoted to the analytic representations of isometries and similarities. We present here equations for these maps in the Gauss plane. Theorems 39.5, 39.7, 39.11, and Corollary 39.13 give the main results. Theorem 39.8 states the interesting result that taking conjugates of the points in the plane is "the basic" opposite isometry.

39.1 Theorem A translation through vector $\vec{OA}$ (mapping Z to Z') has the equation

$$z' = z + a.$$

39.2 Theorem A rotation about point A through angle θ has the representation

$$z' - a = e^{i\theta}(z - a).$$

39.3 Theorem A homothety with ratio k (k real) and center A has the equation

$$z' - a = k(z - a).$$

39.4 Theorem Each equation of the form

$$z' = az + b, \qquad \text{with } a \neq 0,$$

represents
1) a translation if $a = 1$,
2) a rotation if $a \neq 1$ but $|a| = 1$,
3) a homothety if a is real and $a \neq 1$, or
4) the product of a rotation and a homothety if $|a| \neq 1$.

This theorem is readily seen to be true when a is written in the exponential form $re^{i\theta}$, so that

$$z' = zre^{i\theta} + b.$$

In this form it is clear that we have
1) a rotation about O through angle θ,
2) a homothety in center O and ratio r, and then
3) a translation through vector $\vec{OB}$. ▯

39.5 Theorem Each direct similarity may be written in the form

$$z' = az + b, \qquad \text{with } a \neq 0.$$

39.6 Theorem Each direct similarity maps circles into circles and lines into lines.

The direct similarity $z' = \alpha z + \beta$ maps the equation

$$z = \frac{at + b}{ct + d} \qquad \text{into} \qquad z' = \alpha\frac{at + b}{ct + d} + \beta.$$

That is,

$$z' = \frac{(\alpha a + \beta c)t + (\alpha b + \beta d)}{ct + d}.$$

Since this last equation has the same denominator as the original one had, it still represents the same type of curve, circle or line, as the original equation, according to Theorem 38.3. It is left to the reader to show that since $ad - bc \neq 0$ is given, then the corresponding inequality is true in the new equation. ▯

39.7 Theorem Each direct similarity is determined by two given points A and B and their images A' and B', and has the equation

$$\begin{vmatrix} z' & z & 1 \\ a' & a & 1 \\ b' & b & 1 \end{vmatrix} = 0.$$

By Theorem 36.19. ▯

39.8 Theorem A reflection in the x-axis has the equation

$$z' = \bar{z}.$$

39.9 Theorem A reflection in the line m crossing the real axis at point A with angle of inclination θ has the equation

$$z' = a + e^{2i\theta}(\bar{z} - a).$$

Let x denote the real axis. Then $\sigma_m = (\sigma_m \sigma_x)\sigma_x$, a reflection in the x-axis followed by a rotation about A through angle 2θ. The theorem now follows from Theorems 39.8 and 39.2. ▯

39.10 Theorem A reflection in the line m parallel to the real axis and passing through point B has the equation

$$z' = \bar{z} + 2i\operatorname{Im}(b).$$

39.11 Theorem Each opposite similarity has an equation of the form

$$z' = a\bar{z} + b \qquad \text{with } a \neq 0.$$

39.12 Theorem A reflection in the line AB has the equation

$$\frac{a - z'}{b - z'} = \frac{\bar{a} - \bar{z}}{\bar{b} - \bar{z}}.$$

For any given point Z in the plane, triangles ABZ and ABZ' are oppositely similar, so we have

$$\begin{vmatrix} z' & \bar{z} & 1 \\ a & \bar{a} & 1 \\ b & \bar{b} & 1 \end{vmatrix} = 0$$

by Corollary 36.21; that is,

$$\bar{a}z' - \bar{b}z' - a\bar{z} + b\bar{z} + a\bar{b} - \bar{a}b = 0,$$

which may be rewritten in the form

$$\bar{a}b - \bar{a}z' - \bar{z}b + \bar{z}z' = a\bar{b} - a\bar{z} - z'\bar{b} + z'\bar{z}.$$

Now factor this equation to get

$$(\bar{a} - \bar{z})(b - z') = (a - z')(\bar{b} - \bar{z}),$$

from which the theorem follows. ▯

39.13 Corollary An opposite similarity is determined by two given points A and B and their images A' and B', and has the equation

$$\begin{vmatrix} z' & \bar{z} & 1 \\ a' & \bar{a} & 1 \\ b' & \bar{b} & 1 \end{vmatrix} = 0.$$

39.14 Theorem Each opposite similarity maps circles into circles and lines into lines.

Exercise Set 39

1. Prove Theorem 39.1.
2. Prove Theorem 39.2.
3. Prove Theorem 39.3.
4. Answer again the questions posed in 32.3. Compare your current answers with those given earlier.
5. Prove Theorem 39.5.
6. Complete the proof of Theorem 39.6.
7. Verify Theorem 39.9 by writing the desired reflection as the product of a rotation about A through angle $-\theta$, followed by a reflection in the real axis, and then followed by a rotation about A through angle θ.
8. Prove Theorem 39.8.
9. Under what conditions on a and b is the direct similarity of Theorem 39.5 involutoric? Hence deduce the conditions under which it represents a halfturn.
10. Prove Theorem 39.10.
11. Prove Theorem 39.11.
12. Under what conditions on a and b is the opposite similarity of Theorem 39.11 involutoric? Hence deduce the conditions under which it represents a reflection.
13. Solve the equation of Theorem 39.12 for z' and write it in the form of Theorem 39.11.
14. Prove Theorem 39.14.
15. Prove that the product of two translations is a translation.
16. Find the analytic expression for the product of two rotations, and show when this product represents a translation.
17. Write an equation for a reflection in the imaginary axis.
18. Write an equation for a reflection in the line $y = x$.
19. Write an equation for the halfturn about the origin.
20. Write an equation for a halfturn about point A.

5 INVERSION

SECTION 40 MATCHLESS MODERN MATHEMATICS

40.1 During the eighteenth and nineteenth centuries many advances were made in analytic geometry and in the calculus, as well as in synthetic geometry. The methods of the calculus were applied to the study of various curves and problems which defined curves, ushering in a new and fascinating aspect of Euclidean geometry. We shall not, however, consider such work here. Some of the workers that are of interest to us follow.

40.2 Giovanni Ceva (1648–1737), an Italian engineer, studied transversals of geometric figures, publishing the theorem which bears his name (see Theorem 5.2) in 1678 in his *De lineis rectis se invicem secantibus.* It is curious that this theorem should have escaped discovery for so long, since its dual, Menelaus' theorem, was "well known" in A.D. 100, according to Menelaus.

40.3 Matthew Stewart (1717–1785), a professor of mathematics at Edinburgh, extended the theorems of Ceva (see Exercise 2.10).

40.4 Solid analytic geometry was developed by Antoine Parent (1666–1716), and furthered by Alexis Claude Clairaut (1713–1765), one of twenty children of a teacher of mathematics. He and a brother who died of smallpox at age 16 were "two of the most precocious mathematicians of all time," according to Howard Eves.

40.5 In 1706 William Jones (1675–1749) was the first to use the Greek letter π to denote the ratio 3.14159 . . . of a circle's circumference to its diameter.

40.6 Projective geometry was rediscovered by Gaspard Monge (1746–1818), whose outstanding work spurred great interest among mathematicians in this field. At a party once he overheard a scoundrel slandering a certain woman, and immediately leapt to her defense. Some time later he was singularly attracted to a woman, and when they were introduced, he found that she was the very one whose honor he had defended. Later they were married, and a more loyal wife would have been very difficult for anyone to find.

40.7 Constructions using compass alone were investigated by Lorenzo Mascheroni (1750–1800), who proved that all ruler and compass constructions could be accomplished with the compass alone. This assumes, of course, that a line is determined whenever any two points lying on it are located. His method of attack leaned heavily on the reflection transformation.

40.8 After reading Mascheroni's work, Jean-Victor Poncelet (1788–1867) examined straightedge constructions, proving that all ruler and compass constructions could be performed with a straightedge alone, provided that one had available at least one circle of any size along with its center.

40.9 Jacob Steiner (1796–1863) investigated ruler constructions further. Called "the greatest geometrician since the time of Euclid," he extended Pascal's mystic hexagram theorem, which states that the three points of intersection of pairs of opposite sides of a hexagon inscribed in a conic section are collinear (see Exercise 23.2), by showing that if the six vertices are taken in all possible orders, the resulting 60 "Pascal lines" pass three by three through 20 "Steiner points." These Steiner points lie four by four on 15 "Plücker lines." Others showed the Pascal lines also concur three by three in 60 "Kirkman points" which lie three by three upon 20 "Cayley lines" which pass four by four through 15 "Salmon points"!

40.10 Malfatti's problem is to inscribe three circles in a triangle so that each circle is tangent to two sides of the triangle and to the two other circles. Steiner gave a synthetic construction, noting that the problem has 32 solutions.

40.11 Joseph Diaz Gergonne (1771–1859), and Poncelet independently, discovered the principle of duality in projective geometry. Gergonne also solved the problem of Apollonius: to draw with ruler and compass a circle tangent to each of three given circles. The general case has eight solutions.

40.12 His only claim to fame being the theorem that the ninepoint circle is tangent to the four equicircles of a triangle (see Theorem 44.6), Karl Wilhelm Feuerbach (1800–1834) published his theorem when only 22. Having been a political prisoner for a time at age 19, he gradually went mad, spending the last six years of his life locked away from the world.

40.13 During the nineteenth century the three famous problems of antiquity, to trisect an angle, to construct the side of a cube whose volume is twice that of a given cube, and to construct a square equal in area to a given circle, all using only Euclidean tools, were proved impossible. This theorem, which leans heavily on field theory, proves that no one can ever perform these constructions if he uses the ruler and compass correctly. Any construction that claims to have solved these problems either gives only an approximation to the correct result, or uses the tools improperly. This fact has been mathematically proved—just as it is easy to prove, from the rules of the game, that it is impossible to have thirteen kings of one color in a game of checkers. Also, π was proved irrational in 1770, and transcendental in 1882.

40.14 Other contributors to the growth of geometry include Arthur Cayley (1821–1895), Karl von Staudt (1798–1867), Michel Chasles (1793–1880), Johann Plücker (1801–1868), Henri Brocard (1845–1922), and Émile Lemoine (1840–1912). In 1899 David Hilbert (1862–1943) published his famous *Grundlagen der Geometrie* (*Foundations of Geometry*), a careful development of Euclidean geometry from a set of postulates.

40.15 The most significant advance of the nineteenth century was a change in mathematical thinking. Euclidean geometry had been considered the only true geometry that could ever exist. No different type of geometry was conceivable. But mathematicians had been bothered for more than 20 centuries by Euclid's fifth postulate, the "parallel postulate." His first four postulates are clear, concise, and reasonably obvious "truths" about the real world. His fifth postulate, especially in its original form, is none of these. It is not clear without a picture, it is quite wordy, and not at all obvious (see Exercise 40.8).

Since they were *positive* that it was true, mathematicians right from Greek times attempted to show that the fifth postulate could be deduced from the other four. If so, then this troublesome assumption could be simply deleted from the list of postulates. From their many attempts, several statements equivalent to the parallel postulate were discovered, but each attempted proof was shown to have some flaw.

40.16 Girolamo Saccheri (1667–1733) in 1733, Johann Lambert (1728–1777) in 1766, and Adrien-Marie Legendre (1752–1833) in 1794 to 1823, each attempted to prove the parallel postulate, and in so doing, proved many basic and interesting theorems in so-called *absolute geometry*, Euclidean geometry without the parallel postulate. Had one of them recognized and admitted the impossibility of his attempted program, he would have been credited with the discovery of non-Euclidean geometry.

40.17 The non-Euclidean geometry in which through a point not on a given line there is more than one line not intersecting (parallel to) the given line, was discovered independently first by Carl F. Gauss (1777–1855), the greatest mathematician of his age, then by Johann Bolyai (1802–1860), and finally by Nicolai Lobachevsky (1793–1856). Lobachevsky, however, was in 1829 the first to publish his findings, so this geometry generally bears his name.

40.18 Now geometry was free of its traditional mold. Hereafter, a geometry is simply a collection of assumptions about a set of things called points, and the logical results thereof. Many new geometries followed. In 1854 Bernhard Riemann (1826–1866) developed a geometry, realized in the geometry of great circles on a sphere, in which there are no parallel lines whatever.

40.19 Generalization now became the motto in mathematics. Henri Poincaré (1854–1912), who showed Lobachevskian geometry to be as consistent as Euclidean geometry, developed the geometry of continuous functions, namely topology. Maurice Fréchet (1878–   ) in 1906 began the study of very abstract spaces. Today it is quite difficult to state just where geometry leaves off and other branches of mathematics begin.

40.20 In an attempt to obtain order out of the ever-increasing mass of new geometries, Felix Klein (1849–1929) in 1872 defined a *geometry* to be "the study of those properties that are invariant when the elements (points) of a given set are subjected to the transformations of a given transformation group." This codification, while quite convenient and useful even today, is now outdated by geometries that do not fit this classification scheme.

40.21 Thus mathematics has changed from the material axiomatics, the study of the real world, of the Greeks to the *formal axiomatics* of today. Now one is completely free to state a collection of postulates about any set of objects, so long as the postulates do not contradict one another (not always an easy question to answer), and to investigate the consequences thereof. In this way, hundreds of geometries and algebras have been investigated. And the end is not in sight. New mathematics is being constructed daily.

Exercise Set 40

1. Show how to construct, with Euclidean compass alone, each of the points indicated.
   a) Given points A and B, construct point C on line AB so that $d(AB) = d(BC)$.
   b) Given points A, B, X, construct the reflection of X in line AB.
   c) Given points A, B, C, D with C not on line AB, construct the points of intersection of line AB with the circle $C(D)$ (with center C and passing through D).
   d) Look up in Eves' *Survey of Geometry*, Vol. 1, the Mohr-Mascheroni construction theorem, pages 198–202.
2. Given a bisected segment AB and a point P not on line AB, construct a line through P parallel to AB using ruler alone.
3. Given two parallel lines and points A and B on one of them, find the midpoint of segment AB using ruler alone.
4. Look up in Eves' *Survey of Geometry*, Vol. 1, the Poncelet-Steiner construction theorem, pages 204–209.
5. Show that 6 points, no three of which are collinear, are the vertices of 60 different hexagons.
6. a) Trisect a right angle using Euclidean tools; that is, construct a 30° angle.
   b) Explain how it is still impossible to "trisect an angle" in spite of the construction of part (a).
7. Look up in Eves' *Survey of Geometry*, Vol. 2, the proof of the impossibility of the three construction problems of antiquity, pages 30–38. This material is also to be found in the same author's *The Foundations and Fundamental Concepts of Mathematics*, rev. ed., pages 317–325.
8. Euclid's fifth postulate reads, "If a straight line falling on two straight lines makes the interior angles on one side of the straight line together less than two right angles, then the two straight lines, if extended sufficiently, will meet on that side of the given line on which

the two interior angles are together less than two right angles." Interpret this statement and draw an explanatory figure. Do you feel that its "truth" is "obvious"?

9. Referring to Klein's definition of a geometry as given in 40.20, explain why congruence is studied in Chapter 2 (Isometries), and similarity of geometric figures in Chapter 3 (Similarities). What properties should be studied in Chapter 5 (Inversion)?

SECTION 41 | INVERSION

41.1 Although not generally considered a part of high school geometry, inversion should be known to the earnest student of geometry. This transformation does not preserve congruence or even similarity, but it is useful for the quantities it does preserve. Inversion, it will be seen, preserves the set of all circles and lines; that is, a circle maps into a circle or a line, and a line inverts into a circle or a line. It preserves cross ratios and angles between curves. It seems hard to imagine that a transformation not preserving lines, distances, or shapes can be worth studying. Indeed, inversion is most worthwhile, as the applications in Section 44 will demonstrate.

41.2 In order that inversion be a transformation of the plane, it is necessary to append to the plane exactly one ideal point in essentially the same manner as we did for the Gauss plane in Definition 38.4.

41.3 Definition The Euclidean plane with one *ideal point* or *point at infinity* (∞) appended is called the *inversive plane*. This ideal point is considered to lie on every line in the plane.

41.4 Definition Let C be a fixed point in the plane, but not the ideal point, and let r be a fixed positive number. For each point P in the plane we define its *inverse* P' in the circle of center C and radius r by taking P' as that point on ray CP such that

$$(\mathbf{CP})(\mathbf{CP'}) = r^2$$

whenever $P \neq C$. The *inverse* of C is the ideal point, and the *inverse* of the ideal point is C (see Fig. 41.4). This map of the points of the plane is called *inversion* (*in a circle*), or sometimes *reflection in a circle*, point C is the *center* of the inversion, r is its *radius*, and r^2 is its *power*.

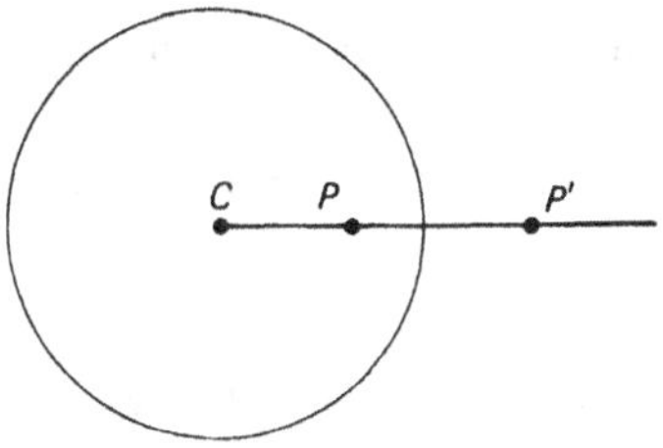

Figure 41.4

41.5 Theorem Inversion is a transformation of the inversive plane and is involutoric.

41.6 Theorem Points inside the circle of inversion map to points outside the circle, and points outside map inside. Points on the circle are invariant and are the only invariant points.

If P lies inside the circle of inversion with center C and radius r, then $CP < r$. Since $\mathbf{CP} \cdot \mathbf{CP'} = r^2$, then CP' cannot be less than or equal to r, for then $\mathbf{CP} \cdot \mathbf{CP'} < r^2$. Hence $CP' > r$, so P' lies outside the circle. The rest of the proof is similar. ▯

41.7 Definition A transformation of the plane is called *conformal* iff it preserves angles between curves. It is *directly* conformal if it preserves the sense of the angle as well, and *inversely* conformal if it reverses the sense.

41.8 Theorem Isometries and similarities are conformal; direct similarities are directly conformal, and opposite similarities are inversely conformal.

41.9 Theorem Let P and Q be two points not collinear with the center of inversion C, and let P' and Q' be their inverses. Then triangles CPQ and $CQ'P'$ are oppositely similar.

Letting r be the radius of inversion, we have, as seen in Fig. 41.9,

$$CP \cdot CP' = r^2 \quad \text{and} \quad CQ \cdot CQ' = r^2.$$

It follows that

$$CP \cdot CP' = CQ \cdot CQ' \quad \text{and} \quad \frac{CP}{CQ} = \frac{CQ'}{CP'}.$$

Since also $\measuredangle QCP = \measuredangle P'CQ'$, then $\triangle CPQ \sim \triangle CQ'P'$ by SAS. Since the similarity can be realized by reflecting triangle CPQ in the bisector of angle C and then applying the homothety $H(C, CQ'/CP)$, it follows that the similarity is opposite. ▯

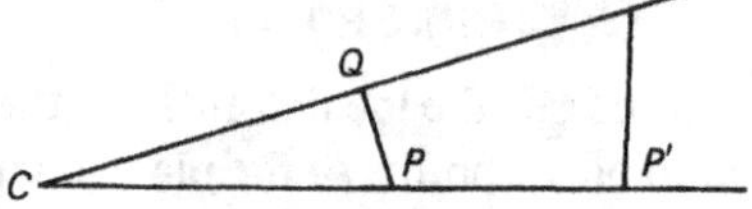

Figure 41.9

41.10 Theorem Inversion is inversely conformal.

Let the curves c and s intersect at point A and suppose that CPQ is a ray different from CA, and intersecting c and s at P and Q, respectively, where C is the center of inversion. Let A', P', Q', c', s' be the inverses of A, P, Q, c, s (see Fig. 41.10). Then $P' \in c'$ and $Q' \in s'$. Now

$$\triangle CAP \sim \triangle CP'A' \quad \text{and} \quad \triangle CAQ \sim \triangle CQ'A'$$

by Theorem 41.9. By subtracting congruent angles in these similar triangles, we obtain

$$\measuredangle PAQ \cong \measuredangle Q'A'P';$$

that is, angles PAQ and $P'A'Q'$ are oppositely congruent. Furthermore, these angles are congruent no matter how small angle ACP (greater than zero) becomes. By taking the limit as $\angle ACP$ approaches 0, angles PAQ and $P'A'Q'$ become the

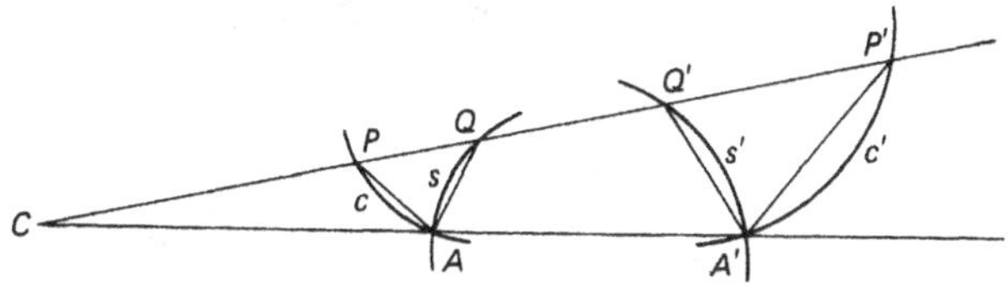

Figure 41.10

angles between the curves c and s, and c' and s'. It follows by the continuity of the inversion transformation that these angles between the two curves and between their inverses are oppositely congruent. ▯

41.11 Theorem A line through the center of inversion is invariant.

41.12 Theorem A line not through the center C of inversion maps into a circle through C.

Drop a perpendicular CP from the center C to the given line m. Let Q be any other point on line m (see Fig. 41.12). Let P' and Q' be the inverses of P and Q. By Theorem 41.9, triangles CPQ and $CQ'P'$ are similar. Hence $\angle CQ'P' \cong \angle CPQ = 90°$, so it follows that Q' lies on the circle of diameter CP'. Since the ideal point maps to C, each point of line m maps to some point on that circle.

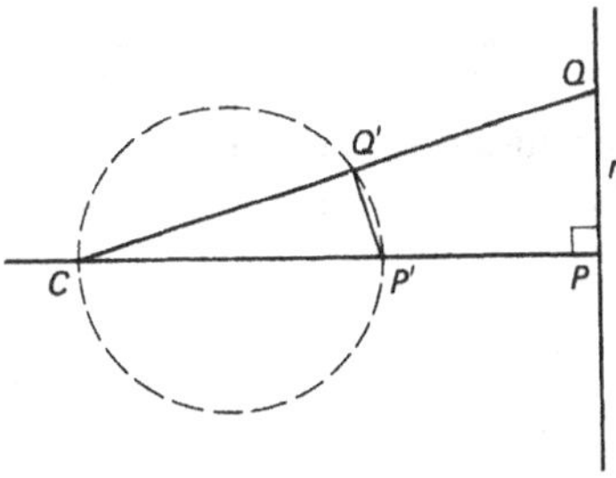

Figure 41.12

Furthermore, for each point Q' (other than C) on that circle, ray CQ' cuts line m in a point Q, and from the first paragraph of this argument, Q inverts to Q'. Hence the image of the line m is the entire circle on CP' as diameter. ▯

41.13 Corollary A circle through the center of inversion maps into a line not through the center of inversion.

41.14 Theorem A circle not through the center of inversion maps into another such circle.

Let that diameter of the given circle s that passes through the center C of inversion cut s in points A and B, and let A' and B' be their inverses as in Fig. 41.14. Take any other P point on s and let its image be P'. By Theorem 41.9,

$$\triangle CAP \sim \triangle CP'A' \qquad \text{and} \qquad \triangle CBP \sim \triangle CP'B'.$$

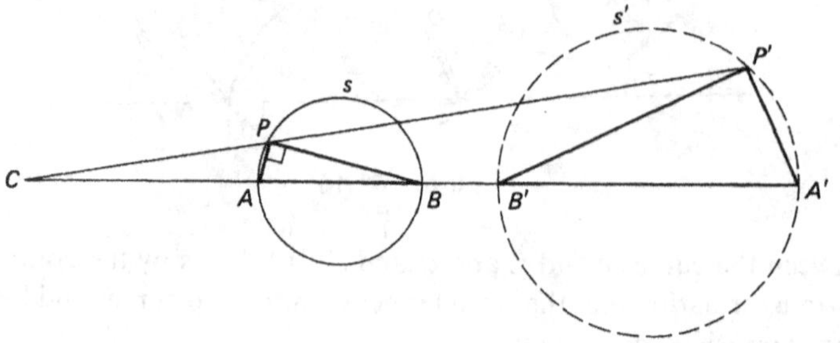

Figure 41.14

From these similar triangles obtain

$$\measuredangle A'P'B' = \measuredangle A'P'C - \measuredangle B'P'C = \measuredangle PAC - \measuredangle PBC = \measuredangle APB = 90^\circ,$$

since PAC is an exterior angle of triangle APB, which triangle is inscribed in a semicircle. Hence P' lies on the circle s' of diameter $A'B'$. It readily follows that the inverse of circle s is the entire circle s'. ▯

▶**41.15 Example** Find the inverse of a right triangle ABC in vertex C, the vertex of the right angle.

Draw the triangle as in Fig. 41.15a. Then assume a convenient radius of inversion, say the length of the longer leg CB. Now lines CB and CA invert into themselves, so draw two perpendicular lines meeting at C, as in Fig. 41.15b. Take B' on one of them so that $CB \cong CB'$. Take A' on the other ray so that $CA \cdot CA' = (CB)^2$. Since

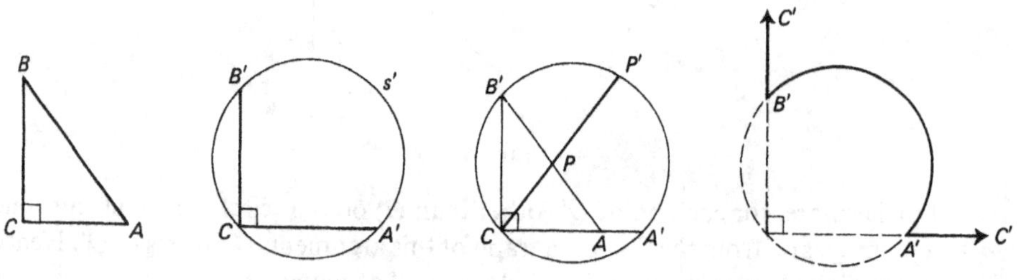

Figure 41.15a Figure 41.15b Figure 41.15c Figure 41.15d

$CA < CB$, then $CA' > CB'$. Finally, line AB inverts into a circle s' through A' and B' and through the center C. Thus we obtain Fig. 41.15b. To see what arcs and segments are the images of the sides AB, BC, CA, superimpose (perhaps mentally) triangle

ABC onto Fig. 41.15b to obtain Fig. 41.15c. Draw to any point P on segment AB, for example, ray CP to cut circle s', the image of side AB, in point P'. Then P' is the inverse of P, and since this is true, it follows that arc $A'P'B'$ is the inverse of segment AB. Since the inverse of C is the ideal point, then the images of CA and CB are the (infinite) segments, the rays, from A' and B' to infinity in the directions away from C. These results are shown in Fig. 41.15d. Here we see the inverse $A'B'C'$ of triangle ABC.

41.16 We see that inversion preserves, but reverses the sense of, angles between curves, and that it maps circles and lines into circles and lines. The circle of inversion is invariant, and circles concentric to it map into other concentric circles. As the radius of such a circle shrinks to zero so that the circle shrinks to the center of inversion, the radius of its image circle increases without bound. Thus, when considering that the point at infinity is the inverse of the center of inversion, one is reminded of the fable of the rider who was so excited that he "jumped up on his horse and rode off madly in all directions".

Exercise Set 41

1. Find the inverse of each of the following points in the circle with radius 10 and centered at the origin.
   a) (6, 8) b) (−6, 8) c) (10, 0)
   d) (0, −10) e) (1, 0) f) (1, −1)
   g) (10, 10) h) (3, 5) i) (2, −7)
   j) (0, 0) k) the ideal point
2. Invert each of the points in Exercise 41.1 in the circle centered at (5, 0) with radius 5.
3. Prove Theorem 41.5.
4. Complete the proof of Theorem 41.6.
5. Prove Theorem 41.8.
6. Prove Theorem 41.10 for the case in which there is no ray CPQ, different from CA, and cutting curves c and s in P and Q.
7. Prove Theorem 41.11.
8. Prove Corollary 41.13.
9. Draw the figure for Theorem 41.14 assuming the center C lies somewhere inside circle s. How does the proof need to be altered for this case?
10. Draw the inverse of a right triangle ABC with the midpoint M of its hypotenuse as center of inversion.
11. Draw the inverse of a square and its diagonals, with its center as center of inversion.
12. Draw the inverse of a square and its diagonals, with a vertex as center of inversion.
13. Invert an equilateral triangle in its centroid.
14. Invert an equilateral triangle in a vertex.
15. Prove that any circle *orthogonal* (perpendicular) to the circle of inversion is invariant.
16. Prove that any circle that is invariant under an inversion is orthogonal to, or coincides with, the circle of inversion.

17. Given that points P and Q are inverted to P' and Q' in a circle of radius r and center C, prove that $P'Q' = (r^2 \cdot PQ)/(CP \cdot CQ)$.

18. a) Prove that when circle s inverts into circle s', then the center of s' is *not* the inverse of the center of s.

    b) Show that when circle s inverts into circle s', then the inverse of the center of s is the inverse of the center of inversion with respect to inversion in s'.

SECTION 42 PROGRESSIONS, RATIOS, AND PEAUCELLIER'S CELL

The theory of inversions is developed further in this section by presenting relations between circles and inversion, progressions and inversion, and cross ratios and inversion. Also defined here is the *Peaucellier cell*, a mechanical tool or linkage of the same sort as the straightedge and the compass. Where the straightedge allows us to draw lines, and the compass permits drawing circles, the Peaucellier cell is designed to draw the inverse of a given curve. Now, we draw circles with the compass, and not by tracing disks. Furthermore, it is not necessary to have a circle in order to build a compass. But we *trace* a given straight line when we use a straightedge. Thus the theoretical question arises as to where the "first" straight line ever came from. One answer to that question is given in Exercise 42.7, for a Peaucellier cell can be used to obtain a line without requiring any previously drawn lines for its construction.

42.1 Definition A transformation of the plane is called *circular* iff it maps circles and lines into circles and lines.

42.2 Theorem Isometries, similarities, and inversions are circular transformations.

42.3 Theorem If P and P' are distinct points inverse in a circle c, then any circle s through P and P' is *orthogonal* (perpendicular) to circle c.

Let C be the center of circle c, and draw the tangent CT to circle s (see Fig. 42.3).

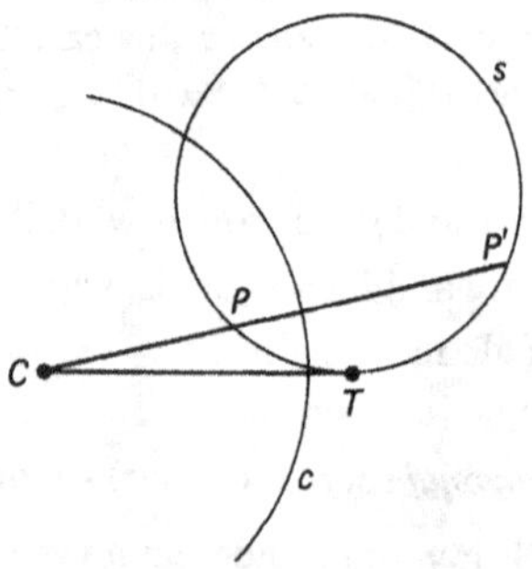

Figure 42.3

Then $\mathbf{CP}\cdot\mathbf{CP}' = (CT)^2$, by Exercise 6.18. But also $CP\cdot CP' = r^2$, where r is the radius of circle c, since P and P' are inverse points in circle c. It follows that $CT = r$, so T lies on circle c, and the two circles are orthogonal. ▯

42.4 Theorem If two circles are orthogonal, let a radius of one circle cut the other circle in two distinct points. Then these two points are inverse with respect to the first circle (see Fig. 42.3).

42.5 Theorem Two points P and P' are inverse in a circle s iff the diameter AB of s that passes through P and P' is divided harmonically by P and P' (see Definition 3.11).

Let C be the center of circle s, as in Fig. 42.5. Then the following statements are equivalent one to the other:

$$(AB, PP') = -1,$$

$$\frac{\mathbf{AP}}{\mathbf{PB}} = -\frac{\mathbf{AP}'}{\mathbf{P'B}},$$

$$\mathbf{AP}\cdot\mathbf{BP}' = -\mathbf{AP}'\cdot\mathbf{BP},$$

$$(\mathbf{AC} + \mathbf{CP})(\mathbf{BC} + \mathbf{CP}') = -(\mathbf{AC} + \mathbf{CP}')(\mathbf{BC} + \mathbf{CP}),$$

$$(\mathbf{AC} + \mathbf{CP})(-\mathbf{AC} + \mathbf{CP}') = -(\mathbf{AC} + \mathbf{CP}')(-\mathbf{AC} + \mathbf{CP}),$$

$$-(AC)^2 + \mathbf{AC}\cdot\mathbf{CP}' - \mathbf{AC}\cdot\mathbf{CP} + \mathbf{CP}\cdot\mathbf{CP}' = (AC)^2 - \mathbf{AC}\cdot\mathbf{CP} + \mathbf{AC}\cdot\mathbf{CP}' - \mathbf{CP}\cdot\mathbf{CP}',$$

and finally

$$\mathbf{CP}\cdot\mathbf{CP}' = (AC)^2 = r^2.$$

The theorem follows. ▯

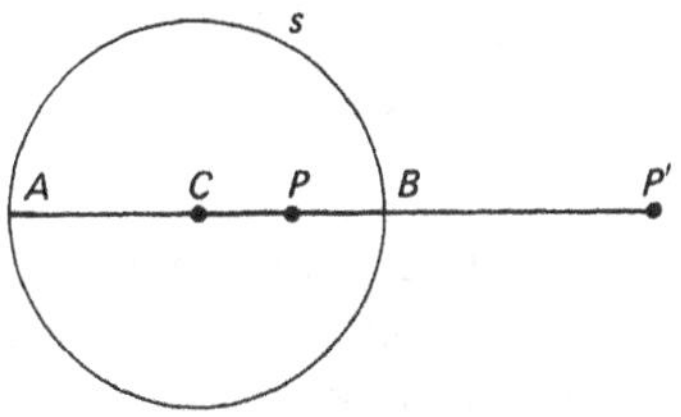

Figure 42.5

42.6 Definition Three numbers a, b, c are said to be in:

1. *Arithmetic progression* iff $c - b = b - a$. We also say that b is the *arithmetic mean* of a and c.
2. *Geometric progression* iff $c/b = b/a$ and $a > 0$, $b > 0$, $c > 0$. We also say that b is the *geometric mean* of a and c.

3. *Harmonic progression* iff $1/a$, $1/b$, $1/c$ are in arithmetic progression, and $a \neq 0$, $b \neq 0$, $c \neq 0$. We also say that b is the *harmonic mean* of a and c.

42.7 Corollary If the two points P and P' of Theorem 42.5 are inverse to each other in circle s, then the segments AP, AB, and AP' (with B lying between P and P') are in harmonic progression, and the segments CP, CB, and CP' are in geometric progression.

We must show that, in Fig. 42.5,

$$\frac{1}{\mathbf{AB}} - \frac{1}{\mathbf{AP}} = \frac{1}{\mathbf{AP'}} - \frac{1}{\mathbf{AB}} \qquad \text{or} \qquad \mathbf{AB} = \frac{2\,\mathbf{AP}\cdot\mathbf{AP'}}{\mathbf{AP}+\mathbf{AP'}}$$

to prove that AP, AB, AP' are in harmonic progression; that is, that AB is the harmonic mean of AP and AP'. It is easy to show that the two displayed equations are equivalent.

Also we must show, to verify the geometric mean property, that

$$\frac{CP}{CB} = \frac{CB}{CP'} \qquad \text{or} \qquad CB = \sqrt{CP\cdot CP'},$$

which is true by the definition of inverse points.

To prove the first relation, start with

$$\mathbf{CP}\cdot\mathbf{CP'} = r^2,$$
$$(\mathbf{CA}+\mathbf{AP})(\mathbf{CA}+\mathbf{AP'}) = r^2 = (CA)^2,$$
$$(CA)^2 + \mathbf{CA}(\mathbf{AP}+\mathbf{AP'}) + \mathbf{AP}\cdot\mathbf{AP'} = (CA)^2,$$
$$-\mathbf{CA}(\mathbf{AP}+\mathbf{AP'}) = \mathbf{AP}\cdot\mathbf{AP'},$$
$$\frac{\mathbf{AB}}{2}(\mathbf{AP}+\mathbf{AP'}) = \mathbf{AP}\cdot\mathbf{AP'},$$

and

$$\mathbf{AB} = \frac{2\,\mathbf{AP}\cdot\mathbf{AP'}}{\mathbf{AP}+\mathbf{AP'}}.$$

The theorem follows. ▯

42.8 Theorem If P and P' are distinct inverse points with respect to a circle s, then the chord joining the points of contact of the two tangents drawn from that point, say P, external to circle s passes through the other point P'.

▶**42.9 Definition** The *Peaucellier cell* is a mechanical linkage defined as follows. Pin together a rhombus $APBP'$ made from four rods (or pieces of heavy cardboard) of equal length that are hinged at each vertex in such a way that the rhombus is free to change its shape. In the same manner pin two congruent segments, longer than the first four, one each to the opposite vertices A and B of the rhombus. Pin the other two ends of these longer segments together and pin them to a fixed point C in the plane (see Fig. 42.9). Now when P traces out a curve c, point P' is said to trace the *Peaucellier image* of c.

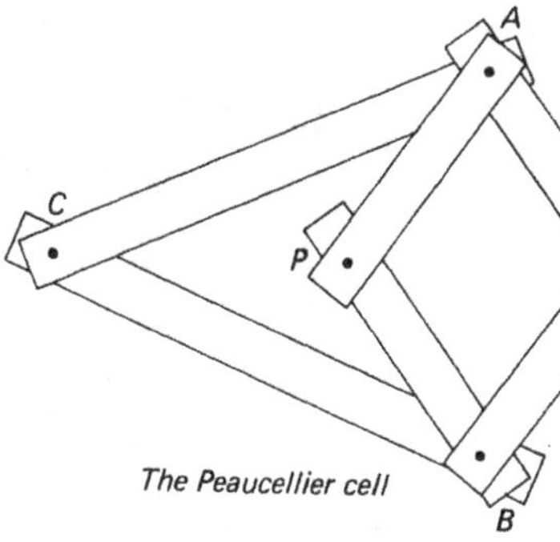

Figure 42.9

▶**42.10 Theorem** The Peaucellier image of a curve is the inverse of the curve in a circle of center C, and with power $(CA)^2 - (AP)^2$.

▶**42.11 Definition** Let A, B, C, D be four points lying on a circle. Define the *cross ratio* (AB, CD) of these four points, taken in that order, by

$$(AB, CD) = \pm\left(\frac{AC}{CB}\right)\Big/\left(\frac{AD}{DB}\right),$$

where AC, CB, AD, DB are chord lengths. The minus sign is taken when points C and D separate points A and B, and the plus sign is taken otherwise. (See also Definition 3.6.)

▶**42.12 Theorem** Inversion preserves the cross ratio of four points on a line or on a circle.

Let O be the center of an inversion that maps the four points A, B, C, D, no two of which are collinear with O and all four of which lie on a circle or a line, to A', B', C', D', which also lie on a line or circle by Theorem 42.2. By Theorem 41.9 we have (Fig. 42.12)

$$\triangle OAC \sim \triangle OC'A', \qquad \text{so} \qquad \frac{AC}{OA} = \frac{C'A'}{OC'},$$

$$\triangle OCB \sim \triangle OB'C', \qquad \text{so} \qquad \frac{CB}{OB} = \frac{B'C'}{OC'},$$

$$\triangle OAD \sim \triangle OD'A', \qquad \text{so} \qquad \frac{AD}{OA} = \frac{D'A'}{OD'},$$

$$\triangle ODB \sim \triangle OB'D', \qquad \text{so} \qquad \frac{DB}{OB} = \frac{B'D'}{OD'}.$$

Now direct substitution into the equation of Definition 42.11 shows that $|(AB, CD)| = |(A'B', C'D')|$. Since A and B separate C and D iff A' and B' separate C' and D', then $(AB, CD) = (A'B', C'D')$. ▯

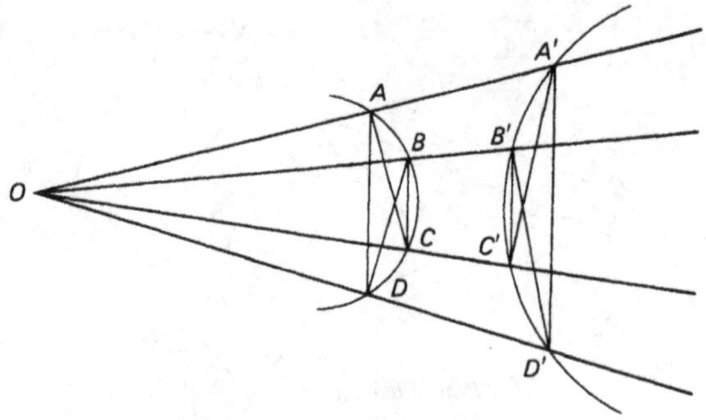

Figure 42.12

Exercise Set 42

1. Prove that two circles are orthogonal iff a radius of one drawn to a point of intersection of the two circles is tangent to the other.
2. Prove Theorem 42.2.
3. Prove that two circles with centers C and D and radii r and s are orthogonal iff $r^2 + s^2 = (CD)^2$.
4. Prove Theorem 42.4.
5. Show that the two equations are equivalent in each displayed pair of equations in the first two paragraphs of the proof of Corollary 42.7.
6. Show that a pair of orthogonal circles will invert into a pair of orthogonal circles, or a circle and one of its diameters, or two perpendicular lines, and locate the center of inversion for each of these cases.
7. To draw a straight line we trace, using a straightedge, a straight line previously constructed by someone. The compass is a mechanical device for constructing circles; we do not draw a circle by tracing someone's previously constructed disc.
   a) Show how to draw the "first" straight line. That is, show how to construct a straight line using a Peaucellier cell, and assuming no previous use of a straight line, not even to construct the Peaucellier cell.
   b) Make a Peaucellier cell out of cardboard without using a straightedge, and use the cell to construct a straight line.
8. Prove Theorem 42.8.
9. Prove that when A, B, C, D are four distinct collinear points, then $(AB, CD) < 0$ iff A and B separate C and D.
10. Prove Theorem 42.10.
11. Given a circle s and a point P, construct the inverse of P with respect to circle s. Do so for P inside, on, and outside the circle s.
12. Prove Theorem 42.12 for the case in which two or more of the points A, B, C, D are collinear with the center O.
13. Given two points A and B lying inside a circle s, prove that there is exactly one line or circle (according as A and B are or are not collinear with the center of s) through A and B and orthogonal to s. Show how to construct this circle or line.

14. Prove that if a radius OR of one circle intersects another circle at two points P and Q so that $\mathbf{OP}\cdot\mathbf{OQ} = (OR)^2$, then each radius OR that cuts the other circle at two points P and Q satisfies that same equation.
15. Assuming that two circles intersect at two points P and Q, and that each circle is orthogonal to a third circle s, prove that P and Q are inverse with respect to circle s.
16. Prove that if two curves are tangent at a point, then their inverses with respect to a given circle are tangent at the inverse of the point. Are there any exceptions?

▶SECTION 43 | INVERSION AND COMPLEX GEOMETRY

43.1 The representation of inversions in the Gauss plane is quite convenient. Since $CZ \cdot CZ' = r^2$ for an inversion with center C and power r^2, we have

$$|z-c||z'-c| = r^2, \qquad \text{and} \qquad \frac{z'-c}{z-c}$$

is real and positive since Z, Z', and C are collinear. This last condition can be written in the form

$$\frac{z'-c}{z-c} = \left|\frac{z'-c}{z-c}\right|.$$

Now we have

$$r^2 = |z-c||z-c|\frac{z'-c}{z-c} = (z-c)(\bar{z}-\bar{c})\frac{z'-c}{z-\bar{c}},$$

so

$$(z'-c)(\bar{z}-\bar{c}) = r^2.$$

Conversely, assuming that this last equation is satisfied, the preceding one follows. Since r^2 is real and $|z-c|^2$ is real, then $(z'-c)/(z-c)$ must also be real. Now the argument reverses completely, to show that Z' is the inverse of Z in the circle with center C and radius r. We have proved the theorem:

43.2 Theorem Inversion in a circle with center C and power r^2 has the complex representation

$$(z'-c)(\bar{z}-\bar{c}) = r^2, \qquad z' = c + \frac{r^2}{\bar{z}-\bar{c}}, \qquad \text{or} \qquad z' = \frac{c\bar{z}-c\bar{c}+r^2}{\bar{z}-\bar{c}}.$$

43.3 Corollary Inversion in the circle of radius r centered at the origin has the complex representation

$$z' = \frac{r^2}{\bar{z}}.$$

43.4 Theorem Inversion is involutoric.

Let the inversion of Theorem 43.2 map Z to Z' and then Z' to Z''. Then we have

$$z'' = \frac{c\bar{z}' - c\bar{c} + r^2}{\bar{z}' - \bar{c}}$$

$$= \frac{c\,\dfrac{\bar{c}z - c\bar{c} + r^2}{z - c} - c\bar{c} + r^2}{\dfrac{\bar{c}z - c\bar{c} + r^2}{z - c} - \bar{c}}$$

$$= \frac{c\bar{c}z - c^2\bar{c} + cr^2 - c\bar{c}z + r^2 z + c^2\bar{c} - cr^2}{\bar{c}z - c\bar{c} + r^2 - \bar{c}z + c\bar{c}}$$

$$= \frac{r^2 z}{r^2} = z.$$

Thus $Z'' = Z$, so inversion is involutoric. ▯

43.5 Theorem Inversion is a circular transformation.

Using the inversion of Theorem 43.2, we show that the image of the circle or line given by the parametric form of Theorem 38.3 is another such circle or line. Thus let the line or circle have the parametric equation, with parameter t,

$$z = \frac{at + b}{dt + e}.$$

Now substitute this value for z into the inversion of Theorem 43.2 to obtain

$$z' = \frac{c\bar{z} - c\bar{c} + r^2}{\bar{z} - \bar{c}}$$

$$= \frac{c(\bar{a}t + \bar{b}) - (c\bar{c} - r^2)(\bar{d}t + \bar{e})}{\bar{a}t + \bar{b} - \bar{c}(\bar{d}t + \bar{e})}$$

$$= \frac{(c\bar{a} - c\bar{c}\bar{d} + r^2\bar{d})t + (c\bar{b} - c\bar{c}\bar{e} + r^2\bar{e})}{(\bar{a} - \bar{c}\bar{d})t + (\bar{b} - \bar{c}\bar{e})},$$

an equation of the same form as given, hence a circle or line. ▯

The parametric form of Theorem 38.3 for a circle or a line,

$$z = \frac{at + b}{ct + d} \qquad \text{with } t \text{ real},$$

maps the real line onto a circle or a line in the Gauss plane. That is, this equation represents a mapping, so if we let t be any complex number, then it is a mapping of the Gauss plane to itself. That it is a transformation of the Gauss plane, and an interesting transformation, is the subject of the rest of this section.

43.6 Definition A *bilinear transformation* or *homography* is that transformation of the complex plane given by

$$z' = \frac{az+b}{cz+d} \qquad \text{where } ad - bc \neq 0.$$

43.7 Theorem The inverse of the bilinear transformation of Definition 43.6 is the bilinear transformation

$$z' = \frac{-dz+b}{cz-a}.$$

43.8 Corollary The bilinear transformation of Definition 43.6 is involutoric iff $a = -d$.

43.9 Theorem The bilinear transformation is a circular transformation.

The bilinear transformation of Definition 43.6 may be written as

$$z' = \frac{a}{c} + \frac{bc-ad}{c^2(z+d/c)}.$$

Thus this transformation is the product of:

a) a translation $z_1 = z + d/c$,
b) a similarity (rotation-homothety) centered at the origin $z_2 = (c^2/(bc-ad))z_1$,
c) a reflection in the x-axis $z_3 = \bar{z}_2$,
d) an inversion in the unit circle centered at the origin $z_4 = 1/\bar{z}_3$, and
e) another translation $z' = z_4 + a/c$.

Since each of these transformations is circular, then so also is the product a circular transformation. ▯

43.10 Theorem 43.9 can also be proved by substituting into the form for a bilinear transformation the equation for a line or circle, as was done in the proof of Theorem 43.5. Although such a proof is slightly shorter, it is far less instructive as to just what a bilinear transformation consists of. Thus a bilinear transformation is direct and is a product of similarities and inversions. In passing, we note without proof that the bilinear transformation and the bilinear transformation preceded by a reflection in the x-axis are the most general circular transformations. They have the equations

$$z' = \frac{az+b}{cz+d} \qquad \text{and} \qquad z' = \frac{a\bar{z}+b}{c\bar{z}+d}$$

with $ad - bc \neq 0$. Clearly both transformations are circular. A proof that they represent all circular transformations is given in Howard Eves' *Functions of a Complex Variable*, Vol. 1, page 39.

43.11 Theorem The set of all bilinear transformations forms a transformation group.

43.12 Theorem A bilinear transformation is determined by three distinct points and their distinct images. If the images of a, b, c are a', b', c', then the bilinear transformation has the form

$$\begin{vmatrix} z'z & z' & z & 1 \\ a'a & a' & a & 1 \\ b'b & b' & b & 1 \\ c'c & c' & c & 1 \end{vmatrix} = 0.$$

Clearly if $z = a$ and $z' = a'$, then the given determinant is zero by Theorem 36.17. Similarly for $z = b$ or for $z = c$. By Definition 36.12 the determinant has the expansion

$$pz'z + qz' + rz + s = 0,$$

where p, q, r, and s are third-order determinants not involving z or z'. Solving for z', we obtain

$$z' = \frac{-rz - s}{pz + q},$$

a bilinear transformation whenever $-rq + ps \neq 0$.

If $-rq + ps = 0$ and $p \neq 0$, then $rq/p = s$, and we have

$$z' = \frac{-rz - s}{pz + q} = \frac{-(r/p)pz - rq/p}{pz + q} = -\frac{r}{p} \cdot \frac{pz + q}{pz + q} = -\frac{r}{p},$$

a constant. This is impossible, since the determinant equation is satisfied by three different values for z'.

Finally assume that $p = 0$ and $ps - rq = 0$. Then we must also have either $r = 0$ or $q = 0$. If $r = 0$, then $z' = -s/q$, a constant, again impossible. If $q = 0$, then we have

$$0 = pa + q = a \begin{vmatrix} a' & a & 1 \\ b' & b & 1 \\ c' & c & 1 \end{vmatrix} - \begin{vmatrix} a'a & a & 1 \\ b'b & b & 1 \\ c'c & c & 1 \end{vmatrix} = (a - b)(c - a)(b' - c').$$

This last statement implies that $a = b$ or $a = c$ or $b' = c'$, contrary to hypothesis.

It follows, then, that $ps - rq \neq 0$, so the given determinant does indeed represent a bilinear transformation. □

Exercise Set 43

1. Prove that the product of two inversions with the same center is a homothety in that center. Find its ratio.
2. Prove Corollary 43.3.
3. Find when a product of two inversions with distinct centers is commutative.
4. Show that the inversion of Theorem 43.2 can be factored into the product of: (a) the translation $z_1 = z - c$, (b) the inversion at the origin $z_2 = r^2/\bar{z}_1$, and (c) another translation $z' = z_2 + c$.

5. Find an equation for the bilinear transformation that maps 0, 1, and ∞ to
   a) $0, 1, \infty$
   b) $0, 1, i$
   c) $0, 1+i, 3-i$
   d) $2, 1+i, 3-i$
   e) $1+i, 1+3i, 2-i$
   f) a, b, c
6. Prove Theorem 43.7.
7. Prove Corollary 43.8.
8. Prove that a homography is indeed a transformation.
9. Prove Theorem 43.11.
10. Find the bilinear transformation that maps 0, 1, and i to each of the sets of points listed in Exercise 43.5.
11. Find the images of each of the four points 0, 1, i, and ∞ under the bilinear transformation
    a) $z' = \dfrac{2z+i}{z-i}$
    b) $z' = \dfrac{(1+i)z+2i}{3z-2-i}$
    c) $z' = \dfrac{1}{z}$
    d) $z' = \dfrac{(2+i)z+3-2i}{(1-i)z}$
12. Show that any bilinear transformation that leaves the real axis invariant can be written with real coefficients.
13. a) Find a bilinear transformation that carries the circle $|z| = 1$ into the circle $|z| = 2$. Is it unique?
    b) Find a bilinear transformation that satisfies part (a) and also maps i to -2. Is it unique?

SECTION 44 | APPLICATIONS OF INVERSION

The usefulness of inversion was alluded to in the introductory paragraph of Section 41. Here we shall demonstrate that usefulness. Observe the transform-solve-transform method (see 18.1) in action: A problem is stated, then transformed into a new problem by an inversion; the new problem is solved by one means or another; and the solution is transformed back to solve the original problem. Theorem 44.2 is one of several theorems whose proof is a clear example of this method: A problem involving circles and lines is inverted into a problem involving a triangle and its altitudes. This latter problem has an immediate solution from our earlier work, and this solves the given problem.

44.1 Theorem Any three circles can be inverted into three circles whose centers are collinear.

Suppose that the centers A, B, C of the circles s, t, u are not collinear. Then (see Exercise 44.2) there is a circle v orthogonal to all three circles (see Fig. 44.1). Choose as center of inversion any point P on circle v but not on any of the other circles. Then v inverts into a line v', and the circles s, t, u invert into circles s', t', u'. Since v is orthogonal to each of s, t, and u, then v' is a diameter of each of the circles s', t', and u'. Now the centers of these circles are collinear on line v'. ▯

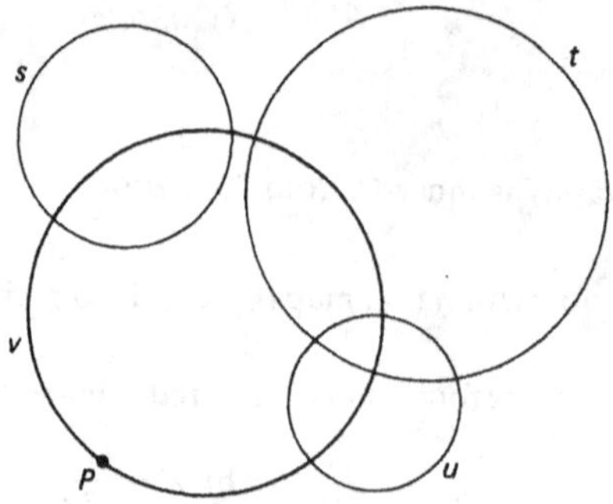

Figure 44.1

44.2 Theorem Let two circles s and t meet at O and P, and let each diameter OS and OT of the two circles cut the other circle at A and B. Then chord OP passes through the center of circle OAB (see Fig. 44.2a).

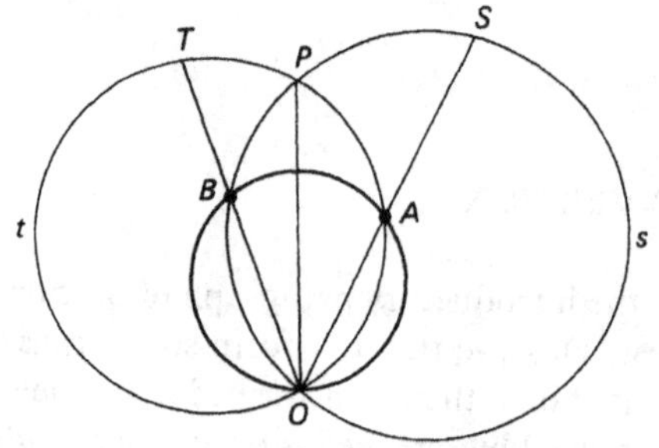

Figure 44.2a

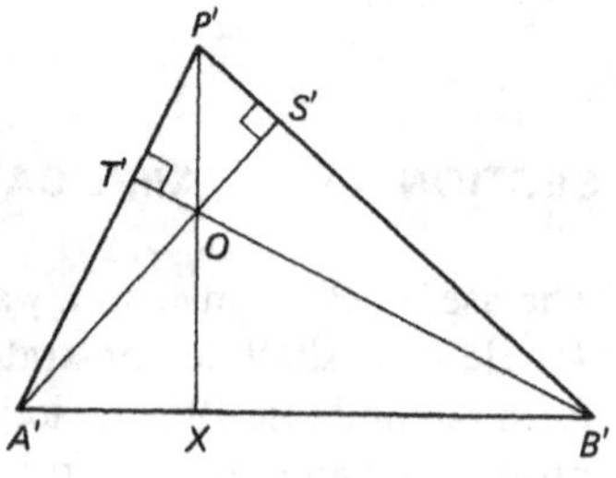

Figure 44.2b

Invert the figure in point O. Lines OP, OBT, and OAS invert into lines OP', $OB'T'$, and $OA'S'$, as in Fig. 44.2b. Circles $OAPT$, $OBPS$, and OAB invert into lines $A'P'T'$, $B'P'S'$, and $A'B'$. Let OP' and $A'B'$ meet at X. Since OS and OT are diameters, then lines OS' and OT' are perpendicular to lines $B'S'P'$ and $A'T'P'$. Hence O is the orthocenter of triangle $A'B'P'$. Now line $P'OX$ is perpendicular to line $A'B'$. Hence, in the original figure, line OP is orthogonal to circle OAB by Theorem 41.10. That is, line OP passes through the center of circle OAB. ▯

44.3 Theorem *Ptolemy's Theorem.* In a convex cyclic quadrilateral the product of the diagonals is equal to the sum of the two products of opposite sides.

Let $ABCD$ be the cyclic quadrilateral and invert in point A (see Fig. 44.3). The inverses B', C', D' of B, C, D lie on a line, by Corollary 41.13, so

$$B'C' + C'D' = B'D'.$$

Letting the power of inversion be r^2, by Exercise 41.17 we have

$$\frac{r^2 \cdot BC}{AB \cdot AC} + \frac{r^2 \cdot CD}{AC \cdot AD} = \frac{r^2 \cdot BD}{AB \cdot AD},$$

from which it follows that

$$AD \cdot BC + AB \cdot CD = AC \cdot BD. \;\square$$

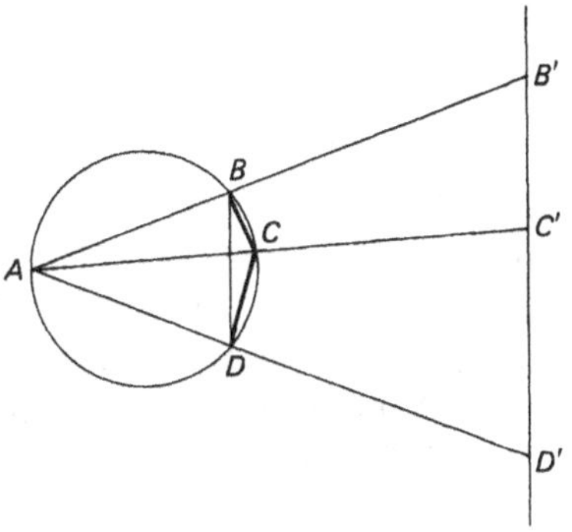

Figure 44.3

44.4 Lemma If a pencil of four lines is cut by two transversals at points A, B, C, D and A', B', C', D', respectively, and neither transversal passes through the vertex of the pencil, then

$$(AB, CD) = (A'B', C'D').$$

See Theorem 3.8. $\square$

44.5 Lemma The tangent at A' to the ninepoint circle of triangle ABC is parallel to that common internal tangent to the equicircles centered at I and I_a that does not pass through A'.

Of course, side BC is one common internal tangent to the equicircles centered at I and I_a. Let the other common internal tangent be PQ, as shown in Fig. 44.5. Then BC and PQ meet at U (which is collinear with A, I, and I_a). Since the ninepoint circle is the circumcircle of triangle $A'B'C'$, which is homothetic to triangle ABC, then the tangent t at A' is parallel to the opposite side DE of the orthic triangle of triangle ABC, by Theorem 7.9. By Corollary 7.10, DE makes the same angle with AB that AC does with BC, namely $\measuredangle C$. But by the symmetry of Fig. $AQCPB$ in mirror AU, PQ makes this same angle with AB. Hence DE is parallel to PQ, so the tangent t is parallel to PQ also. $\square$

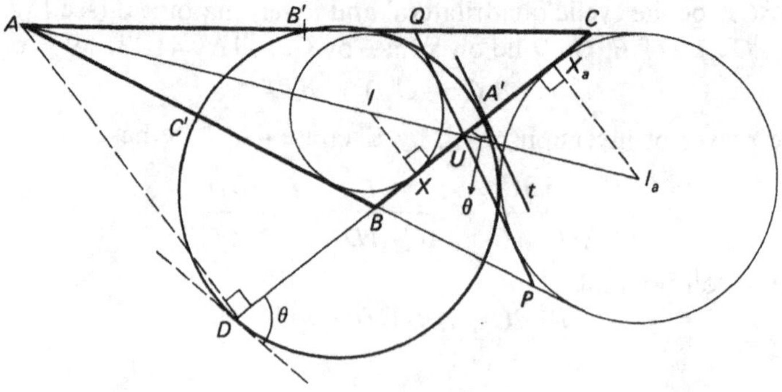

Figure 44.5

44.6 Theorem *Feuerbach's theorem.* The ninepoint circle of a triangle is tangent to each of the equicircles of the triangle.

We show that the ninepoint circle is tangent to the incircle, and to the excircle whose center is I_a. We use Fig. 44.5, and the terminology of Lemma 44.5, adding to that figure the lines shown broken. Thus let X and X_a be the points of contact of the equicircles with side BC, and draw the altitude AD. Now $(II_a, AU) = -1$ by Exercise 24.8, so $(XX_a, DU) = -1$ by Lemma 44.4. Take A' as center of inversion, and $A'X(\cong A'X_a$ by Theorem 6.22) as radius of inversion. By Exercise 41.15 circles I and I_a, the equicircles, invert into themselves. By Theorem 42.5, the inverse of D is U. Line BC is, of course, self-inverse. The ninepoint circle inverts into a line through U by Corollary 41.13. Since angles are preserved (Theorem 41.10), the angle θ made with line BC by the tangent to the ninepoint circle at D is equal and opposite to the angle made by the inverse of the ninepoint circle with line BC. By symmetry, this is the same angle made with line BC by the tangent to the ninepoint circle at A'. By Lemma 44.5, then, the inverse of the ninepoint circle is line PQ. Since PQ is tangent to (the inverses of) both equicircles, then (Theorem 41.10) the ninepoint circle is tangent to both equicircles. The theorem follows. ▯

44.7 The reader is here reminded of the statements following Theorem 7.20, pointing out a total of 32 equicircles associated with a given triangle all 32 of which are tangent to the ninepoint circle.

▶**44.8 Theorem** A homothety with positive ratio can be factored into two inversions.

We prove the theorem, without loss of generality, for the homothety $H(O, k)$ centered at the origin and with ratio $k > 0$. Let α and β denote the inversions given by

$$\alpha(z) = 1/\bar{z} \qquad \text{and} \qquad \beta(z) = k/\bar{z}.$$

Then $\beta\alpha$ is given by

$$(\beta\alpha)(z) = k/\overline{(1/\bar{z})} = kz,$$

an equation for the desired homothety. ▯

▶**44.9 Theorem** Two inversions with positive ratios and in the same center commute iff their radii are equal.

If the circles of inversion have radii r and s, where we are taking the origin as their common center, then, letting α and β represent these transformations, we have

$$\alpha(z) = r^2/\bar{z} \qquad \text{and} \qquad \beta(z) = s^2/\bar{z}.$$

Now the products $\beta\alpha$ and $\alpha\beta$ have the equations

$$(\beta\alpha)(z) = s^2/(\overline{r^2/\bar{z}}) = (s^2/r^2)z \qquad \text{and} \qquad (\alpha\beta)(z) = r^2/(\overline{s^2/\bar{z}}) = (r^2/s^2)z.$$

Thus these two maps are homotheties with ratios s^2/r^2 and r^2/s^2. They are equal iff $r = s$. □

▶**44.10 Theorem** Two inversions with positive ratios and distinct centers commute iff their circles are orthogonal.

Let the circles for the inversions α and β have centers O and A and radii r and s. They are orthogonal, then, iff $r^2 + s^2 = |a|^2 = a\bar{a}$. Equations for α and β are

$$\alpha(z) = r^2/\bar{z} \qquad \text{and} \qquad \beta(z) = a + s^2/(\bar{z} - \bar{a}),$$

so that $\alpha\beta$ and $\beta\alpha$ are given by

$$(\beta\alpha)(z) = a + s^2/((\overline{r^2/\bar{z}}) - \bar{a}) = \frac{ar^2 - a\bar{a}z + s^2 z}{r^2 - \bar{a}z}$$

and

$$(\alpha\beta)(z) = r^2/(\overline{a + s^2/(\bar{z} - \bar{a})}) = \frac{r^2 a - r^2 z}{a\bar{a} - s^2 - \bar{a}z}.$$

These maps will be equal iff their difference is zero. Taking their difference and setting $z = 0$ therein, we obtain

$$(\beta\alpha)(0) - (\alpha\beta)(0) = \frac{ar^2}{r^2} - \frac{r^2 a}{a\bar{a} - s^2}.$$

If this difference is zero, then $a\bar{a} - s^2 = r^2$.

Conversely, if $a\bar{a} = r^2 + s^2$, then we find that

$$\begin{aligned}(\beta\alpha)(z) - (\alpha\beta)(z) &= \frac{ar^2 - (r^2 + s^2)z + s^2 z}{r^2 - \bar{a}z} - \frac{r^2 a - r^2 z}{r^2 + s^2 - s^2 - \bar{a}z}\\ &= \frac{ar^2 - r^2 z}{r^2 - \bar{a}z} - \frac{r^2 a - r^2 z}{r^2 - \bar{a}z}\\ &= 0.\end{aligned}$$

It now follows that $\beta\alpha = \alpha\beta$ iff the circles of inversion are orthogonal. □

Exercise Set 44

1. Explain why, in Theorem 44.1, the circle v could not be taken as the circle through the centers of the three given circles.

2. a) Prove that if two circles intersect, then the center of any circle orthogonal to both of them must lie on the line of their common chord.
   b) Assuming that two circles intersect, prove that any circle orthogonal to one of them and whose center lies on their common chord is orthogonal to the other also.
   c) Show that both parts (a) and (b) remain true when "intersecting" and "common chord" are replaced by "tangent" and "common tangent at their point of tangency."
   d) Prove that the tangents drawn to two intersecting (or tangent) circles from any point on the common chord (or common tangent) are congruent. The locus of all the points from which congruent tangents can be drawn to two nonconcentric circles is called their *radical axis*. Hence the radical axis of two intersecting circles is their common chord, and for two tangent circles is their common tangent drawn at their point of tangency.
   e) Given that two circles s and t do not intersect and are not concentric, then show that it is always possible to draw a circle u intersecting both circles s and t in such a way that the chord common to u and s intersects the chord common to u and t in an ordinary point P.
   f) Show that it is possible to construct infinitely many distinct points P in part (e) by choosing different circles u.
   g) Prove that the locus of all points P of part (e) is the radical axis of circles s and t.
   h) Prove that the radical axis of part (g) is a straight line.
   i) Prove that the radical axis is the locus of the centers of all circles orthogonal to two circles.
   j) Prove that if three circles have noncollinear centers, then the three radical axes of pairs of these circles concur. This point is called their *radical center.*
   k) Show that the radical center of three circles with noncollinear centers is the center of the unique circle orthogonal to each of the three given circles.
3. Show that any three circles with noncollinear centers can be inverted into themselves.
4. Extend Ptolemy's theorem (Theorem 44.3) to the case in which the quadrilateral is not cyclic by proving the general formula

$$AD\cdot BC + AB\cdot CD \geqslant AC\cdot BD.$$

5. Let A and B be inverse points in circle s. Invert in a circle whose center does not lie on s or at A or B to obtain A', B', s'. Then prove that A' and B' are inverse with respect to circle s'.
6. In Exercise 44.5, what happens when the center of inversion lies on circle s? What is the relation of A' and B' to s' then? This explains why reflection in a line is sometimes called *inversion in a line*, and inversion is sometimes called *reflection in a circle.*
7. Find the radius of the circle inverse to a given circle with respect to a given inversion.
8. Show that any two nonconcentric noncongruent circles can be inverted one to another.
9. Show that the inversion with center O and radius r has the Cartesian equations

$$x' = \frac{xr^2}{x^2+y^2} \quad \text{and} \quad y' = \frac{yr^2}{x^2+y^2}.$$

10. Using Exercise 44.9, write an equation for the inverse in the circle of radius 1 centered at the origin of each listed equation:
    a) The line $x = 1$
    b) The line $ax + by + c = 0$
    c) The circle $x^2 + 2x + y^2 = 0$
    d) The circle $x^2 + y^2 + ax + by + c = 0$
    e) The ellipse $b^2x^2 + a^2y^2 = a^2b^2$; is the inverse curve also an ellipse? Sketch its graph.
    f) The parabola $y = x^2$; sketch the graph.
    g) The pair of hyperbolas $x^2y^2 = 1$; sketch the graph.
11. Circles t and u are inverse with respect to circle s. Prove that t and u are mapped into congruent circles by inversion with respect to any point on circle s.
12. Let A, B, C, D be four concyclic points. Show how to invert them into the vertices of a rectangle.
13. Invert the theorem "an angle inscribed in a semicircle is a right angle":
    a) in the center of the semicircle,
    b) in one end of the diameter of the semicircle,
    c) in the vertex of the right angle.
14. What is the image of the line of centers of two given circles under an inversion?
15. Prove Pappus' ancient theorem: In the figure for this exercise, OA, OB, and AB are diameters of circles t, u, and s_0. Then circle s_1 is placed tangent to circles t, u, and s_0, circle s_2 is tangent to circles t, u, and $s_1, \ldots,$ circle s_n is tangent to circles t, u, and s_{n-1}. Prove that the height of the center of circle s_n above line OB is n times the diameter of s_n.

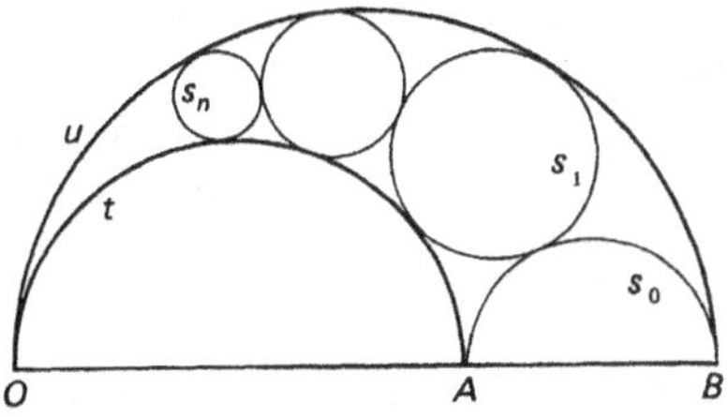

Exercise 44.15

16. Given that three circles concur at a point, prove that there are four distinct circles each of which is tangent to all three circles.
17. When points A and B are inverse in circle s, show that the inverse of B is the center of the inverse of s when A is the center of inversion.

6 ISOMETRIES IN SPACE

SECTION 45 WHAT NEXT?

45.1 The title of this section refers not to possible future trends in the theoretical development of geometry, but rather to the drift of geometry in the high schools. Even more important, perhaps, the question *What next?* may be taken personally, especially by those intending to teach geometry.

45.2 Over the last quarter century, analytic geometry has gained ever-increasing prominence in high school geometry textbooks, while solid geometry has decreased in popularity. These trends are partially due to a lack of emphasis placed on geometry in colleges and normal schools until quite recently. It is easier and requires less skill to prove a theorem by analytic geometry than by synthetic methods. And very few teachers feel comfortable beyond the most elementary work in space.

45.3 These trends will continue, for analytic geometry provides a powerful tool in solid geometry, one that should not be overlooked. Its inclusion in the traditional sophomore geometry course is to be encouraged. Students should be allowed, even encouraged, to use whatever tools are available to solve their problems. This places a heavier burden on the teacher: He must be qualified to evaluate a student's work no matter which tools he uses.

45.4 It is with a sigh of regret that the passing of a semester of solid geometry in the senior year is observed, not as the course generally has been taught, but as it should have been taught. The apparent reason for discontinuing this course was that spatial perception would be taught in the sophomore geometry course simultaneously with plane geometry. It rarely is. But the real reason was the great pressure from teachers who did not have the training to teach solid geometry. Very few high school geometry teachers have significant training in geometry beyond their own sophomore plane geometry courses. This meager training is not enough. At least one significant college-level geometry course should be the minimum requirement for teaching geometry. More than this minimum is strongly recommended.

The effects of inadequate preparation of geometry teachers are felt at the college level. You may recall that students of the calculus have great difficulty working in space. Many cannot visualize surfaces and solids at all, and sketching solids and their plane sections is a difficult task for all but a very few students. Nonetheless, it is doubtful that a course in solid geometry will regain favor in high schools or even in colleges. It is recommended that a unit (of at least six weeks' duration) on spatial perception and sketching graphs of solids be included in the high school program after (or during) sophomore geometry. And it is here that analytic geometry proves to be especially useful—in solid geometry.

45.5 Isometries and similarities are mentioned briefly in current high school geometry texts. And texts on this subject at the college level are appearing in great profusion, this book being one such. As observed throughout this work, these mappings provide another tool of the order of analytic geometry with which to attack problems in geometry. Hence isometries and similarities will gain increasing favor in the high school, and will begin to appear early in textbooks in order that these transformations may be used meaningfully throughout the student's geometry course.

45.6 The understanding of isometries clarifies and rigorizes the hazy method of superposition that was formerly used as a congruence axiom (see Axiom 4 in Appendix A). It was never clear just how one was to pick up one triangle and place it on another. Should you think of the triangle as a piece of cellophane tape to be peeled off the plane? If so, then it could be placed on the lateral surface of a cylinder. On the other hand, it cannot just be slid (translated and rotated) if the congruence is opposite. The problem is, "How do you accurately define just what a rigid motion is?" One answer is, "By isometries." Current geometry texts postulate the SAS congruence condition in order to avoid the necessity of stating considerable background material in isometries before being able to discuss congruence. It seems better pedagogically to be able to prove basic congruence theorems early, and the SAS postulate accomplishes this result. This is not to say that isometries should be dismissed; they should come early in the course. But, as in this text, perhaps isometries should be treated after a preliminary familiarity with basic geometric concepts has been developed.

45.7 Now let us turn to the personal aspect of the question *What next?* What are you planning to do to make geometry meaningful to your students?

In some ways those who teach geometry are more fortunate than teachers of algebra. The student who never really understood arithmetic tends to have insurmountable problems with algebra until his deficiencies are uncovered and corrected. But if we can give him a right *attitude* toward geometry, his earlier difficulties will have a much less detrimental effect upon his progress in geometry.

The experienced teacher is well aware of the average or bright student who scores high on mathematical *ability* tests, but is low on mathematical *achievement* tests. As teachers of geometry, we have a rather unique opportunity to break through the *psychological* barrier of such a student, help to dispel his dread of mathematics, and show him that he *can* find geometry interesting and within his grasp. The observant teacher will use this opportunity to locate and help correct his deficiencies in arithmetic.

But not by simply listing theorem, proof, theorem, proof, theorem, proof, etc., all copied neatly right out of the book.

45.8 A student is most ready for an answer to a question when he feels there really is a *question* and he *wants* to know its answer. If a student looks at an isosceles triangle and immediately *knows* that its base angles are congruent, then a proof of that fact may be simply redundant and meaningless to him.

45.9 In beginning a discussion of isosceles triangles, for example, after defining the term, one might ask what else is true about isosceles triangles, other than that two sides are congruent. When a student answers that the base angles are congruent, one should ask, "Always?" and "How do you know this?" Also, "Is it necessary to prove it? If so, why?"

Such questions and the resulting discussions cause the students to think for themselves, and to develop the curiosity necessary to appreciate deductive mathematics. Of course, in discussing isosceles triangles, properties other than the congruence of the base angles should enter naturally into the discussion, and should be treated as they occur.

45.10 In studying right triangles, tell the students to draw right triangles (for homework, perhaps) having legs of lengths 3 and 4, 8 and 15, 10 and 24, 12 and 35, and 16 and 63, for example. The students should form a table indicating the leg lengths and the corresponding hypotenuse lengths which they carefully measure. They should look for generalizations from the data in the table. Have them form another column for the sum of the squares of the leg lengths for each triangle, and another column for the square of the hypotenuse length. Ask them to look for apparent truths. On their drawings they should locate the midpoint of each hypotenuse and measure and record its distances from the three vertices. Ask what appears to be true. Ask if they can prove it. This is the discovery method.

45.11 The discovery method requires more time than lecturing, and in that sense it might be called inefficient. But certainly the initial parts of a geometry course should be especially well motivated. The student should learn to hunger and thirst after geometric knowledge. Later, he will be more receptive to a formal chain of theorems when time requires that material be covered more rapidly. But class time spent working out original problems together is time wisely spent. It is easy to say,

"Here is the solution." It is far more instructive to say, "How can we obtain a solution? Let us work it out together."

In short, motivate.

45.12 Should this course end your geometric studies? No. Your study of geometry has just begun. Now you are starting to *do* geometry, and your geometric education should be continued, both formally and informally. And it should be reasonably continuous. An occasional course in some aspect of geometry will keep you active and help prevent staleness, which can easily develop, especially in a small school using the same textbook year after year. Specifically, you may wish to study such topics as projective geometry, non-Euclidean geometries, analytic geometry, foundations of geometry, or others. The topics in geometry that are opened to exploration by students completing a basic text such as this one are legion. Working problems regularly from a book such as this one or Horblit and Nielsen's *Plane Geometry Problems with Solutions* (College Outline Series), or Rich's *Principles and Problems of Plane Geometry with Coordinate Geometry* (Schaum's Outline Series) also improves your ability to work original problems. More important, such practice will improve your ability to judge a student's work. Is his original proof valid? If not, where did he go astray? Just because his is not the proof "in the text" does not mean that it is incorrect.

Subscribe to and read journals in your field, such as *The Mathematics Teacher*, *The Arithmetic Teacher*, *Mathematics Magazine*, *The Two-Year College Mathematics Journal*, etc. From time to time you will find interesting items in the literature to bring up in class, or to give to bright students as enrichment material. You will have a better knowledge of what is new in mathematical thinking and will be better able to evaluate new ideas.

45.13 By your love for mathematics and your interest in your students, let your class be one in which this sign (Fig. 45.13) is proudly displayed.

Exercise Set 45

1. Draw the right triangles suggested in 45.10, and note that each hypotenuse is 2 units greater in length than one of the legs. Is this true for all Pythagorean triangles?
2. Draw right triangles with leg lengths 5 and 12, 7 and 24, 20 and 21, and 7 and 10. Measure the length of each hypotenuse. If the legs are of integral length, will the hypotenuse also be of integral length always?
3. Show that when p and q are integers with $p > q$, then $a = 2pq$, $b = p^2 - q^2$, and $c = p^2 + q^2$ form a *Pythagorean triple*; that is, $a^2 + b^2 = c^2$ and a, b, and c are positive integers.
4. Decide if the formulas of Exercise 45.3 give all Pythagorean triples. See Dodge's *Numbers and Mathematics*, pages 243–245.

Figure 45.13

SECTION 46 | INTRODUCTION TO THREE DIMENSIONS

46.1 In Chapter 2 we saw that the reflection in a line is the basic isometry of the plane; all other isometries can be written as products of this basic isometry. The situation in space is quite similar. Here the basic isometry is the *reflection in a plane* (see Fig. 46.1a) which is defined in just the manner that one would expect (see Definition 47.4). Then a *translation* (Definition 48.1) is the product of two reflections in parallel planes, and a *rotation through angle θ about a line m* (Definition 48.8) is the product of reflections in two mirrors through line m, and intersecting at angle $\theta/2$. Figure 46.1b shows the translation, and Fig. 46.1c the rotation, as products of reflections in planes Π and Δ. In each case, point P is mapped to point P'. Compare these figures with their plane counterparts, shown in Figs. 10.6 and 10.8. A rotation about line m through 180° is called a *halfturn about line m*, and is the product of reflections in any two perpendicular mirrors through m, as shown in Fig. 46.1d, where again point P is mapped to point P' by reflections in planes Π and Δ.

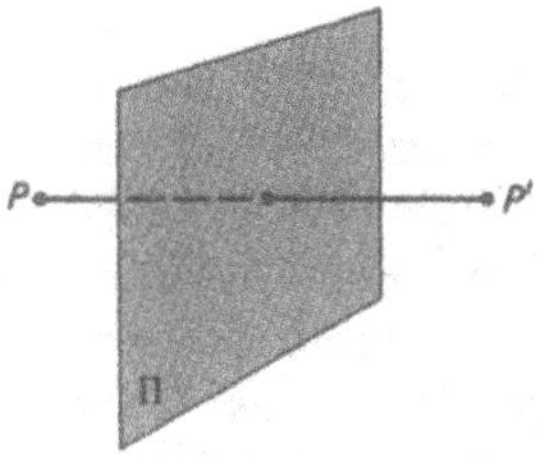

Figure 46.1a

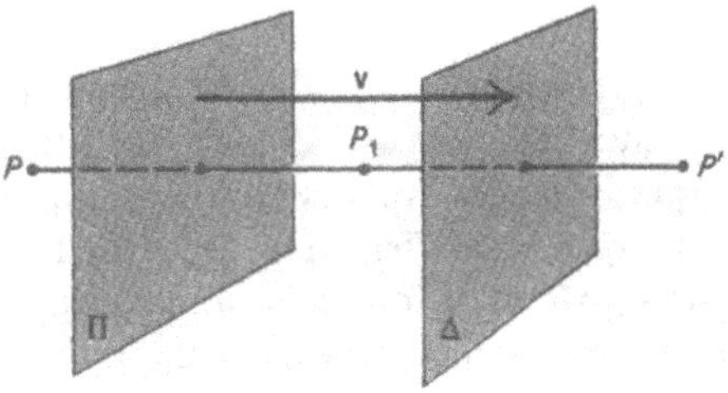

Figure 46.1b

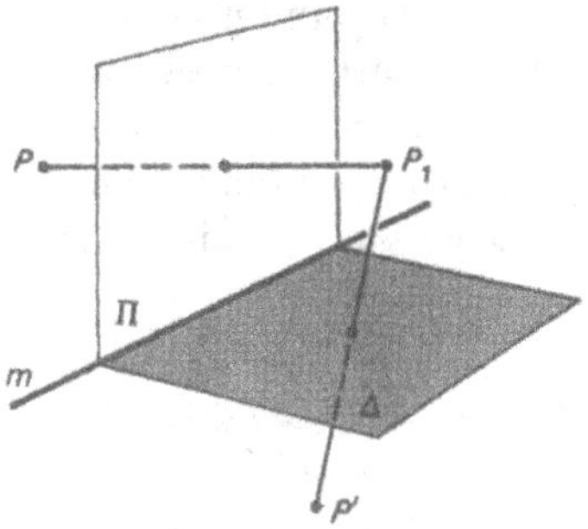

Figure 46.1c

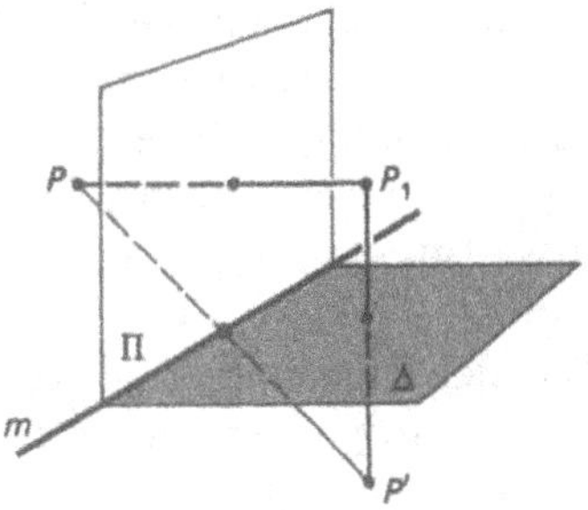

Figure 46.1d

46.2 One more basic isometry occurs in space. A *central inversion* in a point O is that involutoric isometry that maps each point P in space to point P' so that O is the midpoint of segment PP'. It can be factored into the product of three reflections in any three mutually perpendicular planes through O. Figure 46.2 shows a central inversion in point O, mapping point P to P' by reflections in planes Π, Δ, and Γ. Reflections and central inversions are opposite space isometries, reversing the sense of a tetrahedron (see Definition 47.10); translations and rotations are direct.

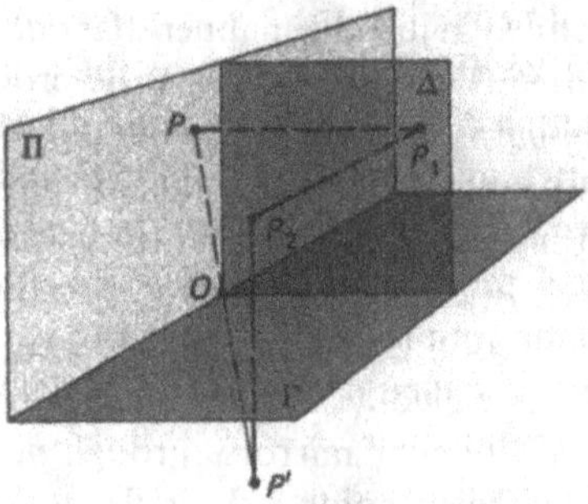

Figure 46.2

46.3 Any isometry in space is completely determined by the four vertices of a tetrahedron and their images, so each such isometry is the product of at most four reflections in planes, three when there is at least one fixed point (see Theorem 47.9 and compare it with Theorem 13.13 for the plane case). Various combinations of products of three or four reflections yield three more new space isometries which we might liken to the glide-reflection in the plane. A *screw displacement* (Definition 49.1) is the product of a rotation about a line m and a translation along line m (see Fig. 49.1). A *glide-reflection* (Definition 49.6) is the product of a reflection in a mirror Π and a translation along any vector lying in plane Π (see Fig, 49.6). It factors into a product of three reflections, two in parallel planes and the third in a plane perpendicular to the first two. Finally, a *rotatory reflection* (Definition 49.7) is the product of a reflection in a plane Π and a rotation about a line m perpendicular to Π (see Fig. 49.7). It is also called a rotatory inversion (see Theorem 49.9), since it can be factored into a product of a central inversion in a point O and a rotation about a line m through O. The screw displacement (Why is it called this?) is direct; the other isometries defined in this paragraph are opposite.

46.4 We shall show in this chapter that each opposite isometry with an invariant point is a rotatory inversion, of which a reflection is a special case. An opposite isometry with no fixed points is a glide-reflection. Each opposite isometry is the product of a reflection and a halfturn.

46.5 Each direct isometry is the product of two halfturns (see Theorem 49.2). If it has a fixed point, the direct isometry is a rotation. If it has no fixed points, then it is a translation or a screw displacement, the former being a special case of the latter.

Exercise Set 46

1. Show that a product of two reflections in parallel planes is a translation through twice the normal vector from the first plane to the second.
2. Show that a product of two reflections in intersecting planes is a rotation through twice the angle from the first plane to the second.
3. Show that reflections in two perpendicular planes commute.
4. There are six different orders in which to write a product of reflections in three planes Γ, Δ, and Π, and two ways to associate each such order. Show that when the three planes are mutually perpendicular, then all twelve such products are equal to the same central inversion.
5. Find a specific isometry in space that can be written as (a product of) no less than the indicated number of reflections, and write it as such a product:
   a) 1 b) 2 c) 3 d) 4
6. State the smallest number of reflections in planes necessary to write each indicated space isometry as a (product of) such reflection(s).
   a) reflection
   b) the identity map ι
   c) translation
   d) central inversion
   e) rotation
   f) screw displacement
   g) halfturn
   h) rotatory reflection
   i) glide-reflection
   j) rotatory inversion
7. State which isometries in Exercise 46.6 are direct and which are opposite.
8. Fashion a suitable definition for *direct* and *opposite* space transformations.
9. State what must be true of the axes of two halfturns in order that their product represents the following:
   a) a rotation
   b) a translation
   c) a screw displacement
   d) the identity map
10. State which points, lines, and planes are fixed under each of the isometries listed in Exercise 46.6.
11. Show that a space isometry preserves:
    a) line segments
    b) lines
    c) planes
    d) angles between lines
    e) angles between planes
12. Show that each isometry can be written as a product of at most four reflections in planes.
13. Show that a product of three halfturns about parallel lines is a halfturn about a line parallel to each of the three lines. Furthermore, show that when any three of these four lines are coplanar, then all four are coplanar.
14. Show that a rotatory reflection and a rotatory inversion are indeed equivalent space isometries.

SECTION 47 | REFLECTION IN A PLANE

47.1 The entire theory of transformations given in Section 12 holds with the word "plane" replaced by "space." Such replacement is assumed for this chapter. The student should now reread Section 12 to verify that all its definitions and theorems do indeed hold in space. Similarly, the definition of isometry is unaltered. In fact, the entire space development parallels quite closely that given for the plane in Chapter 2.

47.2 Definition An *isometry* in space is a map of the points in space to themselves that preserves distance.

47.3 Theorem If $PQRS$ is a tetrahedron, and α and β are isometries such that $\alpha(P)=\beta(P)$, $\alpha(Q)=\beta(Q)$, $\alpha(R)=\beta(R)$, and $\alpha(S)=\beta(S)$, then $\alpha=\beta$.

We need show only that any fifth point in space is uniquely located by its distances from P, Q, R, and S. To that end, let p, q, r, s denote the distances of a point X from points P, Q, R, S. Then there is a sphere of points Y at distance p units from P. Similarly there is a sphere of points Y at distance q from Q. These two spheres intersect in (at most) a circle c of points. Since R does not lie on line PQ, the sphere of points r units from R cuts circle c in (at most) two points Y_1 and Y_2, which are equidistant from plane PQR (see Fig. 47.3). Since S does not lie in plane PRQ, then only

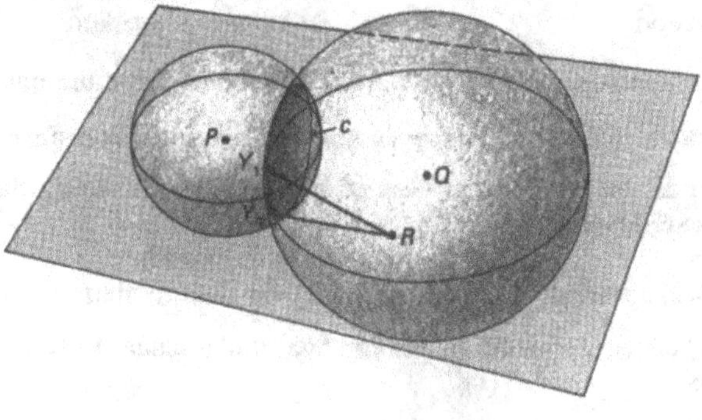

Figure 47.3

one of its distances to Y_1 and Y_2 can be equal to s. Thus point X is uniquely determined by its distances from the four vertices of the tetrahedron. The theorem now follows. ▯

47.4 Definition A space transformation is a *reflection in a plane* Γ, denoted by σ_Γ, iff whenever $B=\sigma_\Gamma(A)$ for points A and B, then Γ is the perpendicular bisector of segment AB, or $A=B$ and $A\in\Gamma$.

47.5 Remember that we have agreed to use upper-case Greek letters $\Gamma, \Delta, \Pi, \ldots$ to denote planes.

47.6 Theorem A reflection in a plane is an isometry.

Let σ_Γ map A to A' and B to B'. Then there is a plane Δ, unique if A, A', B, B' and not collinear, perpendicular to plane Γ and containing A, A', B, B' (see Fig. 47.6). In plane Δ, A' and B' are (by Definition 13.9) the reflections of A and B in the (plane) reflection in line m, the line of intersection of the two planes Γ and Δ. The theorem then follows from Theorem 13.10, the plâne analog of this theorem. ▯

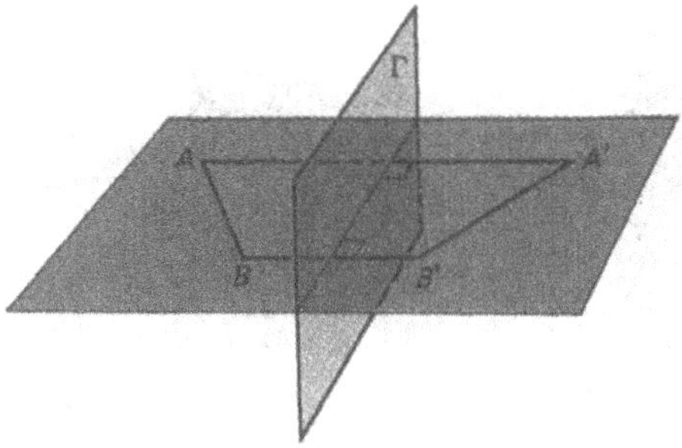

Figure 47.6

47.7 Corollary Each product of reflections is an isometry.

47.8 Theorem For each reflection σ_Γ, $\sigma_\Gamma^{-1} = \sigma_\Gamma$.

47.9 Theorem Each isometry in space is the product of at most four reflections in planes.

Let the isometry α map tetrahedron $ABCD$ into the congruent tetrahedron $A'B'C'D'$. We consider 5 cases.
Case 1. If $A = A'$, $B = B'$, $C = C'$, and $D = D'$, then α is the identity map, which can be written as the square of a reflection in any given plane.
Case 2. If $A = A'$, $B = B'$, $C = C'$, but $D \neq D'$, then plane ABC is the perpendicular bisector of segment DD' (see Exercise 47.11), so α is a reflection in that plane (see Fig. 47.9a).

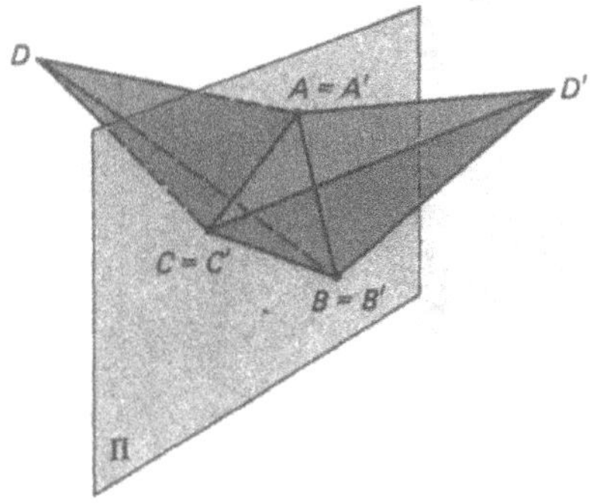

Figure 47.9a

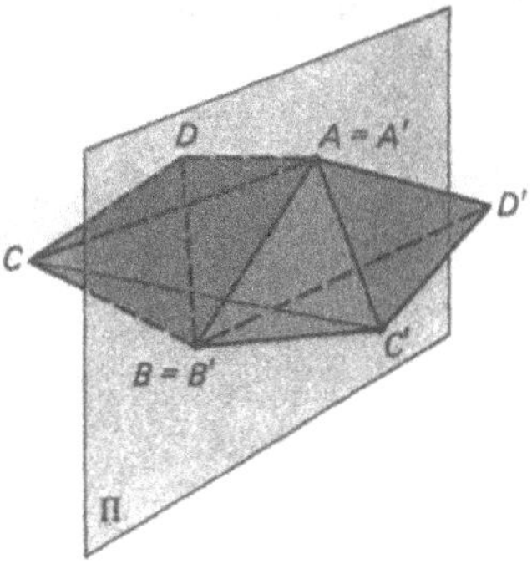

Figure 47.9b

Case 3. If $A = A'$ and $B = B'$, but $C \neq C'$, then A and B both lie on the plane Π, which is the perpendicular bisector of segment CC'. A reflection in plane Π then reduces this case to either Case 1 or Case 2 (see Fig. 47.9b).
Case 4. If $A = A'$ but $B \neq B'$, then A lies on the perpendicular bisector plane Π of segment BB', so a reflection in that plane reduces this case to one of the three previous cases (see Fig. 47.9c).

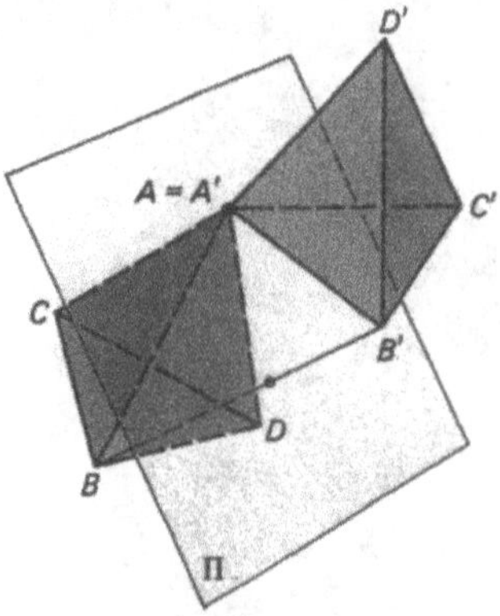

Figure 47.9c

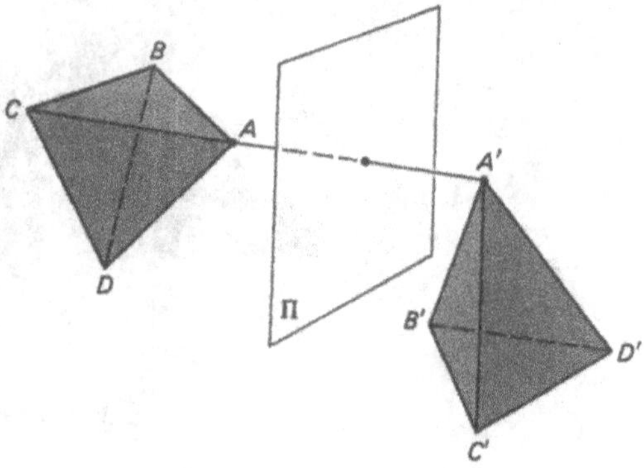

Figure 47.9d

Case 5. Given that $A \neq A'$, reflect tetrahedron $ABCD$ in the perpendicular bisector plane of segment AA', reducing Case 5 to one of the four earlier cases (see Fig. 47.9d).

Of course, if any one of Cases 2 to 5 reduces to Case 1, then a factor of the identity map (a product of two reflections in the same plane) is simply omitted. Thus we see that each isometry can be written as a reflection in a plane or as a product of at most four such reflections. ▯

47.10 Definition The *sense* of a tetrahedron $ABCD$ can be defined as *right* or *left* according as a right-hand screw moves toward or away from vertex D when piercing the plane ABC along the altitude to D, and when rotated in the direction A–B–C–A (see Fig. 47.10). An isometry that maps tetrahedron $ABCD$ to tetrahedron $A'B'C'D'$

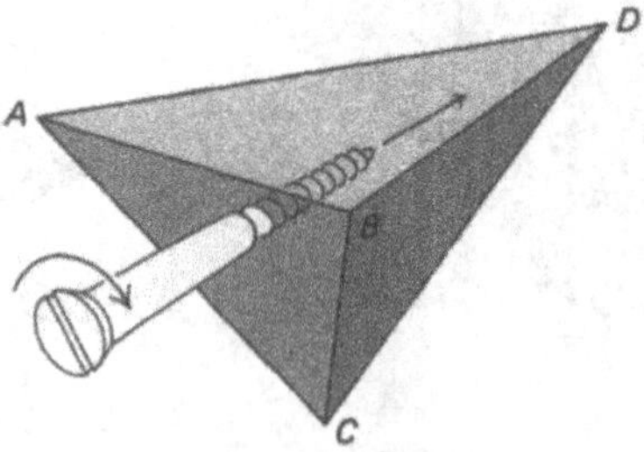

A right-hand tetrahedron

Figure 47.10

is called *direct* or *opposite* according as the two congruent tetrahedrons have the same or different senses.

47.11 We are not as concerned here with memorizing specific details of right and left senses for a tetrahedron as we are with understanding direct and opposite isometries. Of course, it is quite useful to be able to apply this definition of sense to a specific situation in order to ascertain whether the isometry is direct or opposite.

47.12 Theorem A reflection in a plane, or a product of an odd number of such reflections, is an opposite isometry; a product of an even number of reflections is a direct isometry.

47.13 Theorem There are exactly two isometries, one direct and one opposite, that carry a given triangle ABC into a congruent triangle $A'B'C'$.

The method of Theorem 47.9 shows that at most three reflections are needed to map triangle ABC to triangle $A'B'C'$. Let any such isometry be denoted by α. If Π denotes plane $A'B'C'$, then $\sigma_\Pi \alpha$ also is an isometry carrying triangle ABC to $A'B'C'$. And of α and $\sigma_\Pi \alpha$, one is direct and the other is opposite.

Given any fourth point D in space but not in plane Π, there are only two tetrahedrons $A'B'C'D'$ and $A'B'C'D''$ on triangle $A'B'C'$ congruent to tetrahedron $ABCD$ (see Fig. 47.13). Hence there can be no more than two isometries taking triangle ABC to triangle $A'B'C$. ▯

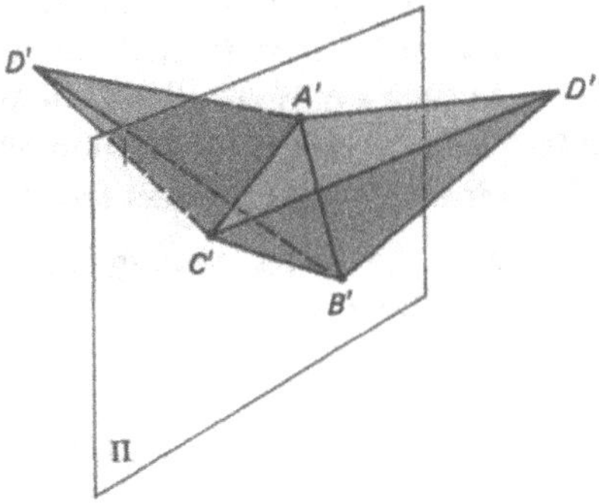

Figure 47.13

Exercise Set 47

1. Reread the transformation theory in Section 12 in terms of space transformations.
2. Prove Corollary 47.7.
3. Prove Theorem 47.8.
4. Make a cardboard model of a right-sensed tetrahedron. Can the vertices be relabeled to change its sense? Explain.
5. Prove Theorem 47.12.
6. Find the two isometries that carry triangle ABC to $A'B'C'$ when the coordinates of the points are $A(0, 0, 0)$, $B(1, 0, 0)$, $C(0, 2, 0)$, $A'(0, 0, 3)$, $B'(1, 0, 3)$, $C'(0, 2, 3)$.

7. Prove that a reflection in a plane is involutoric.
8. Prove that each isometry is either direct or opposite, but not both.
9. How many isometries can be found that map a given segment AB to a congruent segment $A'B'$?
10. Write the coordinates for the vertices of a tetrahedron $A'B'C'D'$, into which tetrahedron $ABCD$, where $A(0, 0, 0)$, $B(1, 0, 0)$, $C(0, 1, 0)$, and $D(0, 0, 1)$, maps under an isometry that is made up of exactly:
    a) one reflection
    b) two reflections
    c) three reflections
    d) four reflections
11. Given that ABC is a triangle and $AD \cong AD'$, $BD \cong BD'$, and $CD \cong CD'$ for two distinct points D and D', prove that plane ABC is the perpendicular bisector of segment DD'.
12. When PAB is a triangle with $PA \cong PB$, prove that P lies on the plane that is the perpendicular bisector of segment AB.

SECTION 48 BASIC SPACE ISOMETRIES

48.1 Definition A space map α is called a *translation through vector* $\overrightarrow{PQ}$ iff for each point A, $\alpha(A) = B$ where $\overrightarrow{AB} = \overrightarrow{PQ}$.

48.2 Theorem Each product of reflections in two parallel planes Π and Δ is a translation through twice the normal vector from plane Π to plane Δ. Conversely, each translation can be factored into such a product of reflections in parallel planes, both of which are perpendicular to the vector of translation and half its length apart. Hence each translation is an isometry and is direct (see Fig. 46.1b).

48.3 Theorem The inverse of a translation through vector $\overrightarrow{PQ}$ is the translation through vector $\overrightarrow{QP}$. If the translation α is given by $\alpha = \sigma_\Pi \sigma_\Delta$, then $\alpha^{-1} = \sigma_\Delta \sigma_\Pi$.

48.4 Theorem Translations commute.

48.5 Theorem The product of two translations is a translation.

48.6 Thus translations behave exactly the same in space as they do in the plane. The proofs of these first theorems in this section are quite analogous to those given for the corresponding plane theorems (see Section 14), so they are not repeated here.

48.7 A rotation about a line in space is also quite similar to a rotation about a point in the plane, as attested by items 48.8 through 48.13. There are some differences, however, in that products of rotations about skew lines do not yield either rotations or translations, as noted in Theorem 49.3.

48.8 Definition Let A be any point and m any line in space, let Π be the plane containing A and perpendicular to m, and let m and Π intersect in point O. A space map α is called a *rotation about line m through angle* θ iff for each such point A, α is the plane rotation of plane Π through angle θ about point O. Line m is called the *axis* of the rotation (see Fig. 48.8).

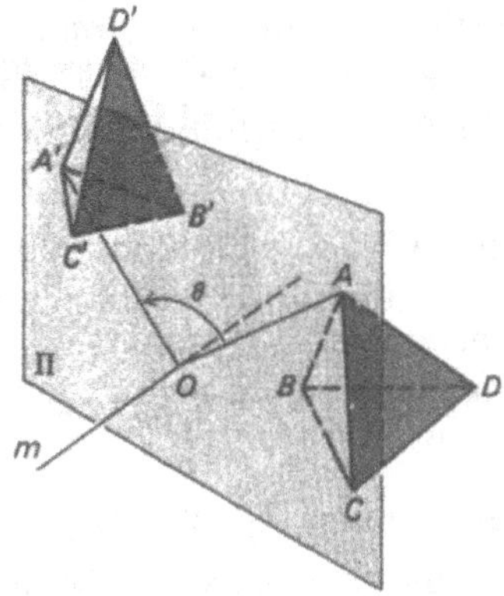

Figure 48.8

48.9 Theorem Each product of reflections in two planes Π and Δ intersecting in a line m at an angle θ is a rotation of angle 2θ about line m. Conversely, each rotation can be factored into such a product of reflections in planes intersecting on its axis at half the angle of the rotation. Hence each rotation is an isometry and is direct (see Fig. 46.1c).

48.10 Theorem The inverse of a rotation about line m through angle θ is the rotation about line m through angle $-\theta$. If the rotation α is given by $\alpha = \sigma_\Pi \sigma_\Delta$, then $\alpha^{-1} = \sigma_\Delta \sigma_\Pi$.

48.11 Theorem Two rotations about the same axis commute.

48.12 Theorem An isometry with an invariant point is a reflection in a plane or a product of at most three reflections.

In the proof of Theorem 47.9, we may assume that point A is the fixed point, so Case 5 is eliminated. Hence the isometry is a reflection or a product of at most three reflections in lines. □

48.13 Theorem A direct isometry with an invariant point is a rotation.

Since the isometry is direct, it is a product of an even number of reflections, hence of just two reflections by Theorem 48.12. But a translation has no fixed points, so the isometry must be a product of reflections in two intersecting planes, a rotation. □

48.14 Definition A rotation of 180° about a line m is called a *halfturn about line m*, denoted by σ_m (see Fig. 46.1d).

48.15 Theorem A halfturn about line m is a product of two reflections in any two perpendicular planes through m.

48.16 Corollary A product of reflections in two perpendicular planes commutes.

48.17 Corollary A halfturn about a line is involutoric.

48.18 Definition A space map α is called a *central inversion* in a point P called its *center* iff $\alpha(P)=P$, and, for each point $A \neq P$, $\alpha(A)=B$, where B is taken so that P is the midpoint of segment AB. This central inversion is denoted by σ_P (see Fig. 46.2).

48.19 Theorem The product of three reflections in mutually perpendicular planes intersecting at a point P is a central inversion in point P. Conversely, each central inversion in a point P can be written as a product of three reflections in any three mutually perpendicular planes passing through P. Hence a central inversion is an isometry and is opposite.

48.20 Theorem A central inversion is involutoric.

The term "central inversion" is somewhat unfortunate, since this map is not to be confused with an inversion in a circle studied in Chapter 5. It is a reflection in a point, and some authors use that term, but then the term "reflection" tends to be overworked. We shall hold to the term "central inversion" here.

Exercise Set 48

1. Indicate the two planes of reflection into which a translation α can be factored if α carries the point (0, 0, 0) into (1, 1, 1).
2. Prove Theorem 48.2.
3. Prove Theorem 48.3.
4. Prove Theorem 48.4.
5. Prove Theorem 48.5.
6. Define a space translation in terms of a plane translation, just as Definition 48.8 does for a space rotation.
7. Show that a product of two rotations about parallel axes is a rotation or a translation.
8. State when a product of two halfturns is a translation.
9. Prove Theorem 48.9.
10. Prove Theorem 48.10.
11. Prove Theorem 48.11.
12. Show that a product of two central inversions is a translation.
13. Prove Theorem 48.15.
14. Prove Corollary 48.16.
15. Prove Corollary 48.17.
16. What isometry is the product of three central inversions?
17. Prove Theorem 48.19.
18. Prove Theorem 48.20.

SECTION 49 | MORE SPACE ISOMETRIES

49.1 Definition A *screw displacement* is the product of a rotation about a line and a translation through a vector parallel to the axis of the rotation (see Fig. 49.1).

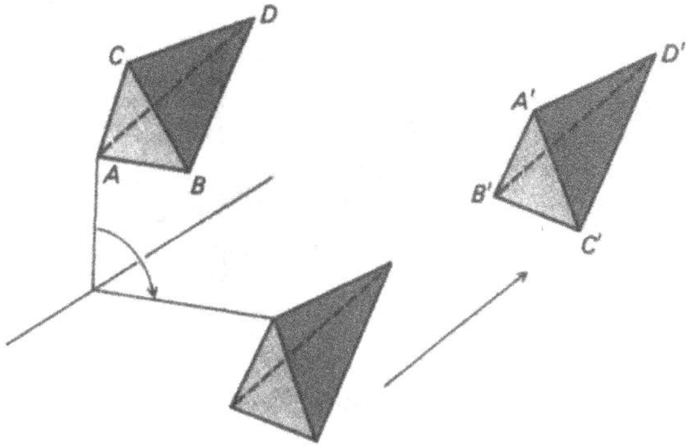

Figure 49.1

49.2 Theorem Each direct isometry can be written as a product of two halfturns.

A translation or rotation α can be factored into a product $\sigma_\Delta \sigma_\Pi$ of two reflections in planes Π and Δ. Take a plane Γ perpendicular to both given planes. Then $\alpha = (\sigma_\Delta \sigma_\Gamma)(\sigma_\Gamma \sigma_\Pi)$, a product of two halfturns.

We need consider only a product α of four reflections. If any three of the mirrors form a pencil, then the isometry reduces to a product of two reflections (Exercise 49.1). Similarly, a product of two translations reduces to a translation (Theorem 48.5). If only the first and second planes or only the third and fourth planes are parallel, then the second and third planes intersect. In any case we may assume that the second and third planes do intersect, so we may rotate them about their line of intersection so that α becomes a product of two rotations. If the rotations are about the same or parallel axes, they may be replaced by a single rotation or translation (Exercise 48.7). Thus we need consider only a product of two rotations in skew or intersecting lines.

Let $\alpha = \sigma_4 \sigma_3 \sigma_2 \sigma_1$ where planes 1 and 2 meet in line m, and planes 3 and 4 meet in line n, and lines m and n are either intersecting or skew. We may assume without loss of generality that plane 3 cuts line m, and that planes 2 and 3 are perpendicular along a line p (see Fig. 49.2a). Now rotate planes 2 and 3 about line p into 2′ and 3′ so that planes 1 and 2′ are perpendicular. Of course, planes 2′ and 3′ remain perpendicular (see Fig. 49.2b). Then we have

$$\sigma_4 \sigma_3 \sigma_2 \sigma_1 = \sigma_4 \sigma_{3'} \sigma_{2'} \sigma_1 = \sigma_4 \sigma_{2'} \sigma_{3'} \sigma_1.$$

Since both planes 1 and 3′ are perpendicular to plane 2, then their line q of intersection is perpendicular to plane 2. Then rotate planes 1 and 3′ about line q into planes 1′ and 3″ so that 3″ is perpendicular to plane 4. Also planes 1′ and 3″ are perpendicular to plane 2′ (since line q is perpendicular to plane 2′). Now we have

$$\alpha = \sigma_4 \sigma_{2'} \sigma_{3'} \sigma_1 = \sigma_4 \sigma_{2'} \sigma_{3''} \sigma_{1'} = (\sigma_4 \sigma_{3''})(\sigma_{2'} \sigma_{1'}),$$

a product of two halfturns. □

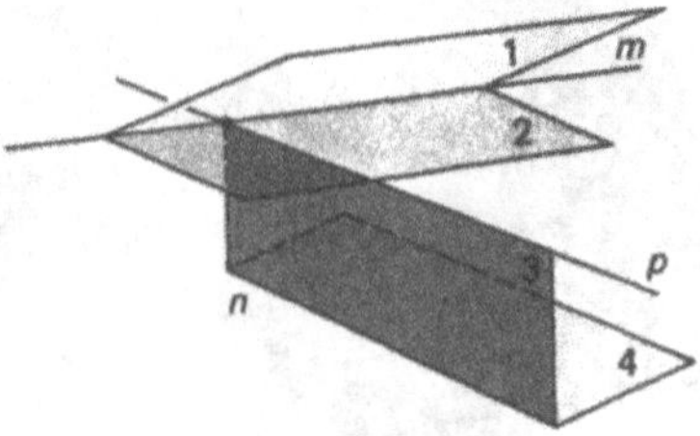

Figure 49.2a

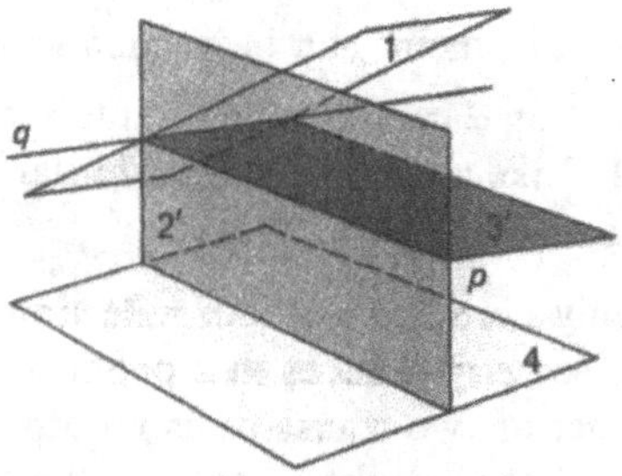

Figure 49.2b

49.3 Theorem A product of two rotations
a) about the same line is a rotation about that line;
b) about parallel lines is a rotation about an axis parallel to the two lines, or a translation whose vector is perpendicular to the direction of the two lines;
c) about two lines intersecting at point O is a rotation about an axis through O;
d) about two skew lines is a screw displacement.

Parts (a) and (b) are analogous to the corresponding plane theorem (Theorem 14.15). Part (c) is established by Theorem 48.13, since point O is a fixed point because it is invariant under both rotations.

For part (d), in view of Theorem 49.2, we need show only that the product of two halfturns about skew lines is a screw displacement. Let $\alpha = \sigma_n \sigma_m = \sigma_4 \sigma_3 \sigma_2 \sigma_1$ be such an isometry, where m and n are skew lines with planes 1 and 2 and planes 3 and 4 perpendicular. We may also assume without loss of generality that plane 1 is parallel

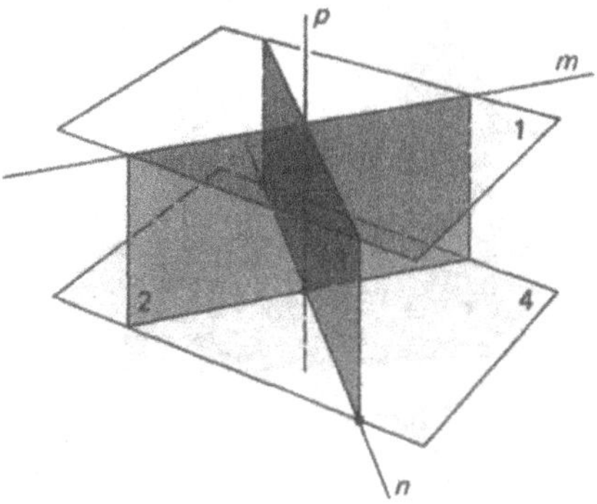

Figure 49.3

to line n and that 4 is parallel to m. Then planes 2 and 3 are each perpendicular to both of planes 1 and 4. Thus

$$\alpha = \sigma_4\sigma_3\sigma_2\sigma_1 = \sigma_4\sigma_3\sigma_1\sigma_2 = \sigma_4\sigma_1\sigma_3\sigma_2,$$

a screw displacement, since planes 1 and 4 are parallel and the line p of intersection of planes 2 and 3 is perpendicular to planes 1 and 4 (see Fig. 49.3). ▯

49.4 Corollary The axes of the two halfturns of a direct isometry with no invariant point are either parallel (a translation) or skew (a screw displacement).

49.5 Corollary Each direct isometry is a rotation, a translation, or a screw displacement.

49.6 Definition A *glide-reflection* is the product of a reflection in a plane and a translation through a vector parallel to the mirror of the reflection (Fig. 49.6).

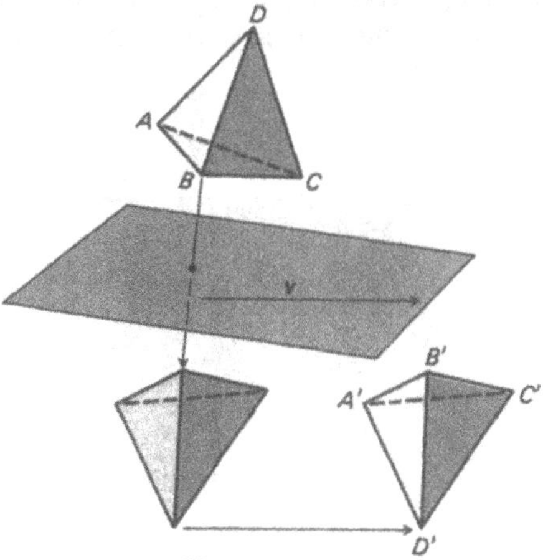

Figure 49.6

49.7 Definition A *rotatory reflection* is the product of a reflection in a plane and a rotation about an axis m perpendicular to the mirror of the reflection (Fig. 49.7).

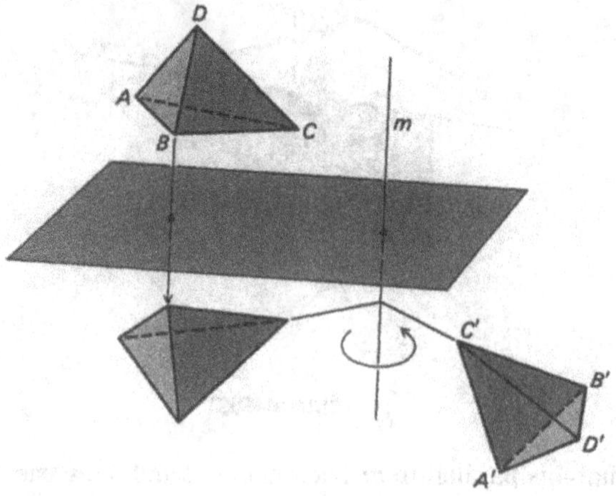

Figure 49.7

49.8 Theorem Each opposite isometry is a reflection, a glide-reflection, or a rotatory reflection.

An opposite isometry α is a reflection or a product of three reflections. So consider the product

$$\alpha = \sigma_\Pi \sigma_\Delta \sigma_\Gamma.$$

This product reduces to a single reflection if the mirrors form a pencil, so we assume that they do not. We may also assume without loss of generality that planes Γ and Δ meet in a line m.

Case 1. If line m is parallel to plane Π (see Fig. 49.8a), then we may assume that Γ is parallel to Π, and it follows that the isometry is a glide-reflection. The proof is analagous to that for Theorem 16.4.

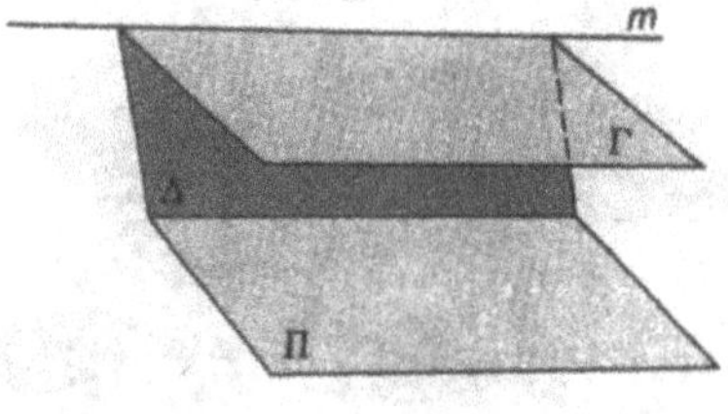

Figure 49.8a

Case 2. Let line m pierce plane Π at a point P (see Fig. 49.8b). First rotate Γ and Δ about line m into Γ' and Δ', so that Δ' is perpendicular to Π (see Fig. 49.8c). Next rotate Δ' and Π about their line n of intersection into Δ'' and Π' so that Δ'' is perpendicular to Γ' (see Fig. 49.8d). Of course, Δ'' is perpendicular to Π'. Then

$$\begin{aligned} \alpha &= \sigma_\Pi \sigma_\Delta \sigma_\Gamma = \sigma_\Pi \sigma_{\Delta'} \sigma_{\Gamma'} \\ &= \sigma_{\Pi'} \sigma_{\Delta''} \sigma_{\Gamma'} = (\sigma_{\Pi'} \sigma_{\Gamma'}) \sigma_{\Delta''}, \end{aligned}$$

a rotatory reflection, since Δ'' is perpendicular to both Π' and Γ', hence to their line of intersection. ▯

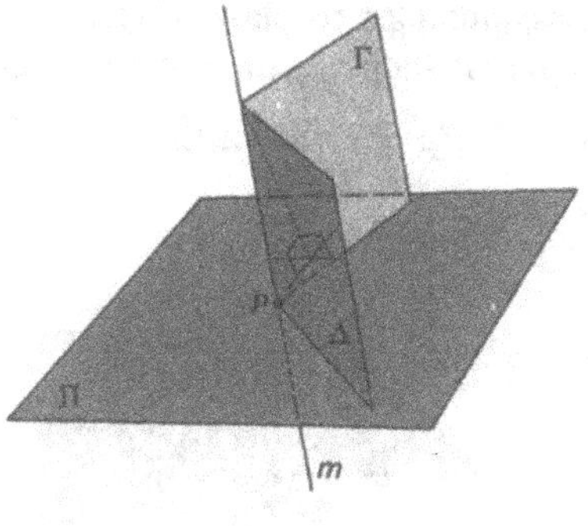

Figure 49.8b

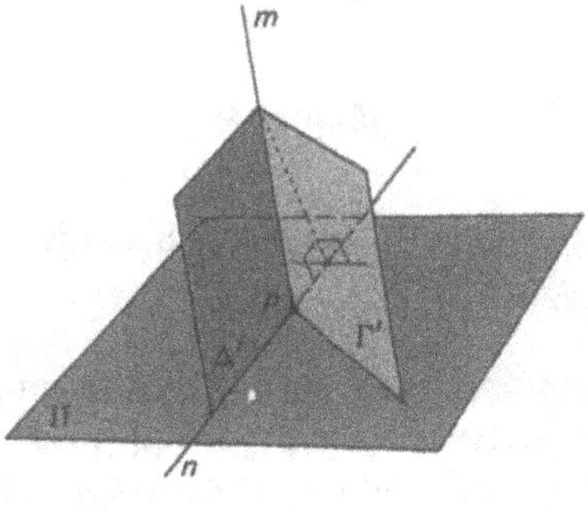

Figure 49.8c

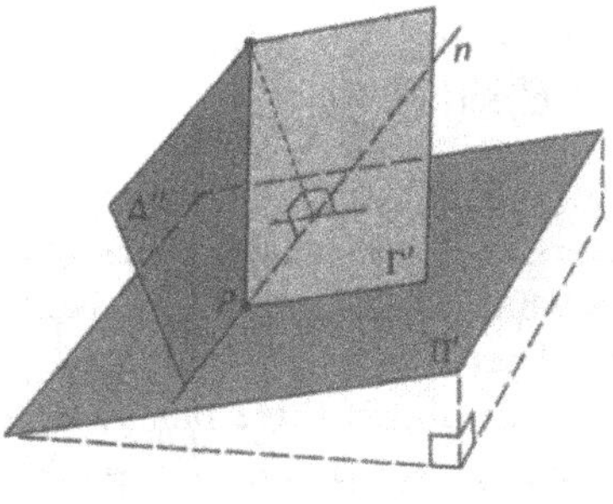

Figure 49.8d

49.9 Theorem A rotatory reflection is the product of a central inversion and a rotation with axis through the center O of the central inversion, hence it is called also a *rotatory inversion.*

Let the rotatory reflection be

$$\alpha = \sigma_\Pi \sigma_\Delta \sigma_\Gamma$$

where Π is perpendicular to the line m of intersection of planes Γ and Δ. Let Ω be the plane through m perpendicular to plane Γ (see Fig. 49.9). Since Π is perpendicular to both planes Δ and Ω, then σ_Π commutes with both σ_Δ and σ_Ω, and we have

$$\alpha = \sigma_\Pi \sigma_\Delta \sigma_\Gamma = \sigma_\Pi \sigma_\Delta \sigma_\Omega^2 \sigma_\Gamma = \sigma_\Delta \sigma_\Omega^2 \sigma_\Pi \sigma_\Gamma = (\sigma_\Delta \sigma_\Omega)(\sigma_\Omega \sigma_\Pi \sigma_\Gamma),$$

a rotatory inversion. ▯

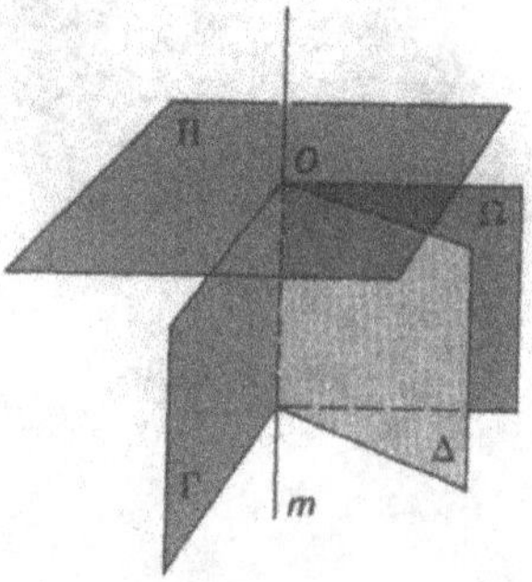

Figure 49.9

49.10 Theorem An isometry with three noncollinear fixed points is either the identity or a reflection in a plane.

49.11 Theorem An opposite isometry with no invariant point is a glide-reflection.

Such an opposite isometry α is the product of three reflections in planes Γ, Δ, and Π that do not form a pencil and do not concur at a point (Exercise 49.9). Thus we assume, without loss of generality, that the three lines of intersection of the pairs of these planes are parallel. By rotating pairs of planes appropriately, as was done for lines in Theorem 16.4, we obtain $\alpha = \sigma_{\Pi'} \sigma_{\Delta'} \sigma_{\Gamma'}$, where Δ' is perpendicular to both Γ' and Π', and these last two planes are parallel (see Fig. 49.8a). Thus $\alpha = (\sigma_{\Pi'} \sigma_{\Gamma'}) \sigma_{\Delta'}$, a glide-reflection. ▯

49.12 Theorem An opposite isometry with an invariant point is a rotatory reflection (of which a central inversion and a reflection are special cases).

The three mirrors which make up such an isometry are parallel, or all pass through its fixed point P (Exercise 49.12a). Now pairs of planes may be rotated about their lines of intersection to obtain three planes Π, Γ, and Δ, with the first two planes both perpendicular to plane Δ (as in Fig. 49.8d). Then $\sigma_\Delta \sigma_\Gamma \sigma_\Pi$ is a rotatory reflection. ▯

49.13 Theorem A direct isometry with no invariant point is a screw displacement (of which a translation is a special case).

49.14 We terminate our development of a theory of space isometry here, having given a sufficient background for our purposes. The last two sections of this chapter contain applications of space isometries to elementary solid geometry and analytic equations for these transformations. Certainly the algebra of isometries could be pursued much further, as it was for plane isometries in Sections 16 and 17. You should now have sufficient preparation to undertake such studies when you feel the need to do so. If so, then our work is complete.

Exercise Set 49

1. Prove that a product of three reflections in planes that form a pencil (they are all parallel or all pass through a line) is a reflection in another plane of the same pencil.
2. Make a model of each of Figs. 49.2a, 49.2b, and 49.3 by gluing or taping together 3-by-5 cards. Label important points, lines, and planes.
3. a) Prove Theorem 49.3, part (a).
   b) Prove Theorem 49.3, part (b).
   c) Prove Theorem 49.3, part (c), by factoring the rotations into appropriately chosen reflections in planes.
4. Prove Corollary 49.4.
5. Prove Corollary 49.5.
6. Make a model of each of Figs. 49.8b, 49.8c, and 49.8d by gluing or taping together 3-by-5 cards. Label important points, lines, and planes.
7. What isometry is the square of a rotatory reflection?
8. Prove Theorem 49.8, Case 1.
9. Show that an opposite isometry with no fixed point is a product of three reflections in planes that do not concur at a point, and do not form a pencil.
10. Prove Theorem 49.10.
11. Complete the proof of Theorem 49.11.
12. a) Prove that the three mirrors of the reflections that make up an opposite isometry with a fixed point are parallel or all pass through the fixed point.
    b) Supply the details in the proof of Theorem 49.12.
13. Prove Theorem 49.13.
14. Show that the translation and reflection of a glide-reflection commute.
15. Show that the rotation and reflection of a rotatory reflection commute.
16. Show that the rotation and central inversion of a rotatory inversion commute.
17. Show that a reflection and a central inversion are both special cases of a rotatory reflection.
18. Show that a translation is a special case of a screw displacement.

SECTION 50 | SOME APPLICATIONS

50.1 Theorem Vertical dihedral angles (angles between planes) are congruent.

50.2 Theorem Parallel planes cut by a transversal plane form congruent corresponding and alternate dihedral angles.

The lines of intersection of the two planes with the transversal plane are parallel, since they lie in the same plane (the transversal) and do not meet. Translate one of the lines of intersection along the transversal plane at right angles to itself into the other line of intersection. The transversal plane maps into itself and one parallel plane maps into the other. Thus this isometry maps one set of vertical dihedral angles into the other. Hence they are congruent. (Compare this proof with that given for Theorem 18.7.) □

50.3 Theorem The opposite faces of a parallelepiped are congruent parallelograms.

50.4 Theorem The diagonals of a parallelepiped concur at a point (called the *center* of the parallelepiped) that bisects each diagonal.

Let AD and BC be a pair of opposite edges. They are parallel and congruent. Thus $ABCD$ is a parallelogram whose diagonals meet at their midpoints by Theorem 18.9. The theorem follows (see Fig. 50.4). □

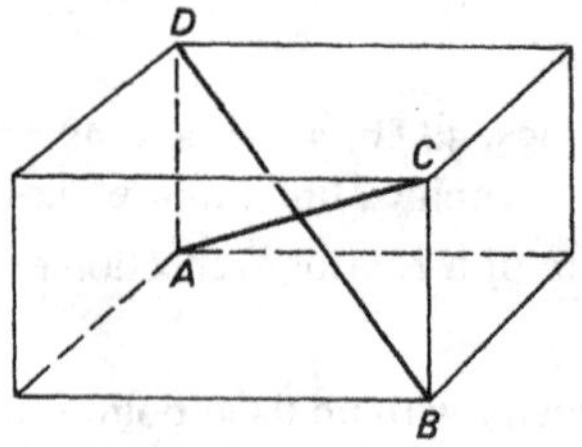

Figure 50.4

50.5 Theorem The diagonals of a rectangular parallelepiped (a box) are congruent.

50.6 Theorem A parallelepiped is symmetric with respect to its center; that is, a parallelepiped is invariant under a central inversion in its center.

50.7 Corollary Any segment through the center of a parallelepiped joining points on opposite faces is bisected by the center.

50.8 Corollary A plane through two opposite edges of a parallelepiped divides the parallelepiped into two congruent triangular prisms.

50.9 Theorem The sides of an isosceles triangle make congruent angles with any plane in which the base lies (see Fig. 50.9).

50.10 Theorem The following statements are equivalent for any three given planes Γ, Δ, and Π:

a) $\Delta = \sigma_\Gamma(\Pi)$,

b) Γ lies midway between Δ and Π (or bisects their angle),

c) $\sigma_\Delta = \sigma_\Gamma \sigma_\Pi \sigma_\Gamma$.

Compare Theorem 50.10 with its plane counterpart, Theorem 17.5.

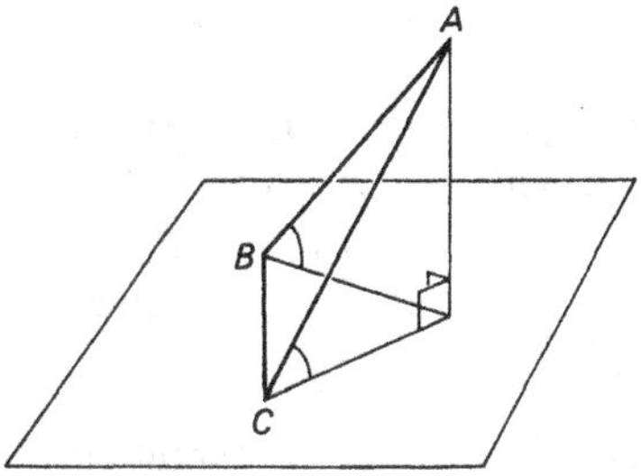

Figure 50.9

50.11 Theorem The three planes that bisect the dihedral angles of a trihedral angle (formed by three planes that concur at a point) concur along a line.

50.12 Definition The *volume* of a rectangular parallelepiped is the product of the lengths of its three edges.

50.13 Theorem The volume of a parallelepiped is the product of the area of any face as base and the length of the altitude to that base.

50.14 Corollary The volume of a triangular prism is the product of the area of its triangular face as base and the length of the altitude to that base.

50.15 Corollary The volume of any prism is the product of the area of its (polygonal) base and the length of the altitude to that base (see Fig. 50.15).

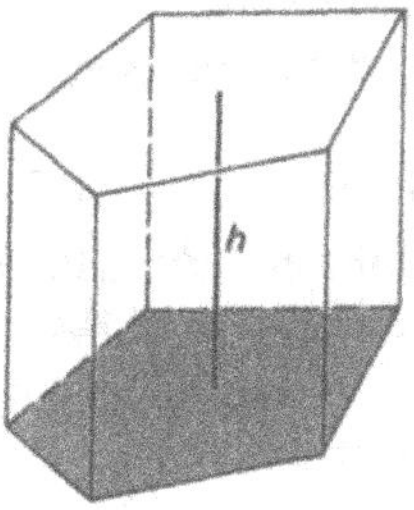

Figure 50.15

50.16 Theorem An oblique prism has the same volume as a right prism whose base is a right section of the oblique prism, and whose altitude is a lateral edge of the oblique prism (see Fig. 50.16).

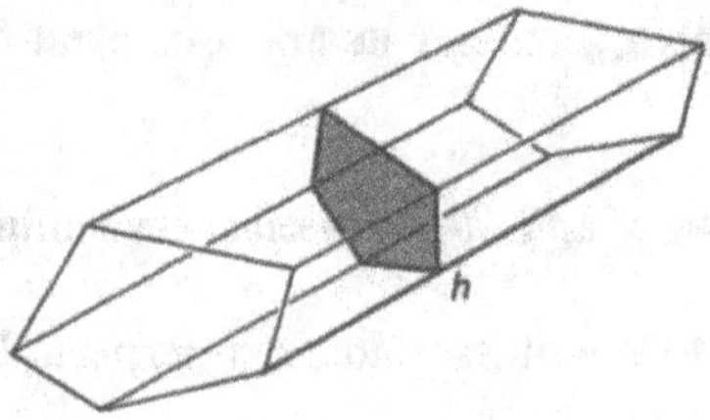

Figure 50.16

50.17 Theorem A sphere is symmetric with respect to its center, with respect to any line through its center, and with respect to any plane through its center.

Because any isometry that leaves the center O of the sphere fixed maps each point P on the sphere into a point P' the same distance (the radius) from O, it follows that P' lies on the sphere. ▯

50.18 Theorem A right circular cylinder is symmetric with respect to
a) the midpoint of its axis,
b) its axis,
c) the normal plane to its axis at its midpoint,
d) any plane through its axis.

50.19 Corollary A right circular cylinder is generated by rotating a rectangle about one side as axis.

50.20 Definition Let P be the centroid of face ABC of tetrahedron $ABCD$. Then segment DP is called a *median* of the tetrahedron (see Fig. 50.20).

50.21 Theorem The four medians of a tetrahedron concur at a point G called the *centroid* of the tetrahedron.

First, central inversions in space behave quite like halfturns in the plane. That is,
1) a central inversion σ_x is involutoric (Theorem 48.20),
2) a product of two central inversions is a translation (Exercise 48.12) and translations commute (Theorem 48.4), and
3) a product $\sigma_C \sigma_B \sigma_A$ of three central inversions is a central inversion (see Exercise 48.16), so $\sigma_C \sigma_B \sigma_A = \sigma_A \sigma_B \sigma_C$.

Let $ABCD$ be the tetrahedron having medians AQ and DP and with A', B', C' the midpoints of the sides of triangle ABC, as shown in Fig. 50.20. By a "shrewd guess," let us place point G three-fourths of the way from D to P along median DP (assume that the theorem is true and apply Menelaus' theorem to triangle $DA'P$ to obtain the "shrewd guess"). Then we have

4) $$(\sigma_G \sigma_D)(\sigma_G \sigma_P)^3 = \iota.$$

Since P and Q are the centroids of triangles ABC and DBC, then P and Q are two-thirds of the way along the triangle medians AA' and DA'. Hence

5) $$(\sigma_Q \sigma_D)(\sigma_Q \sigma_{A'})^2 = \iota \quad \text{and} \quad (\sigma_{A'} \sigma_P)^2(\sigma_A \sigma_P) = \iota.$$

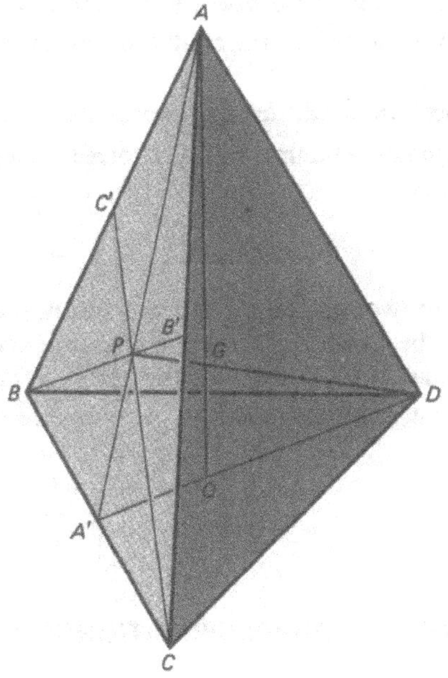

Figure 50.20

To prove the theorem it will suffice to show that G now lies three-fourths of the way from A to Q; that is, we must show that $(\sigma_A \sigma_G)(\sigma_Q \sigma_G)^3 = \iota$. To that end, we take

$$
\begin{aligned}
&(\sigma_A \sigma_G)(\sigma_Q \sigma_G)^3 \\
&= (\sigma_A \sigma_G)(\sigma_Q \sigma_G)^3(\sigma_G \sigma_D)(\sigma_G \sigma_P)^3 && \text{by (4)} \\
&= (\sigma_Q \sigma_G)(\sigma_G \sigma_D)(\sigma_Q \sigma_G)(\sigma_G \sigma_P)(\sigma_Q \sigma_G)(\sigma_G \sigma_P)(\sigma_A \sigma_G)(\sigma_G \sigma_P) && \text{by (2)} \\
&= \sigma_Q \sigma_D \sigma_Q \sigma_P \sigma_Q \sigma_P \sigma_A \sigma_P && \text{by (1)} \\
&= \sigma_Q \sigma_D \sigma_Q \sigma_{A'} \sigma_{A'} \sigma_P \sigma_Q \sigma_P \sigma_A \sigma_P && \text{by (1)} \\
&= \sigma_Q \sigma_D \sigma_Q \sigma_{A'} \sigma_Q \sigma_P \sigma_{A'} \sigma_P \sigma_A \sigma_P && \text{by (3)} \\
&= (\sigma_Q \sigma_D \sigma_Q \sigma_{A'} \sigma_Q \sigma_{A'})(\sigma_{A'} \sigma_P \sigma_{A'} \sigma_P \sigma_A \sigma_P) && \text{by (1)} \\
&= \iota. && \text{by (5)}
\end{aligned}
$$

Similarly, G lies three-fourths of the way along the other two medians. ▯

Exercise Set 50

1. Prove Theorem 50.1.
2. Prove that each product of two halfturns about axes intersecting at an angle θ is a rotation through angle 2θ.
3. Prove Theorem 50.3.
4. Prove that each translation can be written as a product of two central inversions, or two halfturns in parallel axes, or two reflections in parallel mirrors.

5 through 11. Prove Theorems (and Corollaries) 50.5 through 50.11.

12. Prove that each opposite isometry is the product of a reflection in a plane and a halfturn about a line.

13 through 16. Prove Theorems (and Corollaries) 50.13 through 50.16.

17. Prove that each rotation about a line can be factored into a product of two halfturns about intersecting lines.

18. Prove Theorem 50.18.

19. Prove Corollary 50.19.

20. Prove that a product of two halfturns about two skew lines at right angles is a screw displacement, namely the product of a halfturn about the line of shortest distance between the axes of the halfturns and a translation through twice this shortest distance.

21. Name the isometry that maps each point (x, y, z) in space into
   a) $(x, y, -z)$
   b) $(-y, x, z)$
   c) $(x, y, z+1)$
   d) $(-y, x, z+1)$
   e) $(-x, y, z+1)$
   f) $(-y, x, -z)$

▶SECTION 51 | ANALYTIC REPRESENTATIONS

51.1 Let us denote the vector **v** from the origin $O(0, 0, 0)$ to a point $A(h, k, l)$ by $\mathbf{v} = (h, k, l)$.

51.2 Theorem The translation through vector $\mathbf{v} = (h, k, l)$ that carries point $P(x, y, z)$ to point $P'(x', y', z')$ is determined by

$$x' = x + h,$$
$$y' = y + k,$$
$$z' = z + l.$$

51.3 Theorem The rotation about the z-axis through angle θ (see Fig. 51.3) is given by

$$x' = x\cos\theta - y\sin\theta,$$
$$y' = x\sin\theta + y\cos\theta,$$
$$z' = z.$$

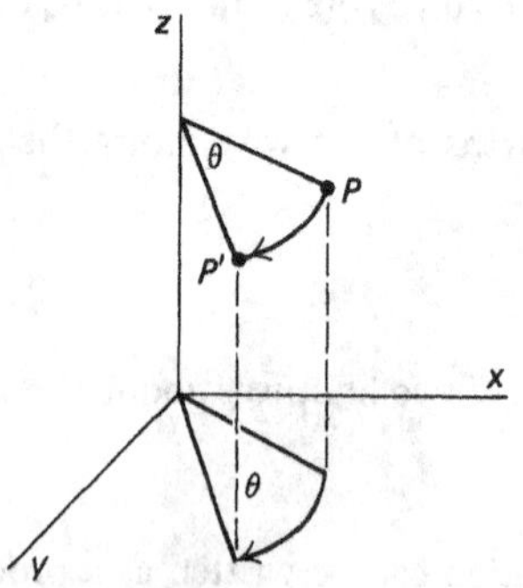

Figure 51.3

51.4 Although we could develop equations for other rotations, we shall be content with equations of rotations about just the coordinate axes. In any case, any rotation can be written as the product of a translation, one to three rotations about the coordinate axes, and then the inverse of the translation. To illustrate, let us perform the translation from $A(h, k, l)$ to $O(0, 0, 0)$, then rotations through angles θ, ϕ, and λ about the z-, y-, and x-axes, then the translation from O back to A, to arrive at the equations

$$\begin{aligned} x' &= (x-h)\cos\theta\cos\phi - (y-k)\sin\theta\cos\phi - (z-l)\sin\phi + h, \\ y' &= (x-h)(\sin\theta\cos\lambda - \cos\theta\sin\phi\sin\lambda) + (y-k)(\cos\theta\cos\lambda \\ &\qquad + \sin\theta\sin\phi\sin\lambda) - (z-l)\cos\phi\sin\lambda + k, \\ z' &= (x-h)(\sin\theta\sin\lambda + \cos\theta\sin\phi\cos\lambda) + (y-k)(\cos\theta\sin\lambda \\ &\qquad - \sin\theta\sin\phi\cos\lambda) + (z-l)\cos\phi\cos\lambda + l. \end{aligned}$$

Clearly these equations are far too complicated to be of practical use to us.

51.5 Theorem The reflection in the xy-plane (see Fig. 51.5) is given by

$$x' = x, \qquad y' = y, \qquad z' = -z.$$

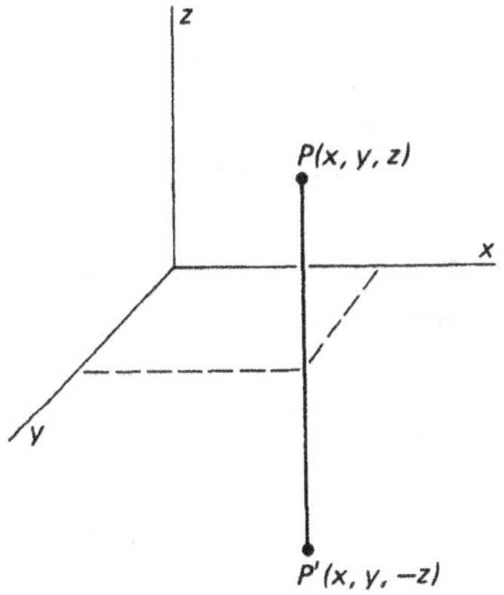

Figure 51.5

51.6 Similar equations determine reflections in the other coordinate planes. Thus we obtain the next two immediate corollaries from Theorems 48.15, 48.19, and 51.5.

51.7 Corollary A halfturn about the z-axis (see Fig. 51.7) is given by

$$x' = -x, \qquad y' = -y, \qquad z' = z.$$

51.8 Corollary A central inversion in the origin (see Fig. 51.8) is given by

$$x' = -x, \qquad y' = -y, \qquad z' = -z.$$

51.9 To develop the equations for a reflection in the general plane Π, let us write the equation of Π in the standard form

$$Ax + By + Cz + D = 0 \qquad \text{where} \qquad A^2 + B^2 + C^2 = 1.$$

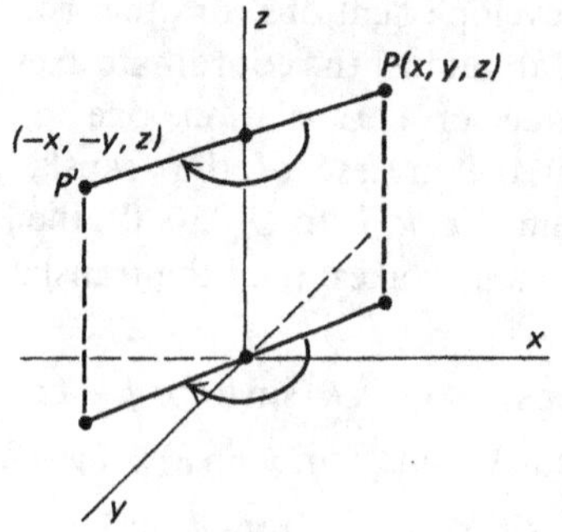

Figure 51.7

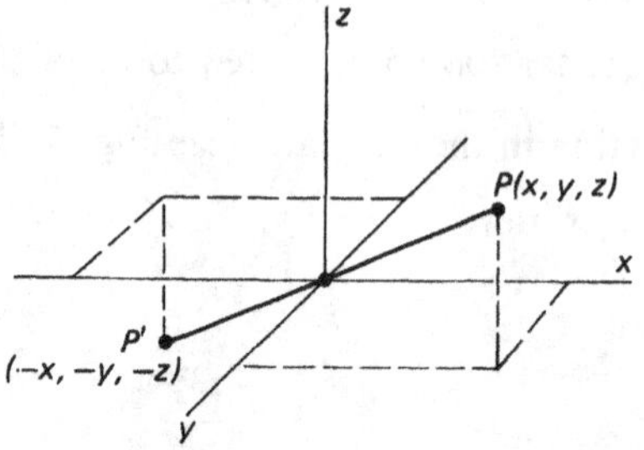

Figure 51.8

Letting σ_Π map $P(x, y, z)$ to $P'(x', y', z')$, we have the line PP' normal to Π, so that the coordinates of P and P' satisfy the equation

$$\frac{x-x'}{A}=\frac{y-y'}{B}=\frac{z-z'}{C}=t,$$

or

$$x'=x+At, \qquad y'=y+Bt, \qquad \text{and} \qquad z'=z+Ct,$$

where t is a real constant.

Now the distance from P to Π is equal to that from P' to Π. Recalling that the distance from P to Π is given by

$$\pm(Ax+By+Cz+D) \qquad \text{when } A^2+B^2+C^2=1,$$

the plus or minus sign being taken according to which side of the plane P lies on, we have, since P and P' lie on opposite sides of the plane Π,

$$Ax+By+Cz+D=-(Ax'+By'+Cz'+D),$$

which we rewrite as

$$A(x+x')+B(y+y')+C(z+z')+2D=0.$$

Substituting the parametric equations of the preceding paragraph into the equation above, we obtain

$$A(2x+At)+B(2y+Bt)+C(2z+Ct)+2D=0,$$

from which

$$t = -2(Ax + By + Cz + D).$$

This value for t may now be substituted into the parametric equations to obtain equations for a reflection in the plane Π. Hence we have proved the following theorem.

51.10 Theorem A reflection in the plane $Ax + By + Cz + D = 0$, in which we have $A^2 + B^2 + C^2 = 1$, has the equations

$$\begin{aligned} x' &= x - 2A(Ax + By + Cz + D), \\ y' &= y - 2B(Ax + By + Cz + D), \\ z' &= z - 2C(Ax + By + Cz + D). \end{aligned}$$

51.11 Since each isometry in space is a product of reflections, it is now possible to write a set of equations for any isometry. The calculations become quite complicated very quickly, so we shall not pursue the matter further.

Exercise Set 51

1. Prove Theorem 51.2.
2. Prove Theorem 51.3.
3. Prove Theorem 51.5.
4. Prove Corollary 51.7.
5. Prove Corollary 51.8.
6. Show that the translation of Theorem 51.2 can be accomplished by a product of two reflections in planes, as given in Theorem 51.10.
7. Show that a product of two reflections in intersecting planes (as given in Theorem 51.10) is a rotation.
8. Find equations for a central inversion about any given point as center.
9. Show that Theorem 51.5 is a special case of Theorem 51.10.
10. Find equations for a halfturn about a given line by taking the product of a reflection in a plane perpendicular to the line and a central inversion in the point of intersection of the line and the plane.

APPENDIXES

APPENDIX A | A SUMMARY OF BOOK I OF EUCLID'S *ELEMENTS*

The Axioms (Common Notions)

1. Things equal to the same thing are themselves equal.
2. When equals are added to equals, then the sums are equal.
3. When equals are subtracted from equals, then the differences are equal.
4. Things which coincide with one another are equal.
5. The whole is greater than any part of the whole.

The Postulates

1. A line segment can be drawn between any two points.
2. A line segment can be extended indefinitely.
3. A circle can be drawn having any point as center and passing through any other point.
4. All right angles are congruent.
5. If a transversal cuts two lines so that the sum of the two interior angles on one side of the transversal is less than two right angles, then the two lines will meet on that same side.

Selections from the 48 Propositions of Book I

1. To construct an equilateral triangle on a given segment.
2. To draw a segment congruent to a given segment from a given point as endpoint.
3. To cut off from a given segment a segment congruent to a smaller given segment.
4. Two triangles that satisfy the SAS condition are congruent.
5. The base angles of an isosceles triangle are congruent.
6. If two angles of a triangle are congruent, then the sides opposite these angles are congruent.
7. Only one triangle directly congruent to a given triangle can be constructed on a given side of a given segment congruent to the base of the given triangle.

8. Two triangles that satisfy the SSS condition are congruent.
9. To bisect a given angle.
10. To bisect a given segment.
11. To erect a perpendicular at a given point on a line.
12. To drop a perpendicular from a point to a line.
13. If a ray emanates from a point on a line, then the two angles formed are right angles or have a sum equal to two right angles.
14. If two rays emanating from opposite sides of a point on a line make their adjacent angles have a sum equal to two right angles, then they lie on a line.
15. Vertical angles formed by intersecting lines are congruent.
16. An exterior angle of a triangle is greater than either opposite interior angle.
17. The sum of any two angles of a triangle is less than two right angles.
18. In a triangle the greater side lies opposite the greater angle.
19. In a triangle the greater angle lies opposite the greater side.
20. In a triangle the sum of any two sides is greater than the third side.
22. To construct a triangle from three given segments.
23. To construct at a point on a line an angle congruent to a given angle.
24. If two triangles have two sides congruent to two sides respectively but the included angle of the first greater than that of the second, then the opposite side of the first is greater than that of the second.
25. If two triangles have two sides congruent to two sides respectively, but the third side of the first is greater than that of the second, then the angle of the first included between the two sides is greater than that of the second.
26. Two triangles that satisfy either the ASA or the AAS condition are congruent.
27. If a transversal cuts two lines so that alternate interior angles are congruent, then the two lines are parallel.
29. A transversal cutting parallel lines makes congruent alternate and corresponding angles.
30. Parallelism of lines is transitive.
31. To draw a line through a point parallel to a given line.
32. An exterior angle of a triangle is equal to the sum of the two opposite interior angles. The sum of the angles of a triangle is equal to two right angles.
33. The segments joining corresponding endpoints of two congruent similarly directed parallel segments are themselves congruent and parallel.
34. The opposite sides and opposite angles of a parallelogram are congruent, and each diagonal bisects the area.
36. Parallelograms on congruent bases and contained within the same parallels have equal areas.
38. Triangles on congruent bases and with congruent altitudes have equal areas.
40. Equal triangles on congruent bases and on the same side of the base line are also within the same parallels.
41. If a parallelogram has the same base and lies between the same parallels as a triangle, then the parallelogram has twice the area of the triangle.

46. To describe a square on a given segment.
47. In a right triangle the square on the hypotenuse is equal to the sum of the squares on the legs.
48. If the square on one side of a triangle is equal to the sum of the squares on the other two sides, then the triangle is a right triangle with the right angle opposite the first side.

APPENDIX B | BASIC RULER AND COMPASS CONSTRUCTIONS

B.1 Notation We denote a circle with center A and radius $r = m(CD)$ by $A(r)$ or $A(CD)$. The circle with center A and passing through point B is denoted by $A(B)$. Unless stated otherwise, throughout this appendix r and s shall denote arbitrary convenient lengths.

B.2 The constructions listed here by no means form an exhaustive list. They are basic constructions that every serious student of geometry should have readily available whenever needed. It is suggested that you practice these constructions, keeping in mind that precision is required. The pencils you use should be always kept fanatically sharp, and an eraser should never be used. Be sure your lines and circles pass precisely through the intended points, not just close to them. And *never* put a large blob on a constructed point to indicate its location! If necessary, draw a small arrow pointing toward that point, but do not obliterate or cover it. Observe how these techniques are followed in the constructions below and do likewise. To be sure your practice constructions are accurate, check your angles with a protractor and measure your distances with a ruler.

B.3 Construction Draw a triangle ABC given the lengths of its three sides a, b, c. From a point A on a line m draw circle $A(c)$ to cut line m at point B. Then draw circles $A(b)$ and $B(a)$ to meet at C (see Fig. B.3).

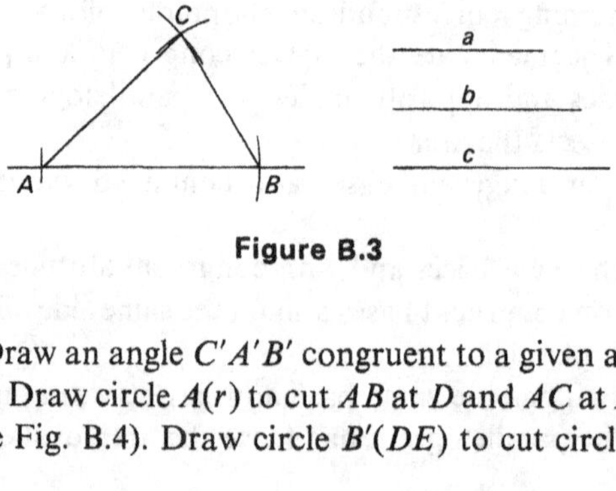

Figure B.3

B.4 Construction Draw an angle $C'A'B'$ congruent to a given angle CAB at a point A' on a given line m. Draw circle $A(r)$ to cut AB at D and AC at E, and circle $A'(r)$ to cut line m at B' (see Fig. B.4). Draw circle $B'(DE)$ to cut circle $A'(r)$ at C'.

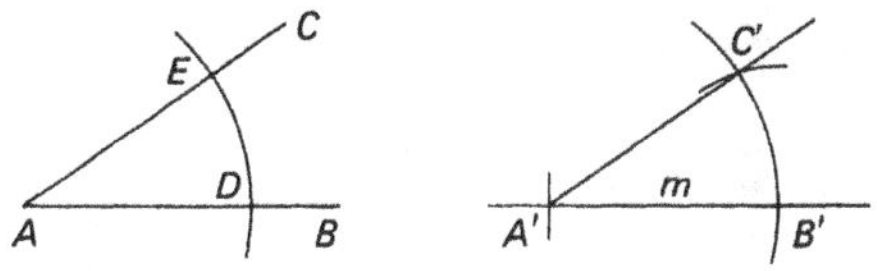

Figure B.4

B.5 Construction Bisect a given angle CAB. Draw circle $A(r)$ to cut AB at D and AC at E. Draw circles $D(s)$ and $E(s)$ to meet at T. Then AT is the desired bisector (see Fig. B.5).

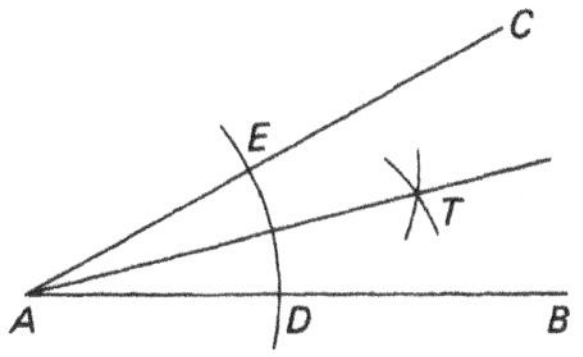

Figure B.5

B.6 Construction Bisect a given segment AB, or erect its perpendicular bisector. Draw circles $A(r)$ and $B(r)$ to meet at C and D. Then CD is the perpendicular bisector of segment AB meeting AB at its midpoint M (see Fig. B.6).

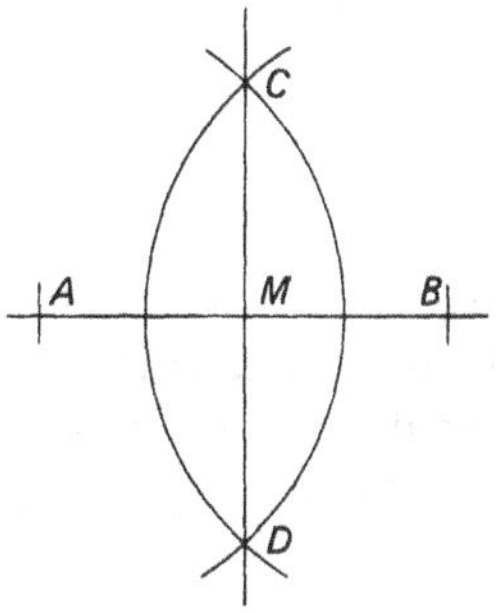

Figure B.6

B.7 Construction Drop a perpendicular from a given point P to a given line m. Draw circle $P(r)$ to cut line m in two points A and B. Draw circles $A(s)$ and $B(s)$ to meet at C as in Fig. B.7. Then line PC is the desired perpendicular.

B.8 Construction Erect a perpendicular at a point P on a given line m.

First method. Draw circle $P(r)$ to cut m at A and B. See Fig. B.8a. With $s > r$, draw circles $A(s)$ and $B(s)$ to meet at C. Then CP is the desired perpendicular.

Second method. Choose any point Q not on m and not on the desired perpendicular, as in Fig. B.8b. Draw circle $Q(P)$ to cut line m again at R and draw diameter RQT. Then PT is the desired perpendicular.

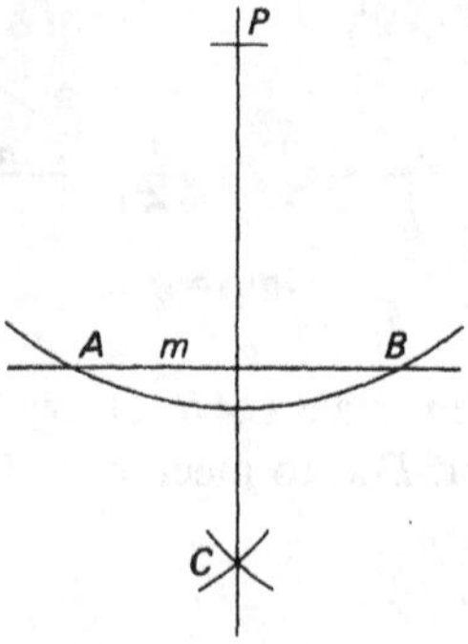

Figure B.7

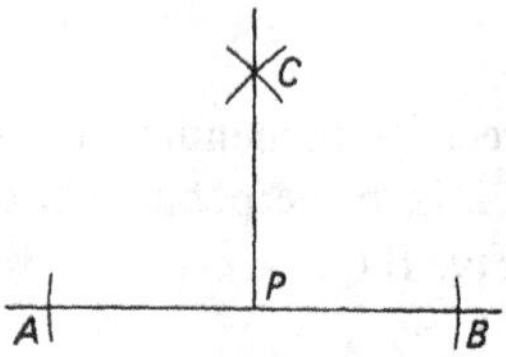

Figure B.8a

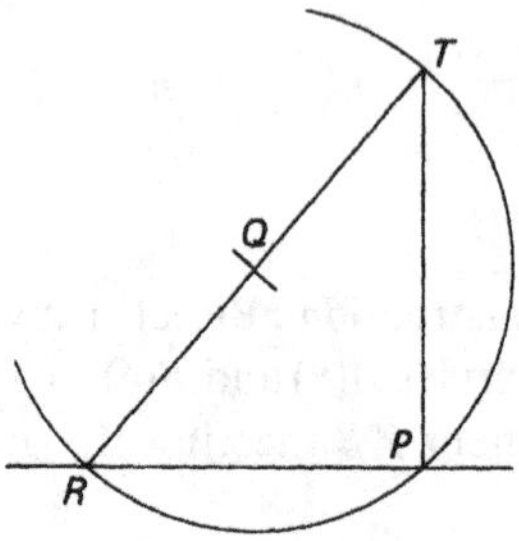

Figure B.8b

B.9 Construction Draw a line *n* parallel to a given line *m* and passing through a given point *P*. Draw any line through *P* to cut line *m* at *A*. Using Construction B.4, draw an angle at *P* congruent to the angle at *A* and on the opposite side of line *PA*. Then line *n* is the terminal side of this constructed angle. (See Fig. B.9.)

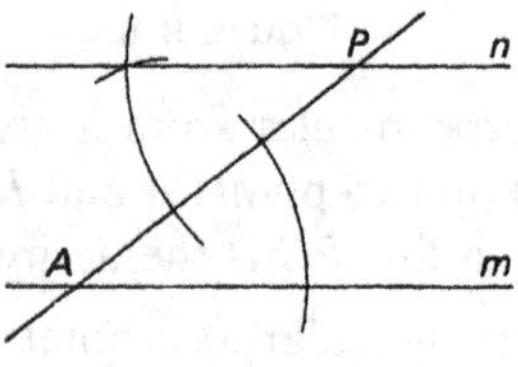

Figure B.9

B.10 Construction Divide a given segment *AB* internally in a given ratio $a:b$. On a convenient ray *AD* not lying on line *AB*, mark *AC′* of length *a* and *C′B′* of length *b* as shown in Fig. B.10. Draw *BB′* and a line through *C′* parallel to *BB′* and cutting *AB* at *C*, the desired point.

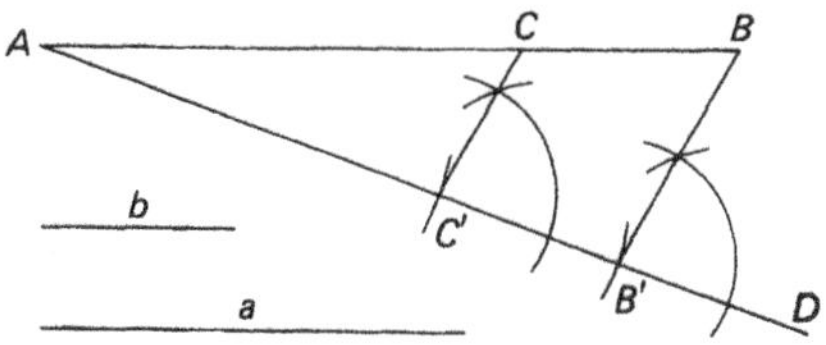

Figure B.10

B.11 Construction B.10 is easily altered to provide the point C which divides AB in the ratio $a{:}b$ externally. If $a > b$, then let point B' lie between A and C' instead of C' between A and B'. The rest of the construction is unaltered. If $a < b$, then interchange the roles played by A and B, and by a and b before starting the construction.

B.12 Construction Locate the center of a given circle s. Draw any chord AB and let its perpendicular bisector (Construction B.6) meet the circle at C and D. The midpoint of CD (Construction B.6 again) is the center O of circle s (Fig. B.12).

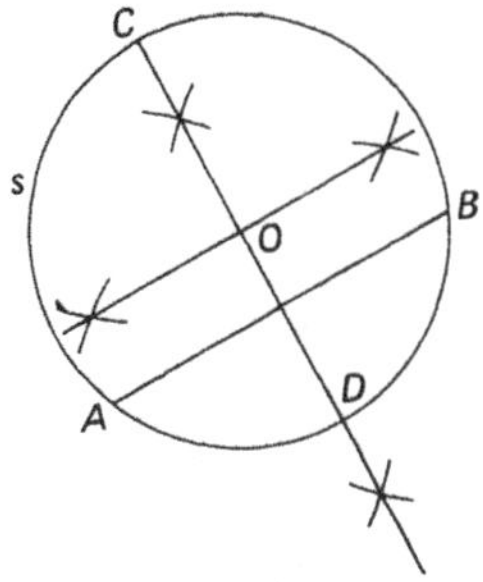

Figure B.12

B.13 Construction Draw a tangent to a circle s at a given point T on the circle. Assuming the center O of the circle is given, draw the radius OT. The line perpendicular to OT at T (Construction B.8) is the desired tangent.

B.14 Construction Draw a tangent to a circle s from a point P external to the circle. Assuming the center O of the circle is given, draw OP and locate its midpoint M (Construction B.6). Draw circle $M(O)$ to cut circle s at point T. Then PT is the desired tangent line (Fig. B.14).

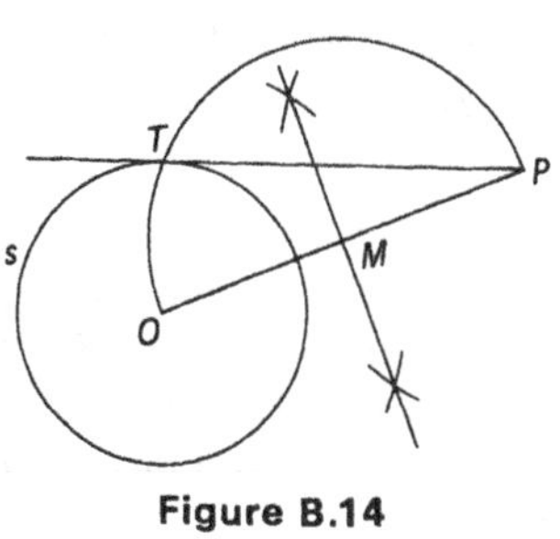

Figure B.14

BIBLIOGRAPHY

Adler, C. F., *Modern Geometry*. New York: McGraw-Hill, 1958

Ahlfors, L. V., *Complex Analysis*. New York: McGraw-Hill, 1953

Anderson, R. D., J. W. Garon, and J. G. Gremillion, *School Mathematics Geometry*. Boston: Houghton Mifflin, 1969

Bachmann, F., *Aufbau der Geometrie aus dem Spiegelungsbegriff*. Berlin: Springer-Verlag, 1959

Ball, W. W. R., *A Short Account of the History of Mathematics* New York: Dover Publications, 1960

Barry, E. H., *Introduction to Geometrical Transformations*. Boston: Prindle, Weber & Schmidt, 1966

Benson, R. V., *Euclidean Geometry and Convexity*. New York: McGraw-Hill, 1966

Bieberbach, L., *Analytische Geometrie*. Leipzig: Teubner, 1930

Blumenthal, L. M., *A Modern View of Geometry*. San Francisco: W. H. Freeman, 1961

Bohuslov, R., *Analytic Geometry*. New York: Macmillan, 1970

Bryant, S. J., G. E. Graham, and K. G. Wiley, *Nonroutine Problems*. New York: McGraw-Hill, 1965

Cajori, F., *A History of Mathematical Notations*, 2 vols. Chicago: Open Court, 1928

Cajori, F., *A History of Mathematics*, 2nd edition. New York: Macmillan, 1919

Choquet, G., *Geometry in a Modern Setting*. Boston: Houghton Mifflin, 1969

Chrestenson, H. E., *Mappings of the Plane*. San Francisco: W. H. Freeman, 1966

Churchill, R. V., *Complex Variables and Applications*, 2nd edition. New York: McGraw-Hill, 1960

Court, N. A., *College Geometry*. Richmond: Johnson, 1925

Coxeter, H. S. M., *Introduction to Geometry*. New York: John Wiley, 1961

Coxeter, H. S. M., *Projective Geometry*. New York: Blaisdell, 1964

Davis, D. R., *Modern College Geometry*. Reading, Mass.: Addison-Wesley, 1949

Deaux, Roland, *Introduction to the Geometry of Complex Numbers*, translated by H. Eves. New York: Frederick Ungar, 1956

Dodge, C. W., *The Circular Functions.* Englewood Cliffs, N.J.: Prentice-Hall, 1966

Dodge, C. W., *Numbers and Mathematics.* Boston: Prindle, Weber & Schmidt, 1969

Dodge, C. W., *Sets, Logic and Numbers.* Boston: Prindle, Weber & Schmidt, 1969

Eves, H. W., *An Introduction to the Foundations and Fundamental Concepts of Mathematics*, revised edition. New York: Holt, Rinehart and Winston, 1965

Eves, H. W., *An Introduction to the History of Mathematics*, 3rd edition. New York: Holt, Rinehart and Winston, 1969

Eves, H. W., *Functions of a Complex Variable*, 2 vols. Boston: Prindle, Weber & Schmidt, 1966

Eves, H. W., *Fundamentals of Geometry.* Boston: Allyn and Bacon, 1969

Eves, H. W., *In Mathematical Circles*, 2 vols. Boston: Prindle, Weber & Schmidt, 1969

Eves, H. W., *Mathematical Circles Revisited.* Boston: Prindle, Weber & Schmidt, 1971

Eves, H. W., *A Survey of Geometry*, 2 vols. Boston: Allyn and Bacon, 1963

Fink, K., *A Brief History of Mathematics*, translated by W. W. Beman and D. E. Smith. Chicago: Open Court, 1900

Forder, H. G., *Geometry.* New York: Harper & Brothers, 1962

Fujii, J. N., *Geometry and Its Methods.* New York: John Wiley, 1969

Gans, D., *Transformations and Geometries.* New York: Appleton-Century-Crofts, 1969

Guggenheimer, H. W., *Plane Geometry and Its Groups.* San Francisco: Holden-Day, 1967

Hall, D. W., and S. Szabo, *Plane Geometry, an Approach Through Isometries.* Englewood Cliffs, N.J.: Prentice-Hall, 1971

Hemmerling, E. M., *Fundamentals of College Geometry.* New York: John Wiley, 1964

Horblit, M., and K. L. Nielsen, *Plane Geometry Problems with Solutions.* College Outline Series. New York: Barnes & Noble, 1947

Hyatt, H. R., and C. C. Carico, *Modern Plane Geometry for College Students.* New York: Macmillan, 1967

Jolly, R. F., *Synthetic Geometry.* New York: Holt, Rinehart and Winston, 1969

Jurgensen, R. C., A. J. Donnelly, and M. P. Dolciani, *Modern School Mathematics: Geometry.* Boston: Houghton Mifflin, 1969

Kay, D. C., *College Geometry.* New York: Holt, Rinehart and Winston, 1969

Kenner, M. R., D. E. Small, and G. N. Williams, *Concepts of Modern Mathematics, Book* 2. New York: American Book, 1963

Lathrop, T. G., and L. A. Stevens, *Geometry, A Contemporary Approach.* Belmont, Cal.: Wadsworth, 1967

Levi, H., *Topics in Geometry.* Boston: Prindle, Weber & Schmidt, 1968

Levy, L. S., *Geometry: Modern Mathematics via the Euclidean Plane.* Boston: Prindle, Weber & Schmidt, 1970

Meder, A. E., Jr., *Topics from Inversive Geometry.* Boston: Houghton Mifflin, 1967

Meserve, B. E., and J. A. Izzo, *Fundamentals of Geometry.* Reading, Mass.: Addison-Wesley, 1969

Miller, L. H., *College Geometry*. New York: Appleton-Century-Crofts, 1957

Moise, E. E., *Elementary Geometry from an Advanced Standpoint*. Reading, Mass.: Addison-Wesley, 1963

Moser, J. M., *Modern Elementary Geometry*, Englewood Cliffs, N. J.: Prentice-Hall, 1971

Newman, J. R., *The World of Mathematics*, 4 vols. New York: Simon and Schuster, 1956

Osgood, W. F., and W. C. Graustein, *Plane and Solid Analytic Geometry*. New York: Macmillan, 1921

Prenowitz, W., and M. Jordan, *Basic Concepts of Geometry*. New York: Blaisdell, 1965

Rees, P. K., *Analytic Geometry*, 3rd edition. Englewood Cliffs, N.J.: Prentice-Hall, 1970

Rich, B., *Principles and Problems of Plane Geometry with Coordinate Geometry*. Schaum's Outline Series. New York: McGraw-Hill, 1963

Rosskopf, M. F., J. L. Levine, and B. R. Vogeli, *Geometry: A Perspective View*. New York: McGraw-Hill, 1969

Shively, L. S., *An Introduction to Modern Geometry*. New York: John Wiley, 1939

Sigley, D. T., and W. T. Stratton, *Solid Geometry*, revised edition. New York: Dryden Press, 1956

Smith, D. E., *A Source Book in Mathematics*. New York: McGraw-Hill, 1929

Spiegel, M. R., *Theory and Problems of Vector Analysis and an Introduction to Tensor Analysis*. Schaum's Outline Series. New York: Schaum, 1959

Tryon, C. W., *Elementary Geometry for College*. New York: Harcourt, Brace & World, 1969

Tuller, A., *A Modern Introduction to Geometries*. Princeton: D. Van Nostrand, 1967

Veblen, O., and J. W. Young, *Projective Geometry*, 2 vols. New York: Blaisdell, 1938

Wells, W., and W. W. Hart, *Modern Solid Geometry*. Boston: D. C. Heath, 1927

Wylie, C. R., Jr., *Foundations of Geometry*. New York: McGraw-Hill, 1964

Wylie, C. R., Jr., *Introduction to Projective Geometry*. New York: McGraw-Hill, 1970

Yaglom, I. M., *Complex Numbers in Geometry*, translated by E. J. F. Primrose. New York: Academic Press, 1968

HINTS FOR SELECTED EXERCISES

In the double-numbering system, the first number refers to the section in which the exercise appears, and the second number refers to the exercise number. Thus 1.3 (below) gives a hint for Exercise 3 of Section 1.

1.1 Drop an altitude from the apex of the isosceles triangle and apply the Pythagorean theorem.
1.2 Set $(\frac{8}{9}d)^2 = \pi(\frac{1}{2}d)^2$ and solve for π.
1.3 Draw diagonals and use Theorem 2.17.
1.4 Let the original pyramid have altitude H. Show that $H/(H-h) = B/b$ and solve this equation for H. Then subtract the volume of the small cut-off pyramid from that of the original pyramid.
1.5 Consider the numerical values of the areas.
1.6 Some persons feel the "molten sea" was oval-shaped, accounting for its circumference being only three times its (maximum) diameter.
1.7 c) Let $y/2 = x$.
d) Let $3y = x$.

2.2 Consider that $|d(AB)| = |d(BA)|$ but AB and BA are oppositely directed.
2.4 Draw the altitude h_c to side AB.
2.5 Use Theorem 2.19.
2.6 Apply Theorem 2.11, using $d(AM) = d(MB)$.
2.9 Replace **AB**, **BC**, **CA** in terms of **DA**, **DB**, **DC**.
2.10 Let E be the foot of the perpendicular from D to line ABC. By Exercise 2.9, the formula holds for A, B, C, and E. Subtract this formula from the desired formula and use the Pythagorean theorem to show that the difference is zero.
2.12 In Stewart's theorem, replace A, B, C, D by B, L, C, A, respectively, in triangle ABC for median AL. Recall that $\mathbf{BL} = \mathbf{LC}$.
2.13 See Hint 2.12. In this case $\mathbf{BL}/\mathbf{LC} = c/b$ by Exercise 2.5.
2.16 Show that the areas represented by the terms of Euler's theorem add together as they should.

3.1 b) Consider the ideal plane.
3.2 b) Consider the ideal line.
d) Let m be the ideal line and P an ordinary point.
3.3 d) Three noncollinear ideal points occur only in extended space in the ideal plane. It cannot be shown in a figure.
3.6 One must violate the word preceding "numerators" in the proof of Theorem 3.5.
3.8 Apply Theorem 2.19 twice to triangle VAB.
3.12 See Theorem 3.9.

4.1 Use 2.18.
4.2 We must agree that, if C is the ideal vertex of triangle ABC, then $0 \cdot CX = 0$ for any point X; if L and M are ordinary points, then $\mathbf{LC} = -\mathbf{CM}$; and if L (or M) is at C, then $\mathbf{LC}$ (or $\mathbf{CM}$) is zero.
4.4 Start with Theorem 4.2 and use Theorem 2.19.
4.6 Use Exercise 4.5 six times.
4.7 Use Exercise 4.2.
4.8 Use the trigonometric form of Menelaus' theorem, showing that, when AL is the external angle bisector of angle BAC, then angles BAL and LAC are supplementary in magnitude and opposite in sense.
4.10 Use Theorem 4.2. Show that $\mathbf{BL} = \mathbf{L'C}$, etc.
4.11 Use Theorem 4.3 and ideas analogous to those of Hint 4.10.
4.12 You must show that, if the four given points are collinear, then the given equation holds. Draw diagonal AC to cut line $LMNO$ at point X. Then apply Menelaus' theorem to the two triangles ABC and ACD.
4.13 Tangents from a point to a circle are congruent.
4.14 Use triangle ABA' cut by line CPQ.
4.16 See Answer 4.15.
4.17 See Answer 4.15.

5.1 See the proof of the converse to Theorem 4.2.
5.2 Assume Exercise 4.1.
5.3 Use Theorem 2.19.
5.4 Show, for example, that $BL/OL = AS/AO$, $CM/MA = BC/AS$, and two similar relations.
5.6 See Theorem 5.9. Note that this construction requires only the straightedge.
5.7 Use Theorem 5.9.
5.10 Let the common ratio $\mathbf{AO/OL} = \mathbf{BO/OM} = \mathbf{CO/ON} = r$. Then use triangle BOC and cevians BN, CM, and OL.
5.11 Use the trigonometric form of Ceva's theorem.
5.12 See Hint 5.11.
5.13 See Exercise 4.11.

5.14 Use Theorem 5.2 and recall that the product of a secant to a circle from an external point P and its external segment is a constant. Thus, for example $AM \cdot AM' = AN \cdot AN'$ (see Exercise 6.18).

6.4 Let O be the midpoint of diagonal AC in parallelogram $ABCD$. Show that triangles ABO and CDO are congruent.
6.6 In Fig. 6.3, show that AA' and $B'C'$ bisect one another.
6.8 See Theorem 6.10.
6.12 See Theorem 6.15.
6.14 Use Theorem 6.20.
6.16 b) Let PT be a tangent and PAB a secant to a given circle and draw line AT. Then $\sphericalangle TAB$ is an external angle for triangle PTA. Now apply part (a).
6.18 See Hint 6.16(b).
6.20 Use Theorems 6.15 and 6.16.
6.21 Use Exercise 6.20, Theorem 6.15, and Corollary 6.19.
6.22 Use Theorem 6.20.

7.2 Use the last paragraph in the proof of Theorem 7.2.
7.3 Use Corollary 7.4.
7.8 See Theorem 29.4.
7.10 Use Theorem 7.17.
7.11 Draw an accurate figure.
7.13 Show they are diagonals of a parallelogram.
7.14 Use right triangle $BA'O$ in Figure 7.16.
7.18 Use Exercise 6.9.
7.20 Start with the right-hand side of the equation and use the indicated theorems.

8.1 a) Draw AX and BX.
b) Use similar triangles.
f) Draw AC, CQ, CP, PD, BD, and DQ.
g) What kind of triangle is triangle PQC?
8.2 Use Problem 8.12.
8.4 Form an isosceles triangle on the given angle as a base angle and the congruent sides the same length as the compass opening. Transfer by Problem 8.19 the base distance to the given point on the given line, etc.
8.5 On a line mark AP and PB of lengths a and 1. On a parallel line draw a semicircle $A'B'$ using the given compass opening as radius. Use proportion.
8.6 The linear dimensions will be two-thirds of the given dimensions.
8.10 Place the isosceles triangle so that one of its congruent sides is the base.
8.11 First draw a figure showing the completed triangle on which the given parts have been marked.
8.14 The first statement in the "proof" is not true.
8.16 h) With these tools one can construct only the intersections of circles and lines.

9.2 See Eves, *Survey of Geometry*, vol. 1, pages 44, 50, 233, and 236, for example.
9.3 b) Start with a hexagon and apply the formula of part (a) four times.
9.4 The formula of part (a) is readily found, but that of part (b) is far too complicated to write explicitly. The perimeter of the circumscribed hexagon being $2\sqrt{3}$, that of the 96-sided polygon is 3.1427.

10.4 Recall vector addition (see Definition 33.6).
10.6 If the corresponding sides of the triangles are not parallel and similarly directed, the center of rotation is the intersection of the perpendicular bisectors of the segments joining corresponding vertices of the triangles.
10.8 Use Exercise 10.1.

11.8 See 11.8.
11.9 See 11.8.
11.10 Let the translation carry point A to point B, and let an arbitrary rotation through angle θ carry A to a point C. Find the rotation through angle $-\theta$ that carries C to B. See Exercise 10.6.
11.11 A rectangle is formed. What is true of its diagonals?

12.3 Try a translation and a reflection or rotation.
12.5 Both these maps are inverse to α^{-1}.
12.6 Both these maps are inverse to $\beta\alpha$.
12.7 b) Let α be involutoric.
12.8 Multiply both sides of each given equation by α^{-1}.
12.9 Multiply both sides of each given equation by γ^{-1}.
12.10 Multiply both sides of the given equation by α^{-1} on the left and by β^{-1} on the right.
12.12 In property (3), let $\beta = \alpha^{-1}$.

13.2 Apply Theorem 13.6.
13.10 See what happens to the sense of a given triangle.
13.12 Consider the circles $P(s)$ and $Q(t)$ for part (a).
13.14 d) See Theorem 13.13.

14.1 See Fig. 10.6.
14.5 Use Theorem 13.13.
14.10 Factor the glide-reflection into a product of reflections in three lines, the third one being perpendicular to the other two lines. Show that if lines m and n are perpendicular, then $\sigma_n\sigma_m = \sigma_m\sigma_n$ (by considering what rotation each of these products represents).
14.18 Find lines a, b, c so that the rotation is $\sigma_b\sigma_a$ and the translation is $\sigma_c\sigma_b$. Their product is $\sigma_c\sigma_a$, etc.

15.2 Look at the geometric picture.
15.4 Use Answer 15.3 as a guide.
15.6 If $ABCD$ is a parallelogram, then $\sigma_D \sigma_C = \sigma_A \sigma_B$ by Theorem 15.11.

16.2 See Hint 14.10.
16.4 Recall Exercise 12.12.
16.6 Eliminate translations and glide-reflections.
16.7 Use Theorem 16.12 and show that no nontrivial rotation has more than one fixed point.
16.11 Consider the location of the image of P.
16.12 Factor σ_A into a product $\sigma_p \sigma_q$ with p parallel to m.
16.13 b) Recall Theorem 14.15.

17.4 See Theorem 16.1.
17.8 See Theorem 15.12.
17.10 See Theorem 15.11.
17.12 See Theorem 15.7.
17.16 This product factors into squares of products of four reflections each, the first one being $(\sigma_a \sigma_b \sigma_c \sigma_d)^2$.

18.1 Use a map similar to that for the proof of Theorem 18.4.
18.2 Map one triangle to the other so that the congruent angles coincide and the triangles lie on the same side of that common side.
18.3 Map one triangle to the other so as to form an isosceles triangle.
18.6 b) This is Theorem 2.3.
18.12 Reflect in the bisector of the apex angle.
18.13 Suppose that median AM is perpendicular to side BC.
18.14 In rhombus $ABCD$, $\sigma_m(\triangle ABC) = \triangle ADC$ where m is line AC.

19.2 Reflect in that diameter.
19.6 This proof is quite like that of Theorem 19.10.
19.8 Consider the sum of pairs of opposite angles.
19.9 Show that $\sigma_{C'} \sigma_C \sigma_{B'} \sigma_B \sigma_{A'} \sigma_A = \iota$ by finding the image of point A under this product of halfturns which is a translation.
19.10 Rotate the triangle a halfturn about the midpoint of its hypotenuse to form a rectangle.
19.14 Place your first coin in the center, then use central symmetry.
19.15 Use symmetry in a line.

20.8 Points L, M, N are the midpoints of the sides of triangle ABC and the midpoints of the segments joining the vertices to the orthocenter of triangle DEF.
20.10 Since the figure is a parallelogram, it has a center of symmetry.
20.12 Triangle ACH is the medial triangle for triangle QDR.
20.14 Reflect either the house or the barn in the river bank. Then solve the problem.

20.15 Apply a halfturn about the midpoint of the side of the triangle.
20.16 Apply Exercise 20.15 for one solution. There is another (trivial?) solution.
20.17 Draw the diameter perpendicular to its base.
20.18 Their altitudes to the common base line are congruent.
20.19 Show that $\sigma_{C''}\sigma_A = \sigma_A\sigma_{B''}$.
20.20 Show that a rotation about the given point leaves the circle fixed.
20.21 Draw lines OA and OB.
20.22 a) Use a 90° rotation about M.
b) Assume that triangle ABC is counterclockwise oriented. Let α and β denote 90° rotations about X and Y, respectively. Then $(\sigma_{B'}\alpha\beta)(C) = C$, so $\sigma_{B'}\alpha\beta = \iota$ because it is a translation. Hence $(\alpha\beta)(B') = B'$. Let $\beta(B') = B''$, so that $\alpha(B'') = B'$. Then triangles $B'XB''$ and $B''YB'$ are each isosceles right triangles, etc.
c) Draw either diagonal of the quadrilateral and use part (b).
d) A triangle can be thought of as a degenerate quadrilateral.

21.6 Recall the formulas for $\sin(\theta + \phi)$ and $\cos(\theta + \phi)$.
21.10 Proceed as in Exercise 21.9.
21.14 Find angle θ so that $\sin\theta = b/(a^2 + b^2)^{1/2}$ and $\cos\theta = a/(a^2 + b^2)^{1/2}$. One method is to set $\theta = 2\ \text{Arctan}\ ([(a^2 + b^2)^{1/2} - a]/b)$.
21.16 For the line use the equations of Exercise 21.9.

22.4 This is similar to Exercise 22.3.
22.6 A reflection is involutoric.

22.8 Since only the translation is reversed, replace r by $-r$ in Theorem 22.9.
22.10 See Hint 22.8.
22.12 Start with the equations of Answer 22.9.
22.14 See Exercises 21.13 to 21.15 and 22.11 to 22.13.
22.16 See Theorems 21.6 and 22.4 and Exercise 22.14.
22.18 Fixed points are (0, 0) and (3, 3).

23.1 Consider the various planes in which each side lies and those in which each pair of corresponding sides lie.
23.2 Extend every other side of the hexagon to form a triangle and apply Menelaus' theorem (Theorem 4.2) several times.
23.4 The values for π given by 1000 and 1001 factors are 3.1400 and 3.1431, respectively.

24.4 See Theorem 25.7.
24.6 Reflect in the bisector of the angle formed by a pair of corresponding sides (extended).
24.8 See Exercise 24.7.

24.10 Its ratio is negative.
24.12 There are none with positive ratios.

25.6 See Theorem 25.5.
25.8 Show that $\mathbf{QB''/B''B'} = j/(1-j)$ and $\mathbf{B'B/BO} = k-1$. Use Menelaus' theorem on triangle OQB' to get $\mathbf{OP/PQ} = (j-1)/(j(k-1))$.
25.12 To find the vector of translation, locate the image of point A under this product of homotheties.
25.16 By Theorem 25.9. One must also show that the product of a translation and a homothety is a homothety. Consider the image of a given segment AB under such a product.
25.20 Use Theorem 25.16.

26.2 See the proof of Corollary 13.7.
26.6 See Theorem 13.20.
26.10 Let n be the other bisector of the angle at Q.
26.12 Reflect figure $OA'B'CD$ in line m.
26.14 This is a corollary to Exercise 26.13.
26.15 a) What isometry leaves a circle fixed?
b) What opposite isometry maps a circle to itself?

27.2 What figure is formed by the union of all four images?
27.4 Remember that areas of similar figures vary as the squares of the linear dimensions.
27.6 Let the midpoints of AC and BD be M and N in Fig. 27.4. The desired result can be obtained by solving algebraically the ratios resulting from the facts that $H(E, \mathbf{EA/EM})$ maps MN to AB and $H(\mathrm{E}, \mathbf{EM/EC})$ maps CD to MN. The algebra is somewhat lengthy.
27.8 See Problem 27.7.
27.10 See Problem 27.7.
27.12 It is opposite.
27.14 See Theorem 27.9.
27.16 Segments AB and $A'B'$ are parallel.

28.2 See Theorem 28.1.
28.3 Let $H(O, 2)$ map triangle ABC to $A'B'C'$.
28.4 See Theorem 27.10.
28.5 Use triangles DEC and FEA, and DEA and GEC.
28.8 Show that if $H(M, k)$ maps XY to $A'C$ and $H(N, j)$ maps XY to BA', then $j = k$. Deduce that M and N are equally distant from BC.
28.9 See Theorem 28.6.
28.10 Use an appropriate rotation and homothety.
28.12 Use Problem 28.10.
28.13 Start with a square that satisfies almost all of the conditions.

28.14 Use the method of Answer 28.13.

28.18 a) Center *any* similar rectangle at the center of the circle.
b) Rest any similar rectangle symmetrically upon the diameter.
d) A diagonal is an axis of symmetry.

29.1 Use Theorems 29.2 and 29.4.

29.2 What does $H(G, -2)$ obviously map into what?

29.7 Reflect one of the circles in the line tangent to it at the given point of intersection.

29.8 Use a homothety of ratio $-r$ instead of the reflection recommended in Hint 29.7.

29.9 Draw a diameter AOB of the smallest circle. See what is true if the desired secant terminates at A.

29.10 Recall Exercise 2.5.

29.11 First use the angle size and the nonparallel side lengths.

29.12 Draw any square first.

29.13 Do not worry about the perimeter at first.

29.14 Show that the ratio of homothety is 1/3.

29.16 Use Exercise 29.14.

29.17 Use Theorem 7.20.

29.18 Using a midpoint of a side of triangle ABC as a center of homothety of ratio 1/3, find the image of the opposite vertex and the orthocenter.

29.21 Use Exercise 29.19.

29.22 Choose an arbitrary point on one of the lines and apply Exercise 29.21.

30.4 Recall Exercise 22.14.

31.4 See Dodge, *Sets, Logic & Numbers*, pages 249–251 or Dodge, *Numbers and Mathematics*, pages 345–347.

32.1 a) Assume $-1 > 0$ is true and apply property (3), then property (1).
b) Use part (a).
c) Use $i^2 = -1$ and the method of part (a).

32.6 See Sections 33 and 34.

32.10 b) $7 = (\sqrt{7})^2$.
c) $2 = -(i\sqrt{2})^2$.
d) Use the method of part (c).
f) See Dodge, *Sets, Logic & Numbers*, Exercises 60-5 or Dodge, *Numbers and Mathematics*, Exercises 90-5.
h) Compare with $a^2 + b^2$, which does not factor *algebraically*, but which is equal to the square *number* 25 when $a = 3$ and $b = 4$.

33.4 Use ordered-pair notation for the vectors.
33.10 This is similar to Exercise 33.9.
33.12 See Answer 33.11.
33.13 Use Exercise 33.11.
33.14 See the proof of Theorem 33.18.
33.15 The vector sum must be zero.
33.16 See Exercise 34.14.

34.2 Use Fig. 34.6 and trigonometry.
34.6 See Theorem 34.10 and Answer 34.5.
34.11 Assume BE and CF meet at O. To locate O, equate expressions for $\overrightarrow{AB} + \overrightarrow{BO}$ and $\overrightarrow{AC} + \overrightarrow{CO}$ in terms of vectors $\mathbf{u} = \overrightarrow{AB}$ and $\mathbf{v} = \overrightarrow{AC}$, and the ratios $\mathbf{BD}/\mathbf{DC} = m/n$, $\mathbf{CE}/\mathbf{EA} = n/p$, and $\mathbf{AF}/\mathbf{FB} = p/m$. Then show that $\overrightarrow{AD}$ is a scalar multiple of $\overrightarrow{AB} + \overrightarrow{BO}$.
34.14 Recall Exercise 2.5 and Exercise 33.12.
34.15 Use Exercise 34.14
34.16 Replace the vectors $\mathbf{u}$ and $\mathbf{v}$ of Exercise 34.14 in terms of $\overrightarrow{PA}$, $\overrightarrow{PB}$, and $\overrightarrow{PC}$.
34.19 If $\mathbf{u}$ and $\mathbf{v}$ are the vectors determined by the sides, then $|\mathbf{u}| = |\mathbf{v}|$.
34.20 Let M be the midpoint of segment AB and let P be any point on the perpendicular bisector of AB. Then $\overrightarrow{PM}$ and $\overrightarrow{MA}$ are orthogonal. Write $|\overrightarrow{PA}|$ and $|\overrightarrow{PB}|$ in terms of $\overrightarrow{PM}$ and $\overrightarrow{MA}$.
34.21 Use Exercise 33.11. If the two congruent medians are BB' and CC', let $\overrightarrow{AB} = \mathbf{u}$ and $\overrightarrow{AC} = \mathbf{v}$ and set $|\overrightarrow{BB'}| = |\overrightarrow{CC'}|$.

35.4 Multiply z by z^{-1} to get 1.
35.6 See Exercise 21.6.
35.7 Use mathematical induction and Exercise 35.6.
35.8 Use Exercise 35.7 and Definition 35.17.
35.9 Use Exercise 35.6.
35.10 Given that n is negative, use Exercise 35.9.
35.14 Use Definition 35.17.
35.16 See Answer 35.15.
35.18 Apply De Moivre's theorem (Exercise 35.7).
35.19 through 35.23 Use Exercise 35.18.
35.24 d) This is a corollary to part (c).
35.25 For a different proof, see Exercises 36.28 through 36.30.
35.26 Eliminate $\bar{z}$ from the two equations $2z + \bar{z} = 5 + i$ and $2\bar{z} + z = 5 - i$.
35.27 through 35.29 Use the method of Exercise 35.26.
35.30 Using the method of Exercise 35.26, from the given equation and its conjugate, obtain $a\bar{a}z + a\bar{b} + b = z$, which has no solution only when the coefficient of z is zero. Observe that if $a\bar{b} + b$ is zero also, then all complex numbers z satisfy the given equation.

36.2 Apply the isometries of Sections 21 and 22.
36.4 Use similar triangles.
36.6 It suffices to prove the theorem for a second-order determinant, so add a multiple of one row or column to the other and evaluate the resulting determinant.
36.7 See Hint 36.6.
36.10 Parts (a) and (b) are true by Theorem 36.17, but note that this exercise is somewhat stronger in stating the proportionality condition as both necessary and sufficient.
36.12 Triangles DEF and $\bar{D}\bar{E}\bar{F}$ are oppositely congruent.
36.14 Use Theorem 36.19 and add appropriate multiples of two of the rows to the third row so as to reduce the element $zb + (1 - z)a$ to zero. Then apply Corollary 36.16.
36.15 Use Exercise 36.5 as a starting point.
36.18 Show that the determinant of Corollary 36.20 is equal to

$$(b - a)(a - c) - (c - b)^2.$$

Adding and subtracting rows so these factors appear as elements of the determinant simplifies the algebra.
36.19 Show that this expression is equivalent to that of Exercise 36.18.
36.20 How can a triangle be both clockwise and counterclockwise?
36.21 Assuming $a = 0$ (for convenience), deduce that $\bar{b}c$ is real, so that $c = kb$ for some scalar k. Now see what must be true of b.
36.22 See Hint 36.6.
36.23 See Hint 36.6.
36.25 Texts on college algebra, linear algebra, or matrix theory generally investigate determinants in detail. Use Exercise 36.22.
36.27 f) The algebraic expression for either product is symmetric in b and c.
36.28 See Answer 36.29.
36.30 Use Exercises 36.28 and 36.29. See also Answer 35.25.

37.2 Show that two points on that line satisfy that linear equation.
37.6 Use Theorem 37.4.
37.10 Start with Theorem 37.10, add the second column to the first, and factor 2 from the first column. Then subtract the first column from the second and factor i from the second column.
37.12 Show that any line concurrent with the first two lines has an equation of the form

$$h(a_1 z + \bar{a}_1 \bar{z} + b_1) + k(a_2 z + \bar{a}_2 \bar{z} + b_2) = 0$$

for some real constants h and k. The determinant formed from the first two equations and this linear combination equation is zero.
37.14 Use Example 34.17.
37.16 Multiply the third row by i and apply Exercise 37.2.

37.18 b) Use Exercise 36.14.
37.20 Apply Exercises 37.18 and 37.19.

38.4 Take absolute values in the equation of Theorem 38.9.
38.5 Use Theorem 38.8. Then let $k = 1$.
38.6 Solve for z and apply Theorem 38.3.
38.13 Show that $a - b = a(1 - \bar{a}b/a\bar{a})$.
38.14 The solutions are quite simple.
38.16 If r is such a root, then $-r$ is the other. Now $z^2 - r^2$ is a factor of the equation. The *other* roots are $(1 \pm i\sqrt{3})/2$.
38.17 Let Q denote the circumcenter and use $|a - q| = |b - q| = |c - q| = R$, the circumradius.

39.14 This is similar to Theorem 39.6.

40.2 Use Ceva's theorem.
40.3 Use Ceva's theorem.
40.8 See Appendix A.

41.6 If there is no ray CP other than CA to curve c, then c coincides with line CA. This case is not difficult. When the rays CP and CQ lie on opposite sides of ray CA, then consider angles PAA' and $P'A'A$ (so long as $A \neq A'$) and angles QAA' and $Q'A'A$.
41.8 See Theorem 41.12.
41.15 Use Exercise 6.18.
41.16 This is a converse to Exercise 41.15.
41.18 a) Locate the inverse S'' of the center S of circle s. Apply Exercise 41.17.
b) Use Exercise 41.17.

42.3 Draw their line of centers and the radii to one of their points of intersection.
42.4 See Exercise 41.15.
42.6 See Theorems 41.10, 41.13, and 41.14.
42.8 Use similar triangles.
42.10 In Fig. 42.9, let line CPP' meet AB at T. Then use right triangles.
42.11 Use Theorem 42.8.
42.12 Modify the given proof to show that the listed ratios are still equal.
42.13 For the construction use Theorem 42.3 and Exercise 42.11.
42.14 See Theorems 42.3 and 42.4.
42.15 Draw the ray from the center of circle s to point P.
42.16 Consider what happens if the curves are tangent at the center of inversion.

43.6 If α and β are inverse, then $\beta\alpha = \iota$.
43.8 Use Theorem 43.9.
43.12 Find the bilinear transformation that maps 0, 1, and -1 to the distinct real images r, s, and t. Consider the two cases $r = \infty$ and $r \neq \infty$.

44.2 b) This is a converse to part (a).
d) Use Exercise 6.18.
h) Let the circles have centers S and T, and let Q be the foot of the perpendicular to ST from any point P on the radical axis. Show that $(SQ)^2 - (TQ)^2$ is equal to the difference of the squares of the circles' radii. Hence the location of Q is independent of the location of P.

44.3 See Theorem 44.1 and Exercise 44.2.

44.4 If the circle on B, C, D does not pass through A, then (in Theorem 44.3) B', C', D' lie on a circle, so that $B'C' + C'D' > B'D'$.

44.5 Use Theorem 42.3.

44.7 Use Exercise 41.17.

44.8 The center is the center of similarity.

44.10 Replace r by 1 in the equations of Exercise 44.9, then solve for x and y in terms of x' and y'. By recalling that inversion is involutoric, much labor is saved.

44.11 Use Exercises 44.5 and 44.6.

44.12 Invert in the point of intersection of the two circles orthogonal to the given circle and each passing through two opposite points.

44.15 Invert in the circle centered at O and orthogonal to circle s_n.

44.16 Invert the circles in their point of concurrence and look at the resulting triangle.

44.17 Draw a circle on AB as diameter.

46.2 See Section 14.

46.8 See Definition 47.10.

46.10 See Theorems 49.10 to 49.13.

46.12 See Theorem 47.9.

46.14 See Theorem 49.9.

47.8 See Theorem 13.19.

47.12 See Exercise 47.11.

48.8 See Theorem 49.2.

48.12 See Theorem 15.11.

48.16 See Theorem 15.12.

49.8 Consider the traces of the three planes in a plane perpendicular to line m and see Theorem 16.4.

49.10 See Theorem 47.9.

50.2 See Theorem 14.9.

50.10 See Theorem 17.5.

50.11 Consider a triangle cut from the sides of the trihedral angle by a plane not through its vertex.

50.12 See Theorems 49.11 and 49.12.
50.17 See Exercise 50.2.
50.20 Factor the halfturns into reflections in appropriate planes.

51.4 See Answer 51.5.
51.6 We have h, k, l proportional to A, B, C.
51.8 See Theorem 21.13.
51.10 Use Exercise 51.8 and Theorem 51.10.

ANSWERS

Answers are included here to alternate parts of all questions having more than one part and to all other odd-numbered questions. In the double-numbering system, the first number refers to the section in which the exercise appears, and the second number refers to the exercise number. Thus Exercise 1.3 is Exercise 3 of Section 1.

1.1 $8\sqrt{6}$.

1.3 Let A, B, C, D denote the vertices between sides d and a, a and b, b and c, c and d, respectively. Since the area of triangle ABC, for example, is given by $\frac{1}{2}ab\sin B$, then the area K of quadrilateral $ABCD$ is half the sum of the four triangles cut from the quadrilateral two at a time by a diagonal. That is,

$$K = \tfrac{1}{4}ab\sin B + \tfrac{1}{4}bc\sin C + \tfrac{1}{4}cd\sin D + \tfrac{1}{4}da\sin A$$
$$\leqslant \tfrac{1}{4}(ab + bc + cd + da) = \tfrac{1}{4}(a + c)(b + d),$$

since $\sin\theta \leqslant 1$ for all θ. Furthermore, equality holds iff

$$\measuredangle A = \measuredangle B = \measuredangle C = \measuredangle D = 90°.$$

1.5 Each right triangle with legs 3 and 4 has area 6, so the total area of this square is $6 \cdot 4 + 1 = 25$, making its side of length 5. That is, the right triangle has sides 3, 4, and 5.

1.7 a) 11 c) 6

2.1 By Theorems 2.10 and 2.11, $d(AB) + d(BC) + d(CA) = d(AC) + d(CA) = 0$.

2.3 Suppose that B is between A and C. Then $d(AB) + d(BC) = d(AC)$, so $d(AB) = -d(BC) + d(AC) = d(CB) - d(CA)$. The other cases are similar.

2.5 Using the notation of Theorem 2.19, let AL be the bisector of angle A. Then $\sin\measuredangle\mathbf{BAL} = \sin\measuredangle\mathbf{LAC}$, so

$$\frac{BL}{LC} = \frac{\mathbf{BL}}{\mathbf{LC}} = \frac{AB\sin\measuredangle\mathbf{BAL}}{CA\sin\measuredangle\mathbf{LAC}} = \frac{AB}{CA}.$$

2.7 By Theorem 2.12, if O is collinear with A, B, C, D, then $2\mathbf{ON} = \mathbf{OC} + \mathbf{OD}$ and $2\mathbf{OM} = \mathbf{OA} + \mathbf{OB}$. Then

$$2\mathbf{MN} = 2\mathbf{MO} + 2\mathbf{ON} = \mathbf{AO} + \mathbf{BO} + \mathbf{OC} + \mathbf{OD} = \mathbf{AC} + \mathbf{BD},$$

etc.

2.9 The given expression is equal to

$$DA^2 \cdot (\mathbf{DC} - \mathbf{DB}) + DB^2 \cdot (\mathbf{DA} - \mathbf{DC}) + DC^2 \cdot (\mathbf{DB} - \mathbf{DA}) + (\mathbf{DB} - \mathbf{DA})(\mathbf{DC} - \mathbf{DB})(\mathbf{DA} - \mathbf{DC}),$$

which reduces to zero when multiplied out.

2.11 This is the converse to Exercise 2.8. Let M and N be the midpoints of AB and CD, respectively. Then $\mathbf{AM} = \mathbf{MB}$ and $\mathbf{CN} = \mathbf{ND}$. Now

$$\mathbf{MN} = \mathbf{MA} + \mathbf{AC} + \mathbf{CN}$$

and also

$$\mathbf{MN} = \mathbf{MB} + \mathbf{BD} + \mathbf{DN} = \mathbf{AM} + \mathbf{CA} + \mathbf{NC} = -(\mathbf{MA} + \mathbf{AC} + \mathbf{CN}) = -\mathbf{MN}.$$

Hence $\mathbf{MN} = 0$, so M and N coincide.

2.13 Let AL denote the bisector of angle A in triangle ABC. Since $BL/LC = AB/CA$ by Exercise 2.5 and $BL + LC = BC$, then

$$BL = AB \cdot BC/(CA + AB) \qquad \text{and} \qquad LC = BC \cdot CA/(CA + AB).$$

By Stewart's theorem,

$$AL^2 \cdot BC + AB^2 \cdot CL + AC^2 \cdot LB + LB \cdot BC \cdot CL = 0,$$

from which we obtain

$$AL^2 = \frac{(AB^2 \cdot CA + AC^2 \cdot AB)(CA + AB) - BC^2 \cdot AB \cdot CA}{(CA + AB)^2}$$

and finally

$$AL = \left[bc\left(1 - \frac{a^2}{(b + c)^2}\right)\right]^{1/2}.$$

2.15 The area $K = ah/2$, so $K^2 = (ah/2)^2$. Using Exercise 2.14 to replace h^2, the resulting expression factors into $s(s - a)(s - b)(s - c)$.

3.1 a) False; an ideal line has more than one ideal point.
c) True
e) False; there is just one.

3.2 a) False; the ideal line has infinitely many ideal points.
c) True
e) False when n is the ideal line.
g) False; there is just one ideal line in the extended plane.

3.3 a) An ordinary triangle having three ordinary vertices.
c) Draw two lines that intersect at ordinary point A. The other two vertices are on the line at infinity in that plane.

3.4 a) $(\frac{1}{2}, 0)$ c) $(0, 0)$
e) The point at infinity
g) $(\frac{1}{3}, 0)$ i) $(\frac{1}{11}, 0)$

3.5 One need only consider the various possible positions for point P.

3.7 a) Let M and N be the points. Then $AM/MB = AN/NC$, so $AM/AB = AN/AC$. Since $\measuredangle BAC = \measuredangle MAN$, then triangles ABC and AMN are similar by SAS.

3.9 For $(AB, CD) = (\mathbf{AC}/\mathbf{CB})(\mathbf{DB}/\mathbf{AD}) = (\mathbf{CA}/\mathbf{AD})(\mathbf{BD}/\mathbf{CB}) = (CD, AB)$, etc.

3.11 a) If $(AB, CD) = (AB, CE)$, then D and E divide AB in the same ratio, so $D = E$ by Theorem 3.5.

3.13 Since the midpoint of a segment divides the segment in the ratio 1, its harmonic conjugate must divide the segment in the ratio -1. The point at infinity does so.

3.15 a) For example,

$$(AB, DC) = (\mathbf{AD}/\mathbf{DB})/(\mathbf{AC}/\mathbf{CB}) = 1/((\mathbf{AC}/\mathbf{CB})/(\mathbf{AD}/\mathbf{DB})) = 1/(AB, CD).$$

c) For example,

$$\begin{aligned}(DB, CA) &= (\mathbf{DC}/\mathbf{CB})/(\mathbf{DA}/\mathbf{AB}) \\ &= \frac{\mathbf{AB}}{\mathbf{CB}} \cdot \frac{\mathbf{DC}}{\mathbf{DA}} = \frac{\mathbf{AC} + \mathbf{CB}}{\mathbf{CB}} \cdot \frac{\mathbf{DC}}{\mathbf{DA}} = \frac{\mathbf{AC}}{\mathbf{CB}} \cdot \frac{\mathbf{DB} + \mathbf{BC}}{-\mathbf{AD}} + \frac{\mathbf{DC}}{\mathbf{DA}} \\ &= -\frac{\mathbf{AC}}{\mathbf{CB}} \cdot \frac{\mathbf{DB}}{\mathbf{AD}} + \frac{\mathbf{AC}}{\mathbf{CB}} \cdot \frac{\mathbf{BC}}{-\mathbf{AD}} + \frac{\mathbf{DC}}{\mathbf{DA}} \\ &= -(\mathbf{AC}/\mathbf{CB})/(\mathbf{AD}/\mathbf{DB}) + 1 = 1 - (AB, CD).\end{aligned}$$

4.1 Suppose Menelaus point $L = B$, for example, and L, M, N are collinear. Then either $N = B$ and M is arbitrary on line AC, or $N \neq B$ and $M = A$. In the first case $\mathbf{BL} = 0$ and $\mathbf{NB} = 0$, so by 2.18 the equation for Menelaus' theorem holds. Other cases are similar.

Conversely, if any factor in the numerator of the Menelaus formula is zero, then a factor in the denominator must also be zero. All such cases yield three collinear Menelaus points.

4.3 Not without some sort of agreement as to how the ratio in which an ideal point divides two ideal points shall be defined. There appears to be no advantage in attempting to do this.

4.5 Apply Menelaus' theorem where PQ cuts side BC at point L. Then, if PQ is parallel to BC, we have $\mathbf{BL}/\mathbf{LC} = -1$.

4.7 By applying Exercise 4.2, the proof given in the text holds for this case also.

4.9 See Hint 4.8.

4.11 We are given $\sin\measuredangle \mathbf{BAL} = \sin\measuredangle \mathbf{L'AC}$, $\sin\measuredangle \mathbf{LAC} = \sin\measuredangle \mathbf{BAL'}$, etc. Then apply the trigonometric form of Menelaus' theorem. Menelaus' formula for L, M, N is equal to the reciprocal of his formula for L', M', N', so if either one is equal to -1, then the other is -1 also.

4.13 By Menelaus' theorem, (**AZ/ZB**)(**BK/KC**)(**CY/YA**) $= -1$. Also, $BZ \cong CX$, $CX \cong CY$, and $AY \cong AZ$, since tangents from an external point to a circle are congruent. Now substitution yields the desired result.

4.15 Let L and N be the points as in Fig. 4.7. Draw lines NA and NA' through N toward L. From another point O, not on either line, draw two lines OAA' and OBB' cutting NA in A and B and NA' in A' and B'. Draw BL and $B'L$. Draw a third line through O to cut BL in C and $B'L$ in C'. Draw AC and $A'C'$ to meet at M. Now triangles ABC and $A'B'C'$ are copolar at O, so they are coaxial. That is, L, M, N are collinear.

4.17 Let m and n be the lines NA and NA' of Answer 4.15, and let the given point P be point L.

5.1 Assume (**BL/LC**)(**CM/MA**)(**AN/NB**) $= 1$ and let BL and CM meet at Q. Let AQ cut BC at L'. The rest of the proof is similar to that for the converse of Theorem 4.2.

One student gave the following proof of this converse. Assuming the given equation, let AL meet BM at O and CN at Q. Then, using Menelaus' theorem on triangles ALB and ALC cut by CQN and BOM, respectively, obtain

$$\frac{\mathbf{BC}}{\mathbf{CL}}\cdot\frac{\mathbf{LQ}}{\mathbf{QA}}\cdot\frac{\mathbf{AN}}{\mathbf{NB}} = -1 \quad \text{and} \quad \frac{\mathbf{LB}}{\mathbf{BC}}\cdot\frac{\mathbf{CM}}{\mathbf{MA}}\cdot\frac{\mathbf{AO}}{\mathbf{OL}} = -1.$$

By multiplying these two equations side for side and simplifying, obtain **AQ/QL** $=$ **AO/OL**, so $O = Q$. The theorem follows.

5.3 Starting with the equation of Theorem 5.2, we may use Theorem 2.19 to make the replacements

$$\frac{\mathbf{BL}}{\mathbf{LC}} = \frac{AB \sin \measuredangle \mathbf{BAL}}{CA \sin \measuredangle \mathbf{LAC}}, \qquad \frac{\mathbf{CM}}{\mathbf{MA}} = \frac{BC \sin \measuredangle \mathbf{CBM}}{AB \sin \measuredangle \mathbf{MBA}},$$

$$\text{and} \quad \frac{\mathbf{AN}}{\mathbf{NB}} = \frac{CA \sin \measuredangle \mathbf{ACN}}{BC \sin \measuredangle \mathbf{NCB}}.$$

The equation of Theorem 5.3 results.

5.5 When $\measuredangle B = 90°$, then $B = D = F$, so the altitudes concur at B.

5.7 By Theorem 5.9, **BL/LC** $= -$**BL′/L′C**, **CM/MA** $= -$**CM′/M′A**, and

$$\mathbf{AN/NB} = -\mathbf{AN'/N'B},$$

so

$$(\mathbf{BL/LC})(\mathbf{CM/MA})(\mathbf{AN/NB}) = -(\mathbf{BL'/L'C})(\mathbf{CM'/M'A})(\mathbf{AN'/N'B}).$$

5.9 If **BL/LC** $=$ **CM/MA** $=$ **AN/NB** $= r$ and the cevians concur, then $r^3 = 1$, so $r = 1$, since r is real.

5.11 Use Theorem 5.3, each fraction in which equals 1 in this case.

5.13 See Answer 4.11. Use this same technique along with Theorem 5.3.

6.3 The medial triangle is just half as large (in linear dimensions) as the given triangle, so their circumcircles also have radii of ratio 1/2. But the ninepoint circle is the circumcircle of the medial triangle.

6.5 In Figure 6.7, let BC cut $B'B''$ at X. Then X bisects CA', so $BX = \frac{3}{4}BC$. Since the altitude of triangle BXB' to side BX is half that of triangle ABC, then triangles ABC and $BB'B''$ have the same altitudes. Hence $K_{BB'B''} = \frac{3}{4}K_{ABC}$.

6.7 By Theorem 6.6, since $BB' \cong CC'$, then $\triangle BGC' \cong \triangle CGB'$ by SAS. Thus $BC' \cong CB'$, from which $AB \cong AC$.

6.9 Let AX and AY be the internal and external bisectors of angle BAC as shown in the accompanying figure. Let α and β be the measures of the angles thus formed. Then $2\alpha + 2\beta = 180°$, so $\alpha + \beta = 90°$. The theorem follows.

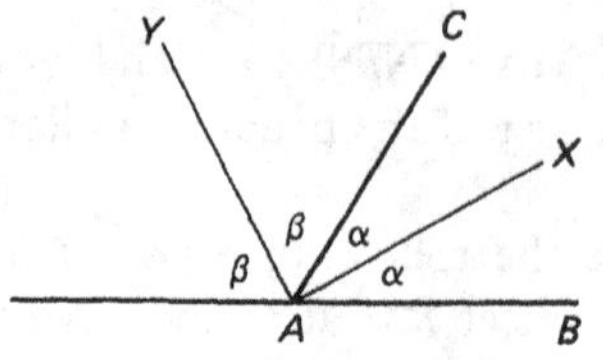

Answer 6.9

6.11 The circumdiameter ST of Theorem 6.12 bisects side BC, since it is perpendicular to chord BC. The theorem follows.

6.13 From Theorems 6.15, 6.16, and 6.17, we have

$$K^2 = s(s-a)(s-b)(s-c) = \frac{K}{r} \cdot \frac{K}{r_a} \cdot \frac{K}{r_b} \cdot \frac{K}{r_c} = \frac{K^4}{rr_ar_br_c},$$

from which the theorem follows.

6.15 In the figure for Exercise 6.15, $OC = (a+b)/2$, $OQ = (a+b-d)/2$, $QD = d/2$, and $OD = (b-a)/2$. Since OQD is a right triangle, then apply the Pythagorean theorem to obtain the desired result.

6.16 a) Let the inscribed angle be ABC, and draw diameter BOD and assume it does not coincide with BA. Then triangle OAB is isosceles, so external angle AOD has twice the measure of either opposite interior angle and specifically angle ABO. Hence $m(\measuredangle ABO) = m(\measuredangle ABD) = \frac{1}{2}m(\measuredangle AOD) = \frac{1}{2}m(\text{arc } AD)$. Similarly $m(\measuredangle OBC) = \frac{1}{2}m(\text{arc } DC)$, whether or not diameter BOD coincides with side BC. Now the angles may be added or subtracted to give the desired result according to whether O lies interior or exterior to angle ABC.

c) Let chords AB and CD meet at E. Since AEC is an exterior angle for triangle ECB, then $\measuredangle AEC = \measuredangle ECB + \measuredangle EBC = (\text{arc } DB + \text{arc } AC)/2$.

6.17 Let the tangents from point P be PT and PU, and let O be the center of the circle. In right triangles POT and POU, $PO = PO$, $OT \cong OU$, and $\measuredangle T \cong \measuredangle U = 90°$. Hence $\triangle POT \cong \triangle POU$ by HL (hypotenuse and leg of one right triangle congruent to the corresponding parts of a second right triangle).

6.19 Let the chords AC and BD meet at E. Since $\measuredangle ADC \cong \measuredangle ABC$ and $\measuredangle DAB \cong \measuredangle DCB$ by Exercise 6.16a, then $\triangle ADE \sim \triangle CBE$, so $AE/DE = CE/BE$, and the theorem follows.

6.21 By Exercise 6.20, Theorem 6.15, and Corollary 6.19,

$$r_a r_b + r_b r_c + r_c r_a = r_a r_b r_c \left(\frac{1}{r_a} + \frac{1}{r_b} + \frac{1}{r_c}\right) = \frac{r_a r_b r_c}{r}$$

$$= \frac{r_a r_b r_c r}{r^2} = \frac{K^2}{K^2/s^2} = s^2.$$

6.22 a) Let BP be the circumdiameter from B. Then $\measuredangle BPC \cong \measuredangle A$, so $\sin\measuredangle A = \sin\measuredangle BPC = a/2R$.

6.23 This is a corollary to Theorem 6.20.

7.1 By Theorem 6.11, each pair of excenters subtends a right angle at each vertex of the triangle not collinear with these excenters. The theorem follows.

7.3 The midpoint of $I_b I_c$ is T by the last paragraph in the proof of Theorem 7.2. Now AI meets the circumdiameter SOT that is the perpendicular bisector of side BC at point S on the circumcircle. Then $A = T$ iff A lies on the perpendicular bisector of BC; that is, iff $AB \cong AC$.

7.5 Since each side (such as BC) intercepts a right angle at each of the two vertices (E and F) of the orthic triangle not lying on that side, the theorem follows.

7.7 Let M be the midpoint of the hypotenuse AB in right triangle ABC, and drop perpendicular MN to leg BC. Since AC is parallel to MN, then $BN \cong CN$. Also $MN = MN$, so $\triangle BMN \cong \triangle CMN$ by SAS. Thus $CM \cong BM \cong AM$.

7.9 Referring to Fig. 7.6, since CA, CB, and CF are perpendicular, respectively, to BH, AH, and AB, then CA, CB, and CF contain the altitudes (AE, BD, and HF) of triangle HAB. Hence C is its orthocenter.

7.11 Point P lies on the circumcircle by Theorem 6.12 and not inside triangle ABC. Now, when $AB < AC$, then the point R lies outside segment AB and Q lies inside segment AC. The stated congruences are all true, but in the last equation, $AB = AR - RB$ and $AC = AQ + QC$, so it is not true that $AB \cong AC$.

7.13 Since AN_a is parallel to OA', and since AH is twice OA' because they are corresponding altitude segments for the given triangle and its medial triangle, then $AN_a \cong OA'$. Thus $AN_aA'O$ is a parallelogram and its diagonals AA' and ON_a bisect each other.

7.15 In Fig. 7.16, OA' is parallel to and half the length of AH by Exercise 7.13. Hence AO and HA' meet at a point P so that $AP = 2OP$ by the similar triangles AHP and $OA'P$. Thus AP is a circumdiameter and P lies on the circumcircle.

7.17 For the orthocentric quadrangle $ABCH$, find its centroids G_a, G_b, G_c, G. The desired quadrangle, say $L_aL_bL_cL$, is to $ABCH$ as $ABCH$ is to $G_aG_bG_cG$. Hence, since $G_aG_bG_cG$ is one-third that of $ABCH$ in linear size, extend each of AG_a, BG_b, CG_c, and HG twice its own length to L_a, L_b, L_c, and L, the vertices of the desired orthocentric quadrangle.

7.19 This is a corollary to Theorem 7.9.

8.1 a) Since AXB is a right angle, as are APX and BPX, since $\measuredangle XAB \cong \measuredangle PAX$ and $\measuredangle XBA \cong \measuredangle PBX$, we have $\triangle ABX \sim \triangle AXP \sim \triangle XBP$ by AA (two angles of one triangle congruent respectively to two angles of the other). From the latter two triangles, $AP/PX = PX/BP$, and the theorem follows.

c) Since $\triangle EHJ \sim \triangle EFG$, then $EJ = 2HJ$, so $5HJ^2 = HJ^2 + (2HJ)^2 = EH^2 = r^2$. Thus $HJ = r/\sqrt{5}$, so $HJ = s/2$.

e) The construction clearly satisfies the given conditions.

g) Since $PCBA$ is a parallelogram, $PC \cong AB$. Triangle PCQ is isosceles since the bisector of angle P is also an altitude of that triangle.

8.3 Let the circle $A(r)$ cut the sides of the angle BAC at points B and C. Draw circles $B(r)$ and $C(r)$ to cut circle $A(r)$ within angle BAC at points D and E. Then $\measuredangle DAE = \measuredangle BAC - 120°$, and the bisector of angle DAE is the bisector of angle BAC also.

An alternative procedure, suggested by a student, is to bisect the supplement of the given angle and then erect a perpendicular to that bisector.

8.5 On a line mark $AP = a$ and $PB = 1$, where $\sqrt{a}$ is desired. On a line parallel to AB mark $A'B' = 2r$ where r is the compass opening. Let AA' and BB' meet at O. Let OP cut $A'B'$ at P'. Construct $P'X' = ((A'P')(P'B'))^{1/2}$ by Problem 8.10. Since $P'X'$ is perpendicular to $A'B'$, let the perpendicular to AB at P cut line OX' at X. See the accompanying figure. By similar triangles, $PX = \sqrt{a}$.

8.7 Compute $(\frac{2}{3})^{1/2}$, and mark points B' and D' on sides AB and AD of trapezoid $ABCD$, so that $AB' = (\frac{2}{3})^{1/2} AB$, and $AD' = (\frac{2}{3})^{1/2} AD$. Draw parallels through B' to side BC and through D' to side CD to meet at C'. Then $AB'C'D'$ is the desired trapezoid.

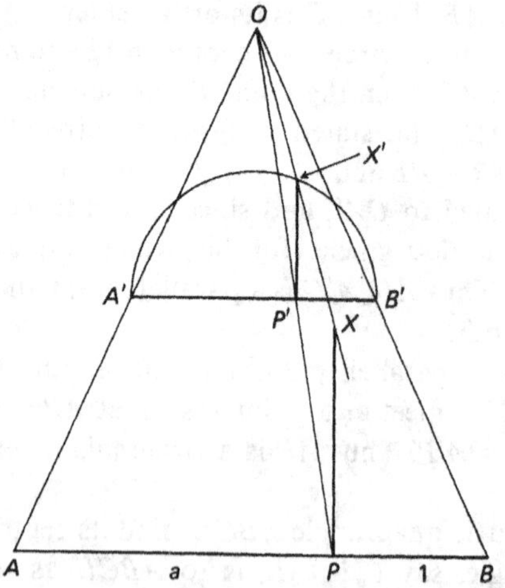

Answer 8.5

8.9 Let b and h be a base and corresponding altitude of the given triangle, and let the given segment have length c. Construct x so that $xc = bh$. On the perpendicular bisector of the given segment, mark a segment of length x from the base line. The point so constructed is the apex of the desired triangle.

8.11 a) At a point E on a base line m erect a perpendicular BE of length h_b. See the accompanying figure. Swing arc $B(t_b)$ to cut m at V. Construct at B, on either side of BV, rays making angles equal to $B/2$ and cutting m at A and C.

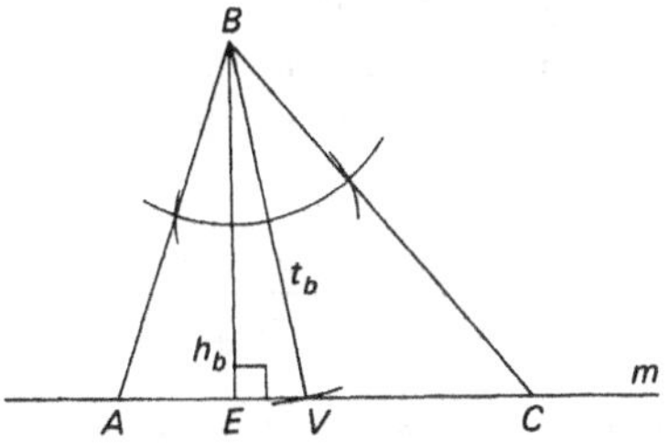

Answer 8.11a

c) At point E on a base line m erect a perpendicular BE of length h_b, as in the accompanying figure. Draw circle $B(c)$ to cut m at A. Let circle $A(b)$ cut m at C_1 and C_2. Then triangles ABC_1 and ABC_2 are solutions.

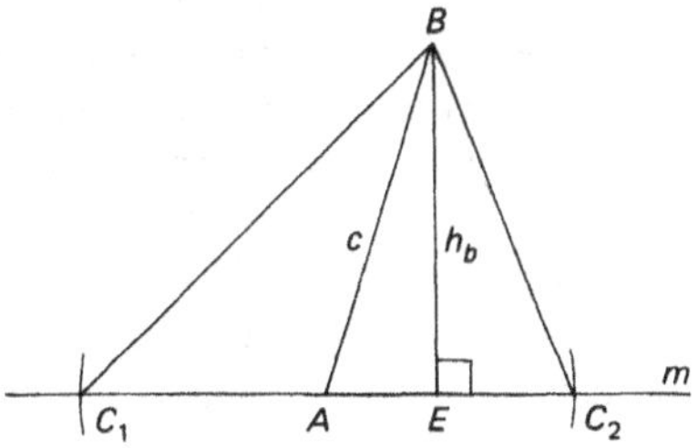

Answer 8.11c

e) At a point C on a line m construct segment CA of length b and making the given angle C with m. Draw circle $A(m_a)$ to cut m at A_1' and A_2'. Locate B_1 and B_2 on m, so that A_1' bisects B_1C and A_2' bisects B_2C, as seen in the accompanying figure. Both triangles AB_1C and AB_2C are solutions.

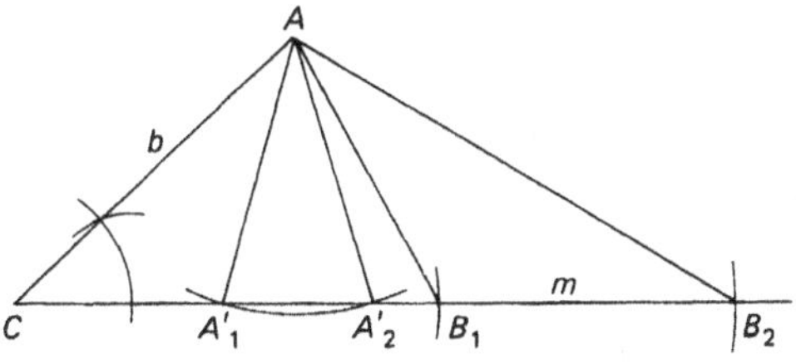

Answer 8.11e

8.13 Since OCA is an exterior angle for triangle COD, then $\sphericalangle OCA = \sphericalangle COD + \sphericalangle ODC = 2\sphericalangle ODC$. Similarly, $\sphericalangle AOB = \sphericalangle AOD + \sphericalangle ADO = \sphericalangle OCA + \sphericalangle ODC = 2\sphericalangle ODC + \sphericalangle ODC = 3\sphericalangle ODC$. Since it requires use of a *marked* straightedge, it does not satisfy the Euclidean restrictions.

8.15 $40°7'$

8.16 c) The line through (a, b) and (c, d) has the equation $y - b = (x - a)(d - b)/(c - a)$.

e) By parts (a) and (d), since only lines and points can be drawn, then only rational numbers can be constructed.

g) Circle $x^2 + y^2 + ax + by + c = 0$ and line $y = vx + u$ meet at the points (p, q) where

$$p = \frac{-(a + bv + 2uv) \pm ((a + bv + 2uv)^2 - 4(1 + v^2)(u^2 + bu + c))^{1/2}}{2(1 + v^2)}$$

and $q = vp + u$. Two circles $x^2 + y^2 + ax + by + c = 0$ and $x^2 + y^2 + dx + ey + f = 0$ meet on the line $(a - d)x + (b - e)y + (c - f) = 0$, if they meet at all, so this case reduces to that of a circle and a line.

i) None of these quantities involves only square roots and rational operations applied to rational numbers.

9.1 Measuring the shadow of the pyramid requires locating the point directly below the apex of the pyramid, which point is inaccessible. For two shadow observations, the distance between the tips of the two shadows of the pyramid is to the height of the pyramid as the corresponding measurements for the stick are to one another.

9.3 a) Let AB be a side of the given polygon in the circle of center O. Let M be the midpoint of arc AB and N the midpoint of chord AB. Then $y = m(AM)$ and $x = m(AB)$. Now use the Pythagorean theorem on triangles OAN and AMN.

10.1 See Theorem 15.1.

10.3 Through the vector from B to A, the negative of the given vector.

10.5 See Theorem 14.11.

10.7 $180°$

10.9 See Theorem 16.1.

11.1 a) (5, 0), (6, 0), (5, 2)

c) (3, −3), (3, −2), (1, −3)

e) (5, −1), (5, −2), (7, −1)

g) (0, 0), (0, 1), (2, 0)

i) (6, 0), (5, 0), (6, −2)

k) (5, 4), (6, 4), (5, 2)

m) (0, 0), (1, 0), (0, 2)

11.2 a) $x = 0,\ x = \frac{5}{2}$.
c) $y = 0,\ y = x - 3$
e) $y = x - 5,\ y = -3$
g) $y = x$
i) $x = 3,\ y = 0$
k) $x = 0,\ x = \frac{5}{2},\ y = 2$
m) $x = 0,\ x = 0$

11.3 A halfturn about its center, and a reflection in either of the two lines through its center and parallel to a pair of sides, and, of course, the identity map ι.

11.4 a) Same as Answer 11.3. Also a 90° or a 270° rotation about its center. Also a reflection in either diagonal.
c) The identity, a halfturn about its center, and reflections in the diagonals.
e) Rotations through multiples of 60°, reflections in perpendicular bisectors of its sides, and reflections in its diagonals that pass through its center.

11.5 a) A translation of 5 units in the negative x-direction.
c) A rotation of 270° about the point (3, 0).
e) A rotation of 90° about Q.
g) This is self-inverse.
i) This is self-inverse.
k) A glide-reflection of 5 units in the negative x-direction with mirror $y = 2$.
m) The identity.

11.6 The isometries of (g) and (i) are involutoric, those of (a), (c), (e), (k), and (m) are not.

11.7 a) There is none.
c) 4
e) 4
g) 2
i) 2
k) There is none.
m) 1

11.8 a) $y = x/2$
c) $y = -x/2$

11.9 a) $y = 1$
c) $y = -1$

11.11 The right triangle ABC with right angle at C and hypotenuse midpoint M and its image under the halfturn about M form a rectangle $ACBC'$ whose diagonals are congruent and bisect each other at M, establishing the theorem.

12.1 For any point A in the plane, let $\alpha(A) = B$ and $\beta(B) = C$. Then $(\beta\alpha)(A) = C$. Since β is one-to-one, no point other than B is mapped to C by β. Similarly, A is the only point mapped to B by α. Hence $\beta\alpha$ is one-to-one.

For any point X in the plane, since β is onto, there is a point Y such that $\beta(Y) = X$. Similarly, there is a point Z such that $\alpha(Z) = Y$. Then $(\beta\alpha)(Z) = X$, so $\beta\alpha$ is onto.

12.3 Let α be a translation of 1 unit in the positive x-direction, and β a reflection in the y-axis. Then $\beta\alpha$ maps the origin to $(-1, 0)$ and $\alpha\beta$ maps the origin to $(1, 0)$. Hence $\beta\alpha \neq \alpha\beta$.

12.5 Since α and $(\alpha^{-1})^{-1}$ are both inverse to α^{-1} by Theorem 12.11, than $\alpha = (\alpha^{-1})^{-1}$ by Theorem 12.12.

12.7 a) We have $\alpha = \alpha\iota = \alpha(\alpha\alpha^{-1}) = \alpha^2\alpha^{-1} = \alpha\alpha^{-1} = \iota$.

12.9 Suppose $\alpha\gamma = \beta\gamma$. Then $\alpha = \alpha\iota = \alpha(\gamma\gamma^{-1}) = (\alpha\gamma)\gamma^{-1} = (\beta\gamma)\gamma^{-1} = \beta(\gamma\gamma^{-1}) = \beta\iota = \beta$. The other case is similar.

12.11 a) Let α, β, γ be a reflection in the y-axis, a reflection in the line $x = 1$, and a translation of 1 unit in the positive x-direction.

c) Yes, but $\alpha = \beta$ if α and β commute.

12.13 By (1), take $a \in S$. By (2*), since $a \in S$ and $a \in S$, then $\iota = a^{-1}a \in S$. By (2*), using a and ι, $a^{-1} = a^{-1}\iota \in S$. Hence condition (2) is satisfied. Now, for given a and b in S, then a^{-1} and b are in S by (2), so $ab = (a^{-1})^{-1}b$ is in S by (2*). Hence (3) is satisfied.

13.1 Theorem 13.3 shows that an isometry α maps any three collinear points into three collinear points. Hence lines map into lines. If P is any point on a circle of radius r and center O, let P' and O' be the images of P and O. Since α is an isometry, $m(OP) = m(O'P')$, so P' lies on the circle of center O' and radius r. Similarly, all such points P' at distance r from O' are images of points on the given circle centered at O.

13.3 If A, B, and A' are collinear, then B' is also collinear with them. Let line $ABA'B'$ cut the mirror at F. By definition of a reflection, $\mathbf{AF} = \mathbf{FA'}$ and $\mathbf{BF} = \mathbf{FB'}$. Then $\mathbf{AB} = \mathbf{AF} + \mathbf{FB} = \mathbf{FA'} + \mathbf{B'F} = \mathbf{B'A'}$.

13.5 For any point A, $\sigma_m(A) = A'$ iff $A = A'$ or m is the perpendicular bisector of AA'. But then $A' = A \in m$ or m is the perpendicular bisector of $A'A$, so $\sigma_m(A') = A$. Thus $\sigma_m^{-1} = \sigma_m$.

13.7 In Fig. 13.16c, segments (such as AB) generally parallel to the mirror m have generally the same senses as their images. That is, AB and $A'B'$ are both sensed generally "downward" along m. Segments (such as CA) roughly perpendicular to m have their senses approximately reversed. That is, CA and $C'A'$ are directed roughly toward the mirror, but on opposite sides of it, so when CA is directed rightward, $C'A'$ is directed leftward. Thus the sense of angle BAC is opposite that of angle $B'A'C'$. It follows that the sense of any triangle is reversed by a reflection.

13.9 Each reflection reverses the sense of a triangle, so a product of two reflections preserves sense. It follows that any direct isometry, being a product of an even number of isometries, preserves the sense of a triangle.

13.11 Corollaries 13.17 and 13.18 establish this theorem for a triangle (a polygon of 3 sides). Suppose the theorem is true for any polygon of k sides, $k \geqslant 3$. Let $A_1A_2 \ldots A_{k+1}$ be a polygon of $k + 1$ sides and let α be an isometry. Then $\alpha(A_1A_2 \ldots A_k) \cong A_1A_2 \ldots A_k$, and $\alpha(\triangle A_kA_{k+1}A_1) \cong \triangle A_kA_{k+1}A_1$ by hypothesis. There must be at least one vertex A_i, $i < k$, such that A_1, A_i, and A_k are not collinear. Since $\alpha(A_i A_{k+1}) \cong A_i A_{k+1}$, it follows that $\alpha(A_{k+1})$ lies on the same side of $\alpha(A_1 A_k)$, relative to the rest of the image polygon, as does A_{k+1}, relative to $A_1 A_k$ and the given polygon. Now we have $\alpha(A_1 A_2 \ldots A_{k+1}) \cong A_1 A_2 \ldots A_{k+1}$ by addition of regions.

13.13 a) $A'(0, 0)$, $B'(1, 0)$, $C'(0, -3)$, $D'(1, -1)$, $E'(2, -3)$, $F'(15, -12)$, $G'(-3, 5)$
c) $A'(0, 4)$, $B'(1, 4)$, $C'(0, 1)$, $D'(1, 3)$, $E'(2, 1)$, $F'(15, -8)$, $G'(-3, 9)$
e) $A'(0, 0)$, $B'(0, 1)$, $C'(3, 0)$, $D'(1, 1)$, $E'(3, 2)$, $F'(12, 15)$, $G'(-5, -3)$

14.1 The proof given in 10.6 is quite sufficient.

14.3 This is a corollary to Theorems 14.2 and 14.4.

14.5 By Theorem 13.13, any isometry α is a product of not more than three reflections. Since it is direct, it is a product of exactly two reflections in mirrors m and n, a rotation if m and n intersect or a translation if m and n are parallel, the identity map being a special case of either a rotation or a translation.

14.7 For $(\beta\alpha)(A) = \beta(A) = (1, -2)$ and $(\alpha\beta)(A) = \alpha(1, -2) = (1, 2)$.

14.9 It is also a translation, but not necessarily the inverse of the translation $\beta\alpha$.

14.11 A reflection is a glide-reflection whose glide is zero. No, a translation is direct and a glide-reflection is opposite.

14.13 Let the glide-reflection be $\alpha = \sigma_g \sigma_b \sigma_a$, in which g is perpendicular to both a and b. Then $\alpha^{-1} = \sigma_a \sigma_b \sigma_g = \sigma_g \sigma_a \sigma_b$ by Hint 14.10, so α^{-1} is the inverse of the given glide, followed by the given reflection.

14.15 No, such a product is a direct isometry.

14.16 a) (0, 0), (0, 1), (−3, 0), (−1, 1), (−3, 2), (−12, 15), (5, −3)
c) (0, 0), (0, −1), (3, 0), (1, −1), (3, −2), (12, −15), (−5, 3)
e) (2, 0), (2, 1), (−1, 0), (1, 1), (−1, 2), (−10, 15), (7, −3)
g) (2, −4), (2, −5), (5, −4), (3, −5), (5, −6), (14, −19), (−3, −1)

14.17 a) (5, 0), (6, 0), (5, 3), (6, 1), (7, 3), (20, 12), (2, −5)
c) (−1, 2), (0, 2), (−1, 5), (0, 3), (1, 5), (14, 14), (−4, −3)

14.19 The center of the rotation $\alpha\beta$ is the point symmetric to the center of the rotation $\beta\alpha$ with the midpoint of AB as center of symmetry. When n and m are parallel, that is, when the angles of the rotations are opposites, the product $\beta\alpha$ is a translation and $\alpha\beta$ is its inverse translation.

15.1 Since, for given angle ABC, any reflection maps triangle ABC to a congruent triangle, it maps angle ABC to a congruent angle. But each isometry is a product of reflections.

15.3 By Theorem 13.12, $\sigma_m^2 = \iota$, but there exist points A and B such that $\sigma_m(A) = B$ and $A \neq B$, so $\sigma_m \neq \iota$. Of course, A can be taken as any point not on m.

15.5 Since b and a are perpendicular, then both $\sigma_b \sigma_a$ and $\sigma_a \sigma_b$ represent 180° rotations about their point P of intersection by Theorem 14.9.

15.7 Let m be the line on A and B, and let a and b be the lines through A and B and perpendicular to m. Let lines c and d be drawn parallel to a and b and spaced so that the directed distance from a to b is equal to that from d to c and with point C on line c. Finally, let line n pass through C perpendicular to c, and let n cut d in D. Then $\sigma_C\sigma_B\sigma_A = \sigma_D\sigma_D\sigma_C\sigma_B\sigma_A = \sigma_D\sigma_d\sigma_n\sigma_n\sigma_c\sigma_b\sigma_m\sigma_m\sigma_a = \sigma_D\sigma_d\sigma_c\sigma_b\sigma_a = \sigma_D$, since $\sigma_d\sigma_c = \sigma_a\sigma_b$.

15.8 a) (1, 0), (2, 0), (1, −3), (2, −1), (3, −3), (16, −12), (−2, 5)
c) (−8, −6), (−9, −6), (−8, −3), (−9, −5), (−10, −3), (−23, 6), (−5, −11)

15.9 a) (0, 0), (−1, 0), (0, −3), (−1, −1), (−2, −3), (−15, −12), (3, 5)
c) (4, 6), (3, 6), (4, 3), (3, 5), (2, 3), (−11, −6), (7, 11)

15.11 If α and β are translations, then there are points A, B, C so that $\alpha = \sigma_B\sigma_A$ and $\beta = \sigma_C\sigma_B$. Then $\beta\alpha = \sigma_C\sigma_B\sigma_B\sigma_A = \sigma_C\sigma_A$. Now, taking D so that $ABCD$ is a parallelogram, then $\alpha = \sigma_B\sigma_A = \sigma_C\sigma_D$ and $\beta = \sigma_C\sigma_B = \sigma_D\sigma_A$, so

$$\alpha\beta = \sigma_C\sigma_D\sigma_D\sigma_A = \sigma_C\sigma_A = \beta\alpha.$$

16.1 If a, b, c are all parallel, then take d so that the directed distance from c to d is equal to that from b to a. If the given lines are concurrent, then replace "distance" by "angle" in the first sentence of this answer.

Conversely, if $\sigma_c\sigma_b\sigma_a = \sigma_d$, then $\sigma_b\sigma_a = \sigma_c\sigma_d$, so $\sigma_c\sigma_d$ is the same translation or rotation as $\sigma_b\sigma_a$. Hence the four lines all pass through the center of the rotation, or all are perpendicular to the direction of translation. In either case, they form a pencil and the directed angles or distances are as stated in the first paragraph of this answer.

16.3 Each direct isometry is a product of two reflections. Each opposite isometry is a glide-reflection by Theorem 16.4, so it can be written as a product $\sigma_g\sigma_b\sigma_a$ where g is perpendicular to both a and b. Now let B be the point of intersection of lines b and g, so that $\sigma_B = \sigma_g\sigma_b$.

16.5 The transformations of each set form a subset of those above and, in each case, the conditions of Exercise 12.12 are satisfied.

16.7 A nontrivial rotation has just its center fixed, for if a rotation α has X as a fixed point, then let $\alpha = \sigma_b\sigma_a$ where line a passes through X. Since $\sigma_a(X) = X$, we have $X = \alpha(X) = (\sigma_b\sigma_a)(X) = \sigma_b(X)$, and X is a fixed point under σ_b, too. Thus X lies on line b, also. This is possible only when X is the center of rotation or when lines a and b coincide so that the rotation is the identity map.

16.9 If a rotation through angle θ is involutoric, then its square, a rotation through angle 2θ, is the identity map. This implies that 2θ is a multiple of 360°, so θ is a multiple of 180°. Since the identity is not called involutoric, then θ is an odd multiple of 180°, and the rotation is a halfturn.

16.11 Since the isometry α maps m to m and n to n, then $\alpha(P)$ must lie on m, since P lies on m and m is a fixed line. Similarly, $\alpha(P)$ must lie on n, since P lies on n and n is a fixed line. Hence $\alpha(P) = P$, the only point on both m and n.

16.13 a) Yes
c) Yes
e) No; the product of two halfturns is a translation.

17.1 Other dual pairs are 17.4 and 17.8, 17.5 and 17.9, and 17.11 and 17.12.

17.3 If $a = b$ or if a is perpendicular to b, then clearly the other conditions follow, mainly by Corollary 15.9.

If $\sigma_b \sigma_a = \sigma_a \sigma_b$, then $(\sigma_b \sigma_a)^2 = \iota$, so apply Exercise 16.11. The other cases are similar or follow from Theorem 16.14.

17.5 Conditions (1), (2), and (3) are equivalent by Theorems 14.4 and 14.10. Conditions (2) and (4) are algebraically equivalent; simply multiply on the right both sides of either equation by σ_b.

17.7 These results are analagous to Theorem 17.5.

17.9 This theorem is dual to Theorem 17.5, so its proof is similar.

17.11 Since $\sigma_C \sigma_b \sigma_A$ is a product of an odd number of reflections, it is a glide-reflection. Then apply Theorem 15.7.

17.13 Apply Theorem 15.12.

17.15 This identity is equivalent to

$$(\sigma_a \sigma_b \sigma_c)^2 (\sigma_c \sigma_a \sigma_b)^2 (\sigma_b \sigma_c \sigma_a)^2 (\sigma_c \sigma_b \sigma_a)^2 (\sigma_a \sigma_c \sigma_b)^2 (\sigma_b \sigma_a \sigma_c)^2 = \iota,$$

which is true, since the first and fourth factors are inverses of one another, as are the second and last, and the third and fifth.

17.17 In Exercise 17.16, each product $(\sigma_A \sigma_B \sigma_C)^2$ is the square of a halfturn, hence it is the identity. In Thomsen's relation, each such product is a translation.

18.1 Let $m(AB) = m(A'B')$, $m(\measuredangle A) = m(\measuredangle A')$, and $m(\measuredangle B) = m(\measuredangle B')$ in triangles ABC and $A'B'C'$. Let an isometry α map $\triangle A'B'C'$ to $\triangle ABC''$ so that C and C'' are on opposite sides of AB, as in Fig. 18.4. Letting m denote line AB, let $\sigma_m(C) = X$. Since AB is the bisector of angle CAC'', X lies on ray AC''. Similarly, X lies on ray BC''. Hence $X = C''$, the point of intersection of the two rays. Now $(\alpha^{-1}\sigma_m)(\triangle ABC) = \alpha^{-1}(\triangle ABC'') = \triangle A'B'C'$, and the theorem follows.

18.3 Let right triangles ABC and $A'B'C'$ have $m(\measuredangle C) = m(\measuredangle C') = 90°$, $m(AC) = m(A'C')$ and $m(AB) = m(A'B')$. Let an isometry α map triangle $A'B'C'$ to triangle $A''B''C$, with B'' on ray CB and A and A'' on opposite sides of BC. Letting m denote line BC, then $\sigma_m(A) = A''$. Thus $m(BA) = m(BA'') = m(B''A'')$, so $B = B''$. Now $\sigma_m(\triangle ABC) = \triangle A''B''C \cong \triangle A'B'C'$. See the accompanying figure.

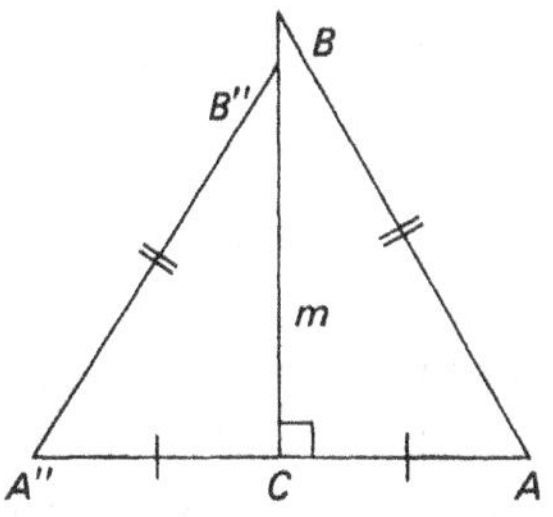

Answer 18.3

18.5 a) If AM is the median, then triangles ABM and ACM are congruent by SSS.
c) If AU is the angle bisector, then triangles ABU and ACU are congruent by SAS.

18.6 a) Let the bisector of apex angle A meet the base BC at U, and let m denote line AU. Now σ_m maps line AB to line AC and, since $m(AB) = m(AC)$, $\sigma_m(B) = C$. So $\sigma_m(\angle ABC) = \angle ACB$.

18.7 By Theorem 17.8, $\sigma_D = \sigma_C \sigma_B \sigma_A$, so $\sigma_D \sigma_N = \sigma_C \sigma_B \sigma_A \sigma_N = \sigma_C \sigma_B \sigma_N \sigma_C = \sigma_N \sigma_B \sigma_C \sigma_C = \sigma_N \sigma_B$, the next to the last equality using Theorem 17.13.

18.9 By Theorem 18.9, $\sigma_N(\triangle ABN) = \triangle CDN$ and $\sigma_N(\triangle BCN) = \triangle DAN$.

18.11 In trapezoid $ABCD$, let $m(\angle A) = m(\angle B)$ and let m be the perpendicular bisector of AB. Now ray $BC = \sigma_m$ (ray AD). Let $\sigma_m(D) = X$. Then X lies on ray BC and, since AB is parallel to CD, then X also lies on CD. Hence $X = C$. That is, $\sigma_m(AD) = BC$, so the trapezoid is isosceles.

18.13 If median AM is perpendicular to side BC, then $\sigma_m(B) = C$, where m is line AM. But then triangle ABC is isosceles.

18.15 Let the perpendicular diagonals AC and BD meet at N. By Theorem 18.9, $m(BN) = m(DN)$. Letting m denote line AC, then $\sigma_m(B) = D$, so $m(AB) = m(AD)$ and $m(CB) = m(CD)$. The theorem follows.

19.1 Let P be any point on a circle of center O, and let m be any diameter. If $\sigma_m(P) = P'$, then $\sigma_m(OP) = OP'$, so $OP \cong OP'$ and P' lies on the circle whenever P does.

19.3 For the converse, suppose chords AB and CD are congruent. Let α be a rotation about the center of the circle carrying AB to CD. That is, assume $\alpha(A) = C$ and $\alpha(B)$ is on the same side of C as is D. Since $\alpha(B)$ is at the intersection of the given circle and circle $C(AB)$, then $\alpha(B) = D$. Since the distance from the center of the circle to a chord is preserved by α, the theorem follows.

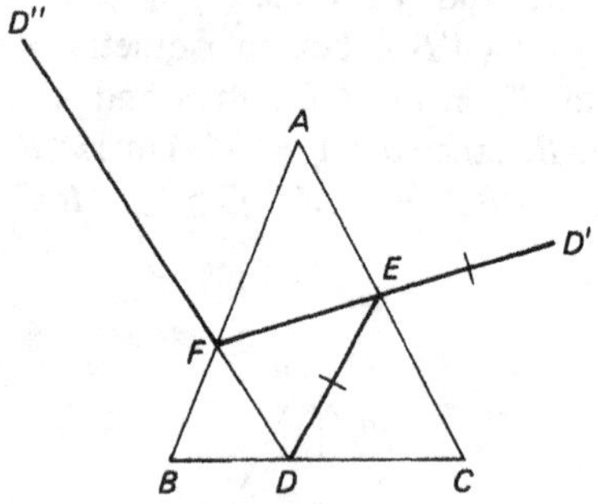

Answer 19.13

19.5 Rotate the triangle a halfturn about the midpoint of any side, forming a parallelogram. Now apply Theorem 19.6.

19.7 For $(\sigma_b \sigma_a \sigma_c \sigma_b \sigma_a)(A) = (\sigma_b \sigma_a \sigma_c \sigma_b)(B) = (\sigma_b \sigma_a \sigma_c)(C) = (\sigma_b \sigma_a)(D) = \sigma_b(E) = F$. But $\sigma_b \sigma_a \sigma_c \sigma_b \sigma_a = \sigma_c$, so $\sigma_c(A) = F$.

19.9 Since $\sigma_{C'}\sigma_C\sigma_{B'}\sigma_B\sigma_{A'}\sigma_A$ is a translation, it is the identity iff it has a fixed point. Now $(\sigma_{C'}\sigma_C\sigma_{B'}\sigma_B\sigma_{A'}\sigma_A)(A) = (\sigma_{C'}\sigma_C\sigma_{B'}\sigma_B\sigma_{A'})(A) = (\sigma_{C'}\sigma_C\sigma_{A'}\sigma_B\sigma_{B'})(A) = (\sigma_{C'}\sigma_C\sigma_{A'}\sigma_B)(C) = (\sigma_{C'}\sigma_B\sigma_{A'}\sigma_C)(C) = (\sigma_{C'}\sigma_B\sigma_{A'})(C) = (\sigma_{C'}\sigma_B)(B) = \sigma_{C'}(B) = A$. The theorem follows.

19.11 One tangent maps into the other by a reflection in that diameter that passes through the given point.

19.13 Line DF is fixed under the glide-reflection $\sigma_c\sigma_b\sigma_a$ (see Theorem 7.6), so DF is the mirror. Point D is fixed under σ_a and maps into a point D' on ray FE under σ_b so that $m(ED') = m(DE)$. Finally, σ_c carries D' to point D'' on ray DF so that $FD'' = FD' = FE + ED' = FE + ED$. Now $DD'' = DF + FD'' = DF + FE + ED$. See the figure on page 262.

19.15 Draw that diameter m of the circumcircle to the polygon that passes through the one coin or bisects the side joining the two coins your opponent takes on his first play. Take the one or two coins at the opposite end of diameter m. From then on, take the image in mirror m of the coin or coins your opponent takes.

20.1 Since $\sigma_d\sigma_g\sigma_f = \sigma_e$, then $\sigma_d\sigma_g = \sigma_e\sigma_f$.

20.3 Since, for example, $CABU$ is a parallelogram, then $\sigma_C\sigma_A\sigma_B\sigma_U = \iota$, so $\sigma_U = \sigma_C\sigma_A\sigma_B$. Since then $\sigma_U\sigma_C = \sigma_C\sigma_V$, then C is the midpoint of UV, etc.

20.5 The condition $\sigma_C(U) = V$ states that C is the midpoint of UV, etc. Now $\sigma_U = \sigma_B\sigma_A\sigma_C$ implies $\sigma_B\sigma_U = \sigma_A\sigma_C$, so $m(BU) = m(AC)$, but $BU = \frac{1}{2}WU$, etc.

20.7 In Theorem 6.20, since triangles ABD and APC are similar, then AD and AP are isogonal conjugates. That is, the altitude and the circumdiameter issuing from a vertex of a triangle are isogonal conjugates. The theorem follows.

20.9 The proof of Theorem 20.15 shows that S trisects median AA'. Similarly S trisects medians BB' and CC'.

20.11 As stated, $(\sigma_f\sigma_e\sigma_d)(Y) = (\sigma_f\sigma_e)(Z) = \sigma_f(X) = Y$, since tangents from a point to a circle are congruent. Thus $\sigma_f\sigma_e\sigma_d$ has a fixed point, so it is not a glide-reflection, but just a reflection in a line through that fixed point and also through the point of intersection of these three mirrors by Theorem 16.1. That is, d, e, and f concur.

20.13 Suppose the midpoints are not all distinct. Then there are distinct points P and S on m with images $\alpha(P) = Q$ and $\alpha(S) = T$ on n, such that some point R is the common midpoint of PQ and ST. Now $\sigma_R(PS) = QT$. Since σ_n leaves the points of n fixed, then $(\sigma_n\sigma_R)(PS) = QT$, too. By Theorem 13.20, then, $\alpha = \sigma_R$ or $\alpha = \sigma_n\sigma_R$. In either case the desired midpoints all coincide at R.

20.15 Let perpendiculars BB_1 and CC_1 be dropped from the vertices B and C onto median AA'. The halfturn $\sigma_{A'}$ carries $A'B$ into $A'C$ and ray $A'B_1$ into ray $A'C_1$. Since there is just one perpendicular from C to line AA', it follows that $\sigma_{A'}(B_1) = C_1$. Hence triangles $BA'B_1$ and $CA'C_1$ are congruent, and the theorem follows.

20.17 Let $ABCD$ be a cyclic trapezoid with sides AB and CD parallel. Reflect in that diameter perpendicular to AB and CD. Then A and D map to points on lines

AB and CD, respectively. But they also map to points on the circle. Hence they map to B and C. Thus $m(AD) = m(BC)$.

20.19 We have $\sigma_{C''} = \sigma_{C'}\sigma_C\sigma_{C'}$ and $\sigma_{B''} = \sigma_{B'}\sigma_B\sigma_{B'}$ by Theorem 17.9. Also,

$$\sigma_{C'}\sigma_A = \sigma_B\sigma_{C'} \qquad \text{and} \qquad \sigma_{B'}\sigma_A = \sigma_C\sigma_{B'}.$$

Now

$$\begin{aligned}\sigma_{C''}\sigma_A\sigma_{B''}\sigma_A &= (\sigma_{C'}\sigma_C\sigma_{C'})\sigma_A(\sigma_{B'}\sigma_B\sigma_{B'})\sigma_A = \sigma_{C'}\sigma_C(\sigma_{C'}\sigma_A)\sigma_{B'}\sigma_B(\sigma_{B'}\sigma_A)\\ &= \sigma_{C'}\sigma_C\sigma_B\sigma_{C'}\sigma_{B'}\sigma_B\sigma_C\sigma_{B'} = \sigma_{C'}(\sigma_{C'}\sigma_B\sigma_C)\sigma_{B'}(\sigma_{B'}\sigma_C\sigma_B)\\ &= \sigma_B\sigma_C\sigma_C\sigma_B = \iota,\end{aligned}$$

using also Theorem 17.13. Hence $\sigma_{C''}\sigma_A = \sigma_A\sigma_{B''}$, so A is the midpoint of $B''C''$ by Theorem 17.9.

20.21 The bisectors a and b of the interior angles at A and B pass through O. Since $\sigma_b\sigma_a$ carries m into n, then $\sigma_b\sigma_a$ is a halfturn (since m and n are parallel). Thus a and b intersect at O in a right angle. Thus the circle on AB as diameter passes through the vertex O of that right angle.

20.22 a) A 90° rotation about M carries one of the triangles AMC and BMD into the other.

c) Let M be the midpoint of either diagonal of the quadrilateral. Then apply part (b). Finally, apply part (a). Note that WY and XZ do not, in general, meet at M.

e) The proofs are the same.

21.1 The equations for the halfturns are: for σ_O, $x' = -x$ and $y' = -y$; for σ_A, $x' = -x + 2h$ and $y' = -y + 2k$; for σ_C, $x' = -x + 2a$ and $y' = -y + 2b$; and for σ_D, $x' = -x + 2a + 2h$ and $y' = -y + 2b + 2k$. Then for $\sigma_C\sigma_D\sigma_A\sigma_O$, $x' = -(-[-(-x) + 2h] + 2a + 2h) + 2a = x$, and $y' = -(-[-(-y) + 2k] + 2b + 2k) + 2b = y$, so $\sigma_C\sigma_D\sigma_A\sigma_O = \iota$.

21.3 Since C is the midpoint of the segment joining $P(x, y)$ and $P'(x', y')$, then $h = (x + x')/2$ and $k = (y + y')/2$, from which the desired equations follow.

21.5 Let α have the equations $x' = x + h$ and $y' = y + k$, and β the equations $x' = x + m$ and $y' = y + n$. Then the equations for $\beta\alpha$ are $x' = (x + h) + m = x + (h + m)$ and $y' = (y + k) + n = y + (k + n)$, equations for a translation.

21.7 Let α and β be the given rotations. Then $\beta\alpha$ has equations

$$\begin{aligned}x' &= (x\cos\theta - y\sin\theta - h)\cos\phi - (x\sin\theta + y\cos\theta - k)\sin\phi + h\\ &= x(\cos\theta\cos\phi - \sin\theta\sin\phi) - y(\sin\theta\cos\phi + \cos\theta\sin\phi) + h(1 - \cos\phi)\\ &\qquad + k\sin\phi\\ &= x\cos(\theta + \phi) - y\sin(\theta + \phi) + h(1 - \cos\phi) + k\sin\phi\end{aligned}$$

and in a similar manner,

$$y' = x\sin(\theta + \phi) + y\cos(\theta + \phi) + k(1 - \cos\phi) - h\sin\phi.$$

For these equations to represent a translation, we must have $\sin(\theta + \phi) = 0$ and $\cos(\theta + \phi) = +1$. This occurs only when $\theta + \phi$ is a multiple of 360°.

21.9 Replace θ by $-\theta$ to get $x' = x\cos\theta + y\sin\theta$ and $y' = -x\sin\theta + y\cos\theta$.
21.11 Since a halfturn is involutoric, use the same equations.
21.13 The translation is $x'' = x' + r$ and $y'' = y' + s$.
21.15 If $a = 1$, then $b = 0$, so $x' = x + c$ and $y' = y + d$.

22.1 The mirror image of point $P(a, b)$ in the x-axis is the point $P'(a, -b)$. Thus $x' = x$ and $y' = -y$ define the reflection in the x-axis. The equations for the reflection in the y-axis are obtained in a similar manner.
22.3 Letting α and β denote these reflections and using the equations as given in Theorem 22.4, we have, for $\beta\alpha$,

$$\begin{aligned} x' &= (x\cos 2\theta + y\sin 2\theta)\cos 2\phi + (x\sin 2\theta - y\cos 2\theta)\sin 2\phi \\ &= x(\cos 2\theta\cos 2\phi + \sin 2\theta\sin 2\phi) + y(\sin 2\theta\cos 2\phi - \sin 2\phi\cos 2\theta) \\ &= x\cos(2\phi - 2\theta) - y\sin(2\phi - 2\theta), \end{aligned}$$

and in a similar manner we obtain

$$y = x\sin(2\phi - 2\theta) + y\cos(2\phi - 2\theta)$$

which are equations for the desired rotation.
22.5 This is straightforward substitution.
22.7 Same as for the given reflection.
22.9 They are $x' = (x - h)\cos 2\theta + (y - k)\sin 2\theta + h + r\cos\theta$ and

$$y' = (x - h)\sin 2\theta - (y - k)\cos 2\theta + k + r\sin\theta.$$

22.11 In the equations of Theorem 22.6, take $a = \cos 2\theta$, $b = \sin 2\theta$, $c = h - h\cos 2\theta - k\sin 2\theta$, and $d = k - h\sin 2\theta + k\cos 2\theta$.
22.13 There is an angle θ such that $a = \cos 2\theta$ and $b = \sin 2\theta$. Then the result follows readily.
22.15 Let us take α with equations $x' = ax - by + c$ and $y' = e(bx + ay) + d$, and β with $x' = fx - gy + h$ and $y' = j(gx + fy) + k$, where $e = \pm 1$, $j = \pm 1$, $a^2 + b^2 = 1$, and $f^2 + g^2 = 1$. Then $\beta\alpha$ has equations

$$\begin{aligned} x' &= f(ax - by + c) - g(ebx + eay + d) + h \\ &= (fa - geb)x - (fb + gea)y + (fc - gd + h) \end{aligned}$$

and

$$\begin{aligned} y' &= j(g(ax - by + c) + f(ebx + eay + d)) + k \\ &= je((gea + fb)x + (fa - geb)y) + (jgc + jfd + k). \end{aligned}$$

These equations have the proper form. All that remains is to show that the sum of the squares of the coefficients of x and y is 1. To that end, noting that $e^2 = 1$, we have

$$\begin{aligned} &(fa - geb)^2 + (fb + gea)^2 \\ &\quad = f^2a^2 - 2fageb + g^2e^2b^2 + f^2b^2 + 2fbgea + g^2e^2a^2 \\ &\quad = f^2(a^2 + b^2) + g^2(a^2 + b^2) = f^2 + g^2 = 1. \end{aligned}$$

22.17 a) We have $x = x'$ and $y = -y'$, and $x = -x'$ and $y = y'$.
c) We have $x = (x' - h)\cos 2\theta + (y' - k)\sin 2\theta + h$ and $y = (x' - h)\sin 2\theta - (y' - k)\cos 2\theta + k$.
e) Now $x = (x' - h)\cos 2\theta + (y' - k)\sin 2\theta + h - r\cos\theta$ and $y = (x' - h)\sin 2\theta - (y' - k)\cos 2\theta + k - r\sin\theta$.

22.19 This is true, since $\sin(\theta + 180°) = -\sin\theta$ and $\cos(\theta + 180°) = -\cos\theta$.

22.21 Let the mirrors m, n, p pass through the points (a, b), (c, d), (e, f), respectively, each with inclination θ. Then, using equations as in Theorem 22.6, we find that $\sigma_p\sigma_n$ has the equations

$$\begin{aligned} x' &= ((x - c)\cos 2\theta + (y - d)\sin 2\theta + c - e)\cos 2\theta \\ &\quad + ((x - c)\sin 2\theta - (y - d)\cos 2\theta + d - f)\sin 2\theta + e \\ &= x + e - c + (c - e)\cos 2\theta + (d - f)\sin 2\theta \end{aligned}$$

and

$$\begin{aligned} y' &= ((x - c)\cos 2\theta + (y - d)\sin 2\theta + c - e)\sin 2\theta \\ &\quad -((x - c)\sin 2\theta - (y - d)\cos 2\theta + d - f)\cos 2\theta + f \\ &= y + f - d + (c - e)\sin 2\theta - (d - f)\cos 2\theta. \end{aligned}$$

It follows that $\sigma_p\sigma_n\sigma_m$ has the equations

$$\begin{aligned} x' &= (x - a)\cos 2\theta + (y - b)\sin 2\theta + a + e - c \\ &\quad + (c - e)\cos 2\theta + (d - f)\sin 2\theta \end{aligned}$$

and

$$\begin{aligned} y' &= (x - a)\sin 2\theta - (y - b)\cos 2\theta + b + f - d \\ &\quad + (c - e)\sin 2\theta - (d - f)\cos 2\theta. \end{aligned}$$

Now translate the point (a, b) by the product $\sigma_p\sigma_n$ to

$$(a + e - c + (c - e)\cos 2\theta + (d - f)\sin 2\theta,$$
$$b + f - d + (c - e)\sin 2\theta - (d - f)\cos 2\theta)$$

$= (g, h)$. Then a bit of algebra shows that $\sigma_p\sigma_n\sigma_m$ is a reflection in a line parallel to the given lines, and has the equations $x' = (x - g)\cos 2\theta + (y - h)\sin 2\theta + g$ and $y' = (x - g)\sin 2\theta - (y - h)\cos 2\theta + h$.

23.1 Let the triangles ABC and $A'B'C'$ be copolar at O. Let the planes of the two triangles intersect on line m. Then AB and $A'B'$ both lie in plane ABO, and it follows that they meet at a point that lies in all three of the planes ABO, ABC, and $A'B'C'$. Hence these lines meet at a point on line m. Similarly, BC and $B'C'$ meet on m, and CA and $C'A'$ meet on m. Hence the triangles are coaxial in line m. The converse is established as in Theorem 4.7.

23.3 Let the five points be A, B, C, D, E. Let AB and DE meet at L. Let any line through L cut BC at M and CD at N. Now ME and NA meet at F, the sixth vertex of the inscribed hexagon $ABCDEF$ by the converse to Pascal's mystic hexagram theorem.

24.1 It suffices to show that a homothety preserves angles. Using Fig. 24.1, let the homothety map angle BAC to $B'A'C'$. Since $B'O/BO = A'O/AO$ and $\measuredangle AOB = \measuredangle A'OB'$, then triangles AOB and $A'OB'$ are similar. Similarly, triangles AOC and $A'OC'$ are similar. By subtracting congruent angles, we have $m(\measuredangle BAC) = m(\measuredangle B'A'C')$.

24.3 The ratio is the product of the two ratios and the center is the common center for the two homotheties.

24.5 Using any center, rotate one triangle so its sides are parallel to the other, then apply Exercise 24.4.

24.7 Letting the circles of radii 3 and 5 be centered at $(0, 0)$ and $(a, 0)$ with $a > 8$, the centers of homothety are at $(3a/8, 0)$ and $(-3a/2, 0)$. The ratios are $\pm 3/5$.

24.9 By similar triangles any such points O and O' divide the line of centers internally and externally in the ratio of their radii. Hence all such points O and O' coincide. Thus these points are the centers of similitude.

24.11 a) $(0, 0), (1, 0), (0, 2)$,
c) $(0, 0), (-1, 0), (0, -2)$,
e) $(-2, 0), (0, 0), (-2, 4)$,
g) $(-4, 0), (-2, 0), (-4, 4)$.

24.13 A homothety with the same center and whose ratio is the reciprocal of the ratio of the given homothety.

24.14 a) Obvious.
c) Since the ratio of a product of homotheties is the product of the ratios, and since $r^2 = 1$ iff $r = \pm 1$, and for $n > 2$, $r^n = 1$ only when $r = 1$ or perhaps -1 (for real r), and the cases for $r = 1$ and $r = -1$ are already covered, then the smallest such n is never greater than 2.

25.1 The result for triangles follows from Theorem 25.2. For the polygon, use the method of Answer 13.11.

25.3 The center is the midpoint of the line of centers and the ratio is -1.

25.5 This is a direct corollary to Theorem 25.5.

25.7 The center lies between A and A' iff the ratio of homothety is negative. Similarly for B and B'.

25.9 If $j \neq 1$ and $k \neq 1$, then the product of the two homotheties in reverse order is a homothety of ratio $(k-1)/k(j-1)$.

25.11 By Definition 25.1, the center is a fixed point. For any other point P, let its image be P'. Then $OP' = k \cdot OP$. Thus $P = P'$ only if $k = 1$. Likewise, if a line is at distance d from O, then its image is at distance kd from O. Hence the only fixed lines are those through O.

25.13 Let the homothety $H(O, k)$ map angle ABC to $A'B'C'$. If the ratio k is positive, then as a point P, interior to $\measuredangle ABC$, moves away from the center O of homothety, its image P' moves in the same direction inside $\measuredangle A'B'C'$. It follows that angles ABC and $A'B'C'$ are similarly oriented, and hence, that triangles ABC and $A'B'C'$ are directly similar.

25.15 Apply Theorems 25.10 and 25.14.
25.17 By Theorem 25.10.
25.19 Apply Theorem 25.9 and Hint 25.16.

26.1 By Definition 13.2.
26.3 This result is similar to that of Theorem 13.4.
26.5 Let similarities α and β each map triangle ABC to $A'B'C'$. Since α and β are transformations (by Theorem 26.4), then $\beta^{-1}\alpha$ is a transformation and, in fact, a similarity. Since $\beta^{-1}\alpha$ maps triangle ABC to itself, it is the identity map. Hence $\alpha = \beta$.
26.7 a) Then AB and $A'B'$ are parallel so point Q is the center of homothety, and the rotation reduces to the identity.
26.9 Point Q will lie between A and A' and between B and B'.
26.11 Follow the hint given in the text.
26.13 Let α and β be similarities of ratios j and k. If $\alpha(AB) = A'B'$ and $\beta(A'B') = A''B''$, then $(\beta\alpha)(AB) = A''B''$, and $A''B'' = k \cdot A'B' = jk \cdot AB$. Thus $\beta\alpha$ is a similarity of ratio jk. The theorem follows.
26.15 a) A circle c is not a segment, and any rotation α about its center carries circle c to itself. So, if β is a similarity, then so also is $\beta\alpha$ a similarity, and $\beta(c) = (\beta\alpha)(c)$. Thus there are infinitely many similarities carrying one given circle to another.
c) Yes, infinitely many similarities.
e) Three of each.
g) Two of each.

27.1 Draw another parallel through the third vertex and then apply Theorem 27.2. Alternatively, if a parallel to side BC of triangle ABC cuts sides AB and AC at M and N, then $H(A, \mathbf{AN/AC})$ maps BC to MN, and $H(A, \mathbf{AM/AB})$ also maps BC to MN, so $AN/AC = AM/AB$, and the theorem follows.
27.3 A line segment parallel to BC and of length $r/(r+1)$ times the length of BC.
27.5 This is a corollary to Theorem 27.4 and the ratio is DC/AB.
27.7 A reflection in the bisector of angle AED followed by the homothety $H(E, ED/EA)$.
27.9 This is similar to Theorem 27.5.
27.11 $\frac{7}{2}\sqrt{21} \approx 16.039$.
27.13 Since the sides of triangles EFA and ADB are directly parallel, there is a similarity mapping one triangle to the other. Thus triangle EFA is isosceles since triangle ADB is isosceles.
27.15 Apply Exercise 27.14 twice.

28.1 One-fourth the area of triangle OAM, or $r^2\sqrt{3}/32$, where r is the radius of the circle.
28.3 Hint 28.3 establishes the theorem.

28.5 Triangles DEC and FEA, and also triangles DEA and GEC are similar by AA. Then $DE/EF = EC/AE$ and $EC/AE = EG/DE$, from which the desired result follows.

28.7 A 90° rotation followed by a homothety carries one triangle to the other.

28.9 Let AU be an external bisector of angle A of triangle ABC, and use the proof of Theorem 28.6.

28.11 A reflection in the bisector of angle B followed by the homothety $H(B, BA/BU)$ maps triangle BAU to triangle BCA. The theorem follows.

28.13 From any point P' on ray CA beyond point A drop a perpendicular $P'Q'$ to BC, and construct square $P'Q'R'S'$ containing point A in its interior. See the accompanying figure. Let CS' cut BA at S. Now S is a vertex of the desired square, homothetic to $P'Q'R'S'$ with point C as center.

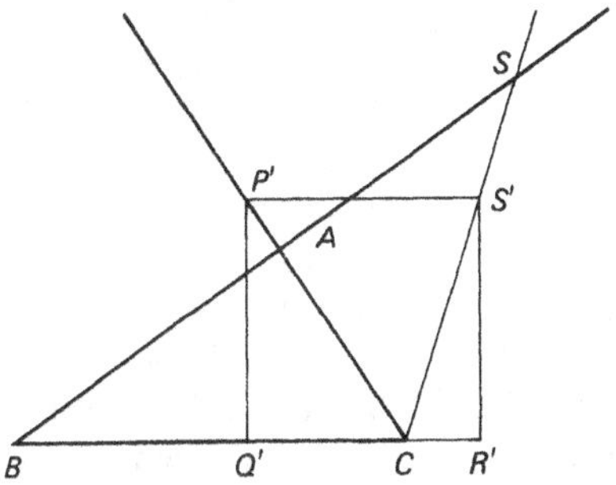

Answer 28.13

28.15 In general, only three vertices of the desired parallelogram will touch the given triangle, so see Answer 28.13 as well as Problem 28.10.

28.17 Use the method of Answer 28.13 and Problem 28.10.

29.1 By Theorem 29.4, H divides NO in the ratio $-1/2$. By Theorem 29.2, $\mathrm{HG}/\mathrm{GO} = 2$. Then

$$\frac{\mathrm{NG}}{\mathrm{GO}} = \frac{\mathrm{NH}+\mathrm{HG}}{\mathrm{GO}} = \frac{\mathrm{NH}}{\mathrm{GO}} + 2 = \frac{\mathrm{NH}}{\mathrm{HO}} \cdot \frac{\mathrm{HO}}{\mathrm{GO}} + 2 = -\frac{1}{2} \cdot \frac{\mathrm{HG}+\mathrm{GO}}{\mathrm{GO}} + 2$$

$$= -\tfrac{1}{2}(2+1) + 2 = \tfrac{1}{2}.$$

29.3 They concur at N.

29.5 First apply Theorem 29.5, then apply a homothety $H(O, OP/OP')$.

29.7 Draw a tangent m to one of the circles at the given point P. Reflect that circle in m, and draw the common chord of the image circle and the other given circle. Note that this reflection can also be effected by $H(P, -1)$.

29.9 Draw diameter AOB of the smallest circle and let circle $B(2s)$, where s is the radius of the second circle, cut the third circle at C. Then AC is the desired secant. To show this fact, let AC cut the second circle at D. Then $OD = s$, so $H(O, AC/AD) = H(O, AB/AO) = H(O, 2)$.

29.11 At angle P, of the given size, construct PQ and PR on its sides of lengths equal to the nonparallel sides. See the accompanying figure. Then draw, joining the sides of angle P and parallel to QR, segments $A'B'$ and $D'C'$ of lengths m and n, where m/n is the given ratio. Trapezoid $A'B'C'D'$ is similar to the desired trapezoid, so use a homothety to get trapezoid $ABCD$ whose nonparallel sides have the proper lengths.

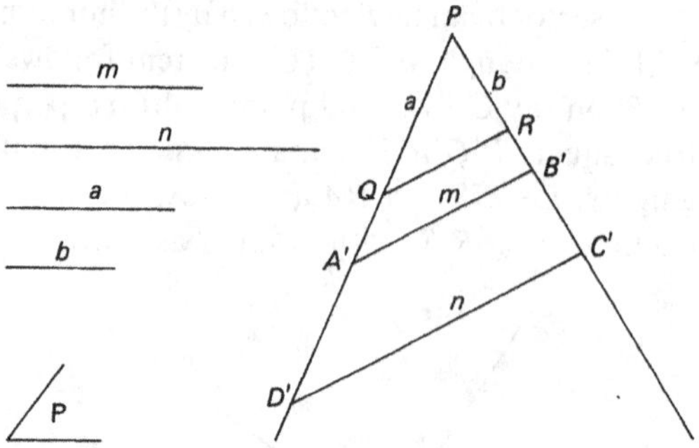

Answer 29.11

29.13 Draw a right triangle having the proper ratio of legs, then use a homothety.

29.15 Since O is the orthocenter of the medial triangle $A'B'C'$ of a triangle ABC under the homothety $H(G, -\frac{1}{2})$, then the ninepoint center N' of the medial triangle lies three-fourths of the way from H to O. By Exercise 29.14, $H(H, \frac{2}{3})$ maps triangle $A'B'C'$ and N' into $G_aG_bG_c$ and its ninepoint center N'', so N'' lies two-thirds of the way from H to N'; that is, N'' lies $(\frac{2}{3})(\frac{3}{4}) = \frac{1}{2}$ way from H to O. Hence $N'' = N$.

29.17 By Theorem 7.20, the orthocentric quadrangle formed by the circumcenters has the same ninepoint center N as, and is congruent to, the given orthocentric quadrangle $ABCH$. Hence the ratio is 1/3 and the center is N for the homothety.

29.19 A similarity consisting of a homothety and a rotation centered at the given point maps the given line to the locus of the third vertex.

29.21 Let the given triangle be ABC and the given point P. Construct any two triangles PQR and PST similar to triangle ABC, with Q and S lying on the first line. Then RT cuts the other line in the desired vertex.

29.23 The homothety $H(A, 2)$ readily establishes the result.

29.25 Referring to Fig. 6.20, rotate triangle ABD about A through $90° - \measuredangle C = \measuredangle DAC$, and apply the homothety $H(A, AC/AD)$. Then AD maps to AC and AB maps to AP, since $m(\measuredangle BAD) = m(\measuredangle PAC)$ and $m(\measuredangle ADB) = m(\measuredangle ACP) = 90°$. The theorem follows from the similar triangles ADB and ACP.

30.1 Since $x' = kx$ and $y' = ky$, then, if $P(x, y)$, we have $OP' = (x'^2 + y'^2)^{1/2} = (k^2x^2 + k^2y^2)^{1/2} = k(x^2 + y^2)^{1/2} = k \cdot OP$.

30.3 Let P be any point, let α be the translation through vector $(a/k - a, b/k - b)$, let $\alpha(P) = Q$, and let $H(O, k)$ map P and Q to P' and Q'. Then the vector $\overrightarrow{P'Q'} = k(a/k - a, b/k - b) = (a - ak, b - bk) = (a(1 - k), b(1 - k))$, the vector of Theorem 30.3 The theorem follows.

30.5 Let $pk = a$ and $qk = b$. Then $x' = k(px - qy + c/k)$, and $y = \pm k(px + qy + d/k)$, an isometry followed by a homothety centered at the origin, into which any similarity can be factored.

30.7 Arccos $\frac{3}{5}$ in the first and fourth quadrants, respectively; that is, 53°8′ and 306°52′.

30.9 See Answer 22.15.

30.10 a) ι

c) We have $x' = (-17x - y + 7)/10$ and $y' = (x - 17y + 19)/10$.

30.11 a) The reflection in the x-axis: $x' = x$ and $y' = -y$.

c) We have $x' = (13x - 11y - 23)/10$ and $y' = (-11x - 13y + 31)/10$.

30.12 a) We have $x' = 2y + 3$ and $y' = 2x$.

c) We have $x' = 3x$ and $y' = 3y$.

e) We have $x' = -x + \frac{3}{2}y - 5$ and $y' = -\frac{3}{2}x - y + 5$.

31.1 a) Multiply the two given equations side for side.

31.3 64 days

31.5 e) The interpretation is $i^2 = -1$.

32.1 a) If $-1 > 0$, then by (3), $1 = (-1)(-1) > 0$. Now (1) is violated since both $-1 > 0$ and $+1 > 0$.

c) If $i > 0$, then by (3), $-1 = i^2 > 0$. But this violates part (a).

e) Since $0^2 \neq -1$, we cannot have $i = 0$.

g) By part (f)

32.2 a) $5 + i$

c) $-5 - i$

e) $12 + 5i$

g) $\frac{1}{2} + 3i/2$

i) -64

k) Yes; the real numbers are a subset of the complex numbers.

32.3 Remember that, when a, b, c, d are real numbers and $a + bi = c + di$, then $a = c$ and $b = d$.

a) $x = \frac{5}{2}$ and $y = -3$

c) $x = 5$ and $y = 10$

e) $x = y = \pm 1/\sqrt{2}$

g) $x = -y = \pm 1/\sqrt{2}$

i) $x = 3s$ and $y = 4s$, where $s = \pm 1$

32.4 a) By definition

c) $i^4 = (i^2)^2 = (-1)^2 = +1$

e) $1/i = i^3/i^4 = i^3 = -i$

32.5 a) i
c) $-i$
e) -1
g) -1
i) 1
k) -1
m) 1
o) $-i$

32.7 Two units east and three units south.

32.9 Yes, but not a real number. It is a complex number.

32.10 a) $\pm\sqrt{5}i$. It cannot be done in the real number system.
c) $(x + \sqrt{2}i)(x - \sqrt{2}i)$. It cannot be done with real factors.
e) This procedure works with 26/65, 19/95, 16/64, and 49/98 only.
g) Two skew lines. It cannot be done in a plane.
i) Three mutually perpendicular lines. They cannot be found in a plane.

33.1 Letting the fourth vertex of the parallelogram of Fig. 33.7 be denoted by D, then from triangles ABC and ADC, we obtain $\mathbf{v} + \mathbf{w} = \overrightarrow{AC} = \mathbf{w} + \mathbf{v}$.
In Fig. 33.8, using triangles ACD and ABD, obtain $(\mathbf{u} + \mathbf{v}) + \mathbf{w} = \overrightarrow{AD} = \mathbf{u} + (\mathbf{v} + \mathbf{w})$.

33.3 In Fig. 33.7, let $\mathbf{v} = (a, b)$ and $\mathbf{w} = (c, d)$. Translating A to the origin, then we have $A(0, 0)$, $B(a, b)$, and by Definition 33.4, $C(a + c, b + d)$ since $\overrightarrow{BC} = (c, d)$. Now $\mathbf{v} + \mathbf{w} = \overrightarrow{AC} = (a + c, b + d)$ by Definition 33.4.

33.5 If $\mathbf{u} = (a, b)$ and $\mathbf{v} = (c, d)$, then $\mathbf{u} - \mathbf{v} = \mathbf{u} + (-1)\mathbf{v} = (a, b) + (-c, -d) = (a - c, b - d)$, a vector. Hence the closure. Now $\mathbf{u} - \mathbf{v} = (a - c, b - d) \neq (c - a, d - b) = \mathbf{v} - \mathbf{u}$. The associativity is similarly disproved.

33.7 Theorem 19.10 establishes this result.

33.9 Let $ABCD$ be the quadrilateral having midpoints M, N, O, P for its sides AB, BC, CD, DA. Let $\overrightarrow{AB} = 2\mathbf{u}$, $\overrightarrow{BC} = 2\mathbf{v}$, and $\overrightarrow{CD} = 2\mathbf{w}$. Then $\overrightarrow{MN} = (\overrightarrow{AB} + \overrightarrow{BC})/2 = \mathbf{u} + \mathbf{v}$ and $\overrightarrow{PO} = (\overrightarrow{AD} + \overrightarrow{DC})/2 = ((\overrightarrow{AB} + \overrightarrow{BC} + \overrightarrow{CD}) - \overrightarrow{CD})/2 = \mathbf{u} + \mathbf{v}$. Since $\overrightarrow{MN} = \overrightarrow{PO}$, then $MNOP$ is a parallelogram. See the accompanying figure.

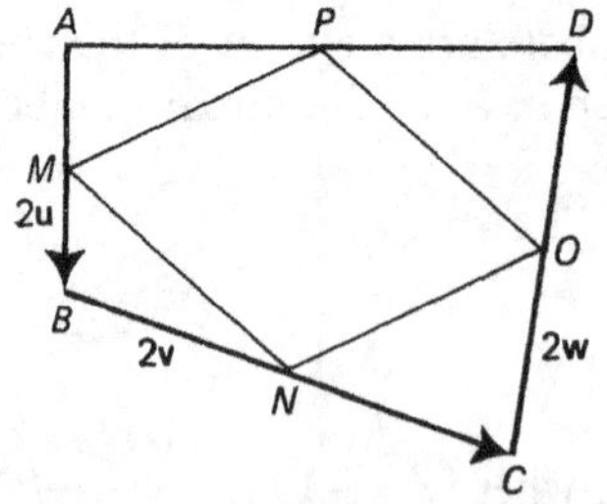

Answer 33.9

33.11 $\overrightarrow{OM} = \overrightarrow{OA} + \overrightarrow{AM} = \overrightarrow{OA} + \overrightarrow{AB}/2 = \overrightarrow{OA} + (\overrightarrow{OB} - \overrightarrow{OA})/2 = (\overrightarrow{OA} + \overrightarrow{OB})/2$.

33.13 By Exercise 33.11, $\overrightarrow{OC'} = (\overrightarrow{OA} + \overrightarrow{OB})/2$, etc.

33.15 Using the usual notation, let $\vec{AB} = \mathbf{u}$ and $\vec{AC} = \mathbf{v}$. Then $\vec{AA'} = (\mathbf{u} + \mathbf{v})/2$, $\vec{BB'} = -\mathbf{u} + \mathbf{v}/2$, and $\vec{CC'} = -\mathbf{v} + \mathbf{u}/2$. Now $\vec{AA'} + \vec{BB'} + \vec{CC'} = \mathbf{0}$, so these three vectors do indeed form a triangle.

33.17 Let $ABCD$ be the trapezoid with $\vec{AB} = \mathbf{u}$, $\vec{DC} = k\mathbf{u}$, and $\vec{BC} = \mathbf{v}$. Let M and N be the midpoints of the nonparallel sides $\vec{DA}$ and $\vec{BC}$. Then $\vec{MN} = \vec{MA} + \vec{AB} + \vec{BN} = \vec{DA}/2 + \vec{AB} + \vec{BC}/2 = (\vec{DC} + \vec{CB} + \vec{BA})/2 + \vec{AB} + \vec{BC}/2 = k\mathbf{u}/2 - \mathbf{v}/2 - \mathbf{u}/2 + \mathbf{u} + \mathbf{v}/2 = (\mathbf{u} + k\mathbf{u})/2$, establishing the theorem.

34.1 Since $|\mathbf{v}| = (a^2 + b^2)^{1/2}$, then $|\mathbf{v}| \geqslant 0$. Hence (1). For (2), clearly $|\mathbf{0}| = 0$. If $0 = |\mathbf{v}| = (a^2 + b^2)^{1/2}$, then $a^2 + b^2 = 0$, so $a = b = 0$, and $\mathbf{v} = \mathbf{0}$. For (3), $|c\mathbf{v}| = |c(a, b)| = |(ca, cb)| = ((ca)^2 + (cb)^2)^{1/2} = |c|(a^2 + b^2)^{1/2} = |c||\mathbf{v}|$. For (4), from $0 \leqslant (bc + ad)^2$ obtain $2abcd \leqslant b^2c^2 + a^2d^2$, so

$$a^2c^2 + 2abcd + b^2d^2 \leqslant a^2c^2 + b^2c^2 + a^2d^2 + b^2d^2,$$

$$ac + bd \leqslant (a^2 + b^2)^{1/2}(c^2 + d^2)^{1/2},$$

$$(a + c)^2 + (b + d)^2 \leqslant a^2 + b^2 + 2(a^2 + b^2)^{1/2}(c^2 + d^2)^{1/2} + c^2 + d^2,$$

$$((a + c)^2 + (b + d)^2)^{1/2} \leqslant (a^2 + b^2)^{1/2} + (c^2 + d^2)^{1/2},$$

and finally, $|\mathbf{u} + \mathbf{v}| \leqslant |\mathbf{u}| + |\mathbf{v}|$. Geometrically, this part is trivial; it simply states that one side of a triangle is always less than or equal to the sum of the other two sides. Hence it is often called the *triangle inequality*. For (5), replace $\mathbf{u}$ in (4) by $\mathbf{u} - \mathbf{v}$, obtaining

$$|\mathbf{u}| = |\mathbf{u} - \mathbf{v} + \mathbf{v}| \leqslant |\mathbf{u} - \mathbf{v}| + |\mathbf{v}|,$$

so $|\mathbf{u}| - |\mathbf{v}| \leqslant |\mathbf{u} - \mathbf{v}|$.

34.3 $|-\mathbf{v}| = |-(a, b)| = |(-a, -b)| = ((-a)^2 + (-b)^2)^{1/2} = (a^2 + b^2)^{1/2} = |\mathbf{v}|$.

34.5 For the distributivity, let $\mathbf{u} = (a, b)$, $\mathbf{v} = (c, d)$, and $\mathbf{w} = (e, f)$. Then

$$\begin{aligned}\mathbf{u}(\mathbf{v} + \mathbf{w}) &= (a, b)(c + e, d + f)\\ &= (ac + ae - bd - bf, ad + af + bc + be)\\ &= (ac - bd, ad + bc) + (ae - bf, af + be)\\ &= (a, b)(c, d) + (a, b)(e, f) = \mathbf{uv} + \mathbf{uw}.\end{aligned}$$

34.7 If $\mathbf{u} = (a, b)$ and $\mathbf{v} = (c, d)$, then $|\mathbf{uv}| = |(ac - bd, ad + bc)| = ((ac - bd)^2 + (ad + bc)^2)^{1/2} = ((a^2 + b^2)(c^2 + d^2))^{1/2} = |\mathbf{u}||\mathbf{v}|$.

34.9 If $\mathbf{v} = (a, b)$, then $|\mathbf{v}\bar{\mathbf{v}}| = |(a, b)(a, -b)| = |(a^2 + b^2, 0)| = a^2 + b^2 = |\mathbf{v}|^2$.

34.11 Using Fig. 34.24, let Cevians AD, BE, CF for triangle ABC satisfy $(\mathbf{BD}/\mathbf{DC})(\mathbf{CE}/\mathbf{EA})(\mathbf{AF}/\mathbf{FB}) = +1$, and let $\mathbf{BD}/\mathbf{DC} = m/n$ and $\mathbf{CE}/\mathbf{EA} = n/p$, so that $\mathbf{AF}/\mathbf{FB} = p/m$. Let $\mathbf{u} = \vec{AB}$ and $\mathbf{v} = \vec{AC}$. If BE and CF meet at O, then there are scalars s and t such that $\vec{BO} = s(p(\mathbf{v} - \mathbf{u}) + n(-\mathbf{u}))$, and $\vec{CO} = t(p(\mathbf{u} - \mathbf{v}) + m(-\mathbf{v}))$. Equate $\vec{AB} + \vec{BO} = \mathbf{u} + \vec{BO}$ and $\vec{AC} + \vec{CO} = \mathbf{v} + \vec{CO}$, and set the coefficients of $\mathbf{u}$ equal to each other. Do the same for the coefficients of $\mathbf{v}$, and solve for s and t, obtaining $s = m/(np + pm + mn)$ and $t = n/(np + pm + mn)$. Now $\vec{AB} + \vec{BO} = (np\mathbf{u} + mp\mathbf{v})/(np + pm + mn)$, a multiple of $n\mathbf{u} + m\mathbf{v}$. Since $\vec{AD} = (n\mathbf{u} + m\mathbf{v})/(m + n)$, also a multiple of $n\mathbf{u} + m\mathbf{v}$, then $\vec{AD}$ does pass through O.

34.13 Let M be the midpoint of the hypotenuse AB of the right triangle ABC with $\mathbf{u} = \overrightarrow{CA}$ and $\mathbf{v} = \overrightarrow{CB}$. Then $\overrightarrow{CM} = (\mathbf{u} + \mathbf{v})/2$ by Exercise 33.11, so

$$\begin{aligned}|\overrightarrow{CM}|^2 &= |\mathbf{u} + \mathbf{v}|^2/4 = |\mathbf{u} + \mathbf{v}||\bar{\mathbf{u}} + \bar{\mathbf{v}}|/4 = |(\mathbf{u} + \mathbf{v})(\bar{\mathbf{u}} + \bar{\mathbf{v}})|/4 \\ &= |\mathbf{u}\bar{\mathbf{u}} + \mathbf{u}\bar{\mathbf{v}} + \bar{\mathbf{u}}\mathbf{v} + \mathbf{v}\bar{\mathbf{v}}|/4 = |\mathbf{u}\bar{\mathbf{u}} + \mathbf{v}\bar{\mathbf{v}}|/4 = (|\mathbf{u}\bar{\mathbf{u}}| + |\mathbf{v}\bar{\mathbf{v}}|)/4 \\ &= (|\mathbf{u}|^2 + |\mathbf{v}|^2)/4 = |\overrightarrow{AB}|^2/4\end{aligned}$$

by Theorems 34.21 and 34.22, Example 34.17, and the Pythagorean theorem, so $|\overrightarrow{CM}| = |\overrightarrow{AB}|/2$.

34.15 Either apply Ceva's theorem (Example 34.24), or the method used in that example to establish this theorem.

34.17 Let $\overrightarrow{AB} = \overrightarrow{CD} = \mathbf{u}$ and $\overrightarrow{BC} = \overrightarrow{AD} = \mathbf{v}$. Then

$$\begin{aligned}AC^2 + BD^2 &= |\mathbf{u} + \mathbf{v}|^2 + |\mathbf{v} - \mathbf{u}|^2 = |(\mathbf{u} + \mathbf{v})(\bar{\mathbf{u}} + \bar{\mathbf{v}})| + |(\mathbf{v} - \mathbf{u})(\bar{\mathbf{v}} - \bar{\mathbf{u}})| \\ &= |(\mathbf{u} + \mathbf{v})(\bar{\mathbf{u}} + \bar{\mathbf{v}}) + (\mathbf{v} - \mathbf{u})(\bar{\mathbf{v}} - \bar{\mathbf{u}})| = |\mathbf{u}\bar{\mathbf{u}} + \mathbf{v}\bar{\mathbf{v}} + \mathbf{v}\bar{\mathbf{v}} + \mathbf{u}\bar{\mathbf{u}}| \\ &= |\mathbf{u}\bar{\mathbf{u}}| + |\mathbf{v}\bar{\mathbf{v}}| + |\mathbf{v}\bar{\mathbf{v}}| + |\mathbf{u}\bar{\mathbf{u}}| = AB^2 + BC^2 + DA^2 + CD^2\end{aligned}$$

by Theorems 34.20 and 34.22.

34.19 Let $\mathbf{u}$ and $\mathbf{v}$ be vectors representing adjacent sides of the rhombus. Then $|\mathbf{u}| = |\mathbf{v}|$ and the diagonals are $\mathbf{u} + \mathbf{v}$ and $\mathbf{u} - \mathbf{v}$. From Example 34.17, $(\mathbf{u} + \mathbf{v})(\bar{\mathbf{u}} - \bar{\mathbf{v}}) + (\bar{\mathbf{u}} + \bar{\mathbf{v}})(\mathbf{u} - \mathbf{v}) = 2\mathbf{u}\bar{\mathbf{u}} - 2\mathbf{v}\bar{\mathbf{v}} = 0$.

34.21 Let $\overrightarrow{AB} = \mathbf{u}$ and $\overrightarrow{AC} = \mathbf{v}$ in triangle ABC. Then medians

$$\overrightarrow{BB'} = (-\mathbf{u} + (\mathbf{v} - \mathbf{u}))/2 = (\mathbf{v} - 2\mathbf{u})/2 \qquad \text{and} \qquad \overrightarrow{CC'} = (\mathbf{u} - 2\mathbf{v})/2.$$

Assume $m(BB') = m(CC')$, so that $|\mathbf{v} - 2\mathbf{u}|^2 = |\mathbf{u} - 2\mathbf{v}|^2$. Then

$$|(\mathbf{v} - 2\mathbf{u})(\bar{\mathbf{v}} - 2\bar{\mathbf{u}})| = |(\mathbf{u} - 2\mathbf{v})(\bar{\mathbf{u}} - 2\bar{\mathbf{v}})|,$$
$$(\mathbf{v} - 2\mathbf{u})(\bar{\mathbf{v}} - 2\bar{\mathbf{u}}) = (\mathbf{u} - 2\mathbf{v})(\bar{\mathbf{u}} - 2\bar{\mathbf{v}}),$$

from which we obtain $3\mathbf{u}\bar{\mathbf{u}} = 3\mathbf{v}\bar{\mathbf{v}}$; that is, $|\mathbf{u}| = |\mathbf{v}|$.

35.1 Either product gives $(-7, 22) = -7 + 22i$.

35.5 Use 35.16 and Exercise 35.4.

35.7 By definition, $(\text{cis}\,\theta)^1 = \text{cis}\,1\theta$. Now suppose $(\text{cis}\,\theta)^n = \text{cis}\,n\theta$. Then $(\text{cis}\,\theta)^{n+1} = (\text{cis}\,\theta)^n \text{cis}\,\theta = \text{cis}\,n\theta\,\text{cis}\,\theta = \text{cis}\,(n\theta + \theta) = \text{cis}\,((n+1)\theta)$ by Exercise 35.6. The theorem follows by mathematical induction.

35.9 Since $\text{cis}\,\alpha\,\text{cis}\,\beta = \text{cis}\,(\alpha + \beta)$ by Exercise 35.6, let $\alpha = \phi$ and $\beta = \theta - \phi$. Then $\alpha + \beta = \theta$, and the theorem follows.

35.11 We have $e^{i\theta}/e^{i\phi} = e^{i(\theta - \phi)}$.

35.13 $(1/2 + i\sqrt{3}/2)^{27} = (\text{cis}\,60°)^{27} = \text{cis}\,1620° = -1$.

35.15 $a + bi = a - bi$ iff $2bi = 0$ iff $b = 0$.

35.17 For $z + \bar{z} = 2a$, $z\bar{z} = a^2 + b^2$, and $(z - \bar{z})/i = 2b$ when $z = a + bi$.

35.19 $2, 2\,\text{cis}\,120°, 2\,\text{cis}\,240°$.

35.21 We have $\text{cis}\,45° = (\sqrt{2}/2)(1 + i)$ and $\text{cis}\,225° = -(\sqrt{2}/2)(1 + i)$.

35.23 $\text{cis}\,45°, \text{cis}\,135°, \text{cis}\,225°, \text{cis}\,315°$; that is, $\pm\sqrt{2}/2 \pm \sqrt{2}i/2$

35.24 a) $|z| = (a^2 + b^2)^{1/2} = (a^2 + (-b)^2)^{1/2} = |\bar{z}|$.

c) $\overline{zw} = \overline{(a + bi)(c + di)} = \overline{ac - bd + (ad + bc)i} = ac - bd - (ad + bc)i = (a - bi)(c - di) - \bar{z}\bar{w}$.

e) $\overline{\overline{(a + bi)}} = \overline{(a - bi)} = a + bi$.

35.25 $|z + w|^2 + |z - w|^2 = (z + w)(\bar{z} + \bar{w}) + (z - w)(\bar{z} - \bar{w}) = 2z\bar{z} + 2w\bar{w} = 2(|z|^2 + |w|^2)$.

35.27 From $(5 - 2i)z + \bar{z} = 6 - 16i$ and $(5 + 2i)\bar{z} + z = 6 + 16i$, eliminate $\bar{z}$ to obtain $z = 2 - 3i$.

35.29 Since $|z|$ is real, then $\text{Im}(|z| + z) = \text{Im}(z) = \text{Im}(1 + 5i) = 5$. Then $|z| = -z + 1 + 5i = -\text{Re}(z) + 1$. Squaring this equation, get $(\text{Re}(z))^2 + 25 = (\text{Re}(z))^2 + (\text{Im}(z))^2 = |z|^2 = (\text{Re}(z))^2 - 2\,\text{Re}(z) + 1$, so $25 = -\text{Re}(z) + 1$, and $\text{Re}(z) = -12$. Hence $z = -12 + 5i$.

36.1 For $\text{Re}(z) = x = ((x + yi) + (x - yi))/2 = (z + \bar{z})/2$. Also $(z - \bar{z})/2i = ((x + yi) - (x - yi))/2i = 2yi/2i = y = \text{Im}(z)$.

36.3 For then $d + z = w$, so $d = w - z$.

36.5 By definition, $e^{i\theta} = \cos\theta + i\sin\theta$, so $z = r\cos\theta + ir\sin\theta$. Then $\cos\theta = (\text{Re}(z))/r$, etc.

36.7 It suffices to prove the theorem for a second-order determinant. Thus

$$\begin{vmatrix} ka & kb \\ c & d \end{vmatrix} = kad - ckb = k(ad - cb) = k \begin{vmatrix} a & b \\ c & d \end{vmatrix}$$

and

$$\begin{vmatrix} ka & b \\ kc & d \end{vmatrix} = kad - kcb = k(ad - cb) = k \begin{vmatrix} a & b \\ c & d \end{vmatrix}.$$

36.9 Then adding an appropriate multiple of one of the rows or columns to the other will produce a row or column of zeros, and Corollary 36.16 applies.

36.10 a) Theorem 36.17 proves half of parts (a) and (b). If the determinant is zero, then $ad - bc = 0$, so $ad = bc$ and $a/c = b/d$. A similar result holds even when $c = 0$ or $d = 0$.

c) The determinant of the matrix whose rows are 1, 2, 3, and 1, 1, 1, and 2, 3, 4 is zero, but no row or column is a multiple of any other row or column.

36.11 By Theorem 36.19, the determinant equation of Corollary 36.20 holds iff triangles ABC and BCA are directly congruent, which implies $m(\measuredangle A) = m(\measuredangle B)$ and $m(\measuredangle B) = m(\measuredangle C)$ and $m(\measuredangle C) = m(\measuredangle A)$, and these equations are true iff triangle ABC is equilateral. Of course, if triangle ABC is equilateral, then triangles ABC and BCA are directly congruent.

36.13 By Corollary 36.21, the given condition determines whether triangles ABC and ACB are oppositely congruent; that is, whether $m(\measuredangle B) = m(\measuredangle C)$. (See Theorem 2.3.)

36.15 Let $b - a = z$ and $c - a = w$. Then $\cos \measuredangle BAC = \cos \measuredangle ZOW$. Furthermore, if the affix of U is 1 and if $v = z/w$, then

$$\cos \measuredangle VOU = \cos \measuredangle ZOW = \frac{\operatorname{Re}(v)}{|v|} = \frac{v + \bar{v}}{2\sqrt{v\bar{v}}} = \frac{z/w + \bar{z}/\bar{w}}{2\sqrt{z\bar{z}/w\bar{w}}} = \frac{z\bar{w} + \bar{z}w}{2\sqrt{z\bar{z}w\bar{w}}}$$
$$= \frac{(b-a)(\bar{c}-\bar{a}) + (\bar{b}-\bar{a})(c-a)}{2\sqrt{(b-a)(\bar{b}-\bar{a})}\sqrt{(c-a)(\bar{c}-\bar{a})}}.$$

36.16 a) We have

$$\sin \measuredangle BAC = \frac{(b-a)(\bar{c}-\bar{a}) - (\bar{b}-\bar{a})(c-a)}{2i\sqrt{(b-a)(\bar{b}-\bar{a})}\sqrt{(c-a)(\bar{c}-\bar{a})}}$$

36.17 If $a - c = (c - b)\operatorname{cis} 60°$, then angle ACB is 120°, etc.

36.19 Since $-(c - b) = (b - a) + (a - c)$, then $1/(c - b) + 1/(b - a) + 1/(a - c) = 0$ iff, clearing of fractions,

$$\begin{aligned} 0 &= (b-a)(a-c) + (c-b)(a-c) + (c-b)(b-a) \\ &= (b-a)(a-c) + (c-b)((a-c) + (b-a)) \\ &= (b-a)(a-c) - (c-b)^2, \end{aligned}$$

and the theorem follows from Exercise 36.18.

36.21 We may assume, without loss of generality, that $a = 0$. But if not, then subtract a times the last column from the first column and $\bar{a}$ times the last column from the second. Thus we assume that

$$0 = \begin{vmatrix} 0 & \bar{b} & 1 \\ b & \bar{c} & 1 \\ c & 0 & 1 \end{vmatrix} = \bar{b}c - c\bar{c} - b\bar{b}.$$

Since $b\bar{b}$ and $c\bar{c}$ are real, then $\bar{b}c$ is real, so there is a real number k, such that $\bar{b}c = kb\bar{b}$ (since $b\bar{b}$ is real), whence $c = kb$. Then $0 = \bar{b}c - c\bar{c} - b\bar{b} = bk\bar{b} - k^2 b\bar{b} - b\bar{b}$, so $k^2 - k + 1 = 0$ if $b \neq 0$. Now k is real, but $k^2 - k + 1$ has no real roots. Thus $b = 0$, so $c = 0$, too. That is, the given determinant condition requires that $a = b = c$, so triangle ABC degenerates to a single point.

36.23 See the comment in Answer 36.7. Then

$$\begin{vmatrix} a & b+e \\ c & d+f \end{vmatrix} = a(d+f) - c(b+e) = (ad - cb) + (af - ce) = \begin{vmatrix} a & b \\ c & d \end{vmatrix} + \begin{vmatrix} a & e \\ c & f \end{vmatrix}.$$

36.25 Denote by a_{ij} the element in the ith row and jth column of the determinant. Denote by A_{ij} the determinant formed from the given determinant A by deleting the ith row and also the jth column. Then

$$A = \sum_{i=1}^{n} (-1)^{i+j} a_{ij} A_{ij} \quad \text{and} \quad A = \sum_{j=1}^{n} (-1)^{i+j} a_{ij} A_{ij},$$

where the determinant is of order n.

36.27 a) By Exercise 36.14, $a*b = zb + (1 - z)a$.

c) For $(0*1)*1 = z*1 = z + (1 - z)z$, and $0*(1*1) = 0*1 = z$.

e) For $c = zb + (1 - z)a$ can be solved for a or b, so long as $z \neq 0$ and $z \neq 1$; that is, when 0, 1, z form a nondegenerate triangle.
g) $A*(B*C) = (A*A)*(B*C) = (A*B)*(A*C)$ by parts (d) and (f).

36.29 $|z - w|^2 = (z - w)(\bar{z} - \bar{w}) = z\bar{z} - z\bar{w} - w\bar{z} + w\bar{w} = z\bar{z} - \overline{(z\bar{w})} - (z\bar{w}) + w\bar{w} = |z|^2 - 2\,\mathrm{Re}(z\bar{w}) + |w|^2$.

37.1 By Theorem 36.21, triangle ABC is oppositely congruent to itself. The only way triangle ABC can be both clockwise and counterclockwise oriented is for that triangle to degenerate into three collinear points.

37.3 This corollary follows immediately from the second proof of Theorem 37.2.

37.5 Since A_2 lies on line OA_1 iff a_1/a_2 is real, the theorem follows from Theorem 37.4.

37.7 Clearly, line OB has the parametric form $z = tb$, where t is real, since then $\mathbf{OZ} = t \cdot \mathbf{OB}$. Then $z = a + tb$ is obtained from $z = tb$ by the translation $z' = z + a$.

37.9 Lines OB and OD are perpendicular iff $b = rid$ for some real number r; that is, iff b/d is pure imaginary. Then apply Answer 37.7.

37.11 In the proof of Theorem 37.10, if triangle ABC is clockwise oriented, then the equation "$\beta = (|\beta|/|\alpha|)e^{iC}\alpha$" becomes $\beta = (|\beta|/|\alpha|)e^{-iC}\alpha$, in which C is the nondirected measure of angle BCA. Then the sign of $\sin C$ is changed, and the theorem follows.

37.13 $z = (\bar{a}_1 b_2 - \bar{a}_2 b_1)/(a_1 \bar{a}_2 - \bar{a}_1 a_2)$.

37.15 Apply Exercise 37.14.

37.17 a) 1 c) 3

37.18 a) From Theorem 33.18, $g = a + ((c - a) + (b - a))/3 = (a + b + c)/3$.
c) By Corollary 36.20, since triangle ABC is equilateral, we have the determinant equation shown at the left below. The corresponding determinant for triangle $A'B'C'$ is shown at the right. We have

$$\begin{vmatrix} a & b & 1 \\ b & c & 1 \\ c & a & 1 \end{vmatrix} = 0 \quad \text{and} \quad \begin{vmatrix} zb + (1 - z)a & zc + (1 - z)b & 1 \\ zc + (1 - z)b & za + (1 - z)c & 1 \\ za + (1 - z)c & zb + (1 - z)a & 1 \end{vmatrix}$$

We can show this second determinant to be zero by applying Exercise 36.23 several times.

37.19 We have $(c - a)/(b - c) = r$.

38.1 For $r^2 = |z - c|^2 = |(z - c)(\bar{z} - \bar{c})| = z\bar{z} - c\bar{z} - \bar{c}z + c\bar{c}$.

38.3 Let z_1, z_2, z_3 be the values for t_1, t_2, t_3, Then we have three equations $z_k = (at_k + b)/(ct_k + d)$, which reduce to $t_k a + b - t_k z_k c - z_k d = 0$ for $k = 1, 2, 3$. These three equations in the four unknowns a, b, c, d may be solved to yield values of three of the unknowns in terms of the fourth unknown. The algebra for the general case is somewhat messy.

38.5 Theorem 38.8 takes care of the case when $k \neq 1$. For $k = 1$, we have $|z - a| = |z - b|$; that is, $m(AZ) = m(BZ)$, so Z lies on the perpendicular bisector of AB.

38.7 $(a\bar{d} - b\bar{c})(c\bar{d} - \bar{c}d)$ and $|(ad - bc)/(c\bar{d} - \bar{c}d)|$.

38.9 For the circle with center E and radius r, r real, use Theorem 38.5, letting $z_0 = e + r$, $z_\infty = e - r$, and $z_1 = e + ir$. Then let $d = i$. Now $c = 1$, $b = i(e + r)$, and $a = e - r$.

38.10 a) $z = t$

c) $z = (4 + 2i)t/((1 + i)t + 2 - 2i)$

e) $z = ((2 + 4i)t + 5 - 3i)/(2it + 1 - 4i)$

38.11 When $d \neq 0$, the equation becomes $(ct + d)z = (b/d)(ct + d)$, since $a = bc/d$. If also $c \neq 0$, then all points in the plane satisfy when $t = -d/c$. If $c = 0$, then $z = b/d$ is the only point that satisfies the equation. A similar situation exists when $d = 0$.

38.13 If $a \neq 0$, then $a - b = a(1 - b/a) = a(1 - \bar{a}b/a\bar{a})$. So, when $|a| = 1$, then $|a - b| = |a||1 - \bar{a}b/a\bar{a}| = |1 - \bar{a}b|$ and the theorem follows. The situation is similar when $|b| = 1$.

38.15 $c = (1 + i)a - bi$ and $d = (1 - i)b + ia$, $c = (1 - i)a + bi$ and $d = (1 + i)b - ia$, or $c = (a + b + bi - ai)/2$ and $d = (a + b - bi + ai)/2$.

38.17 The circumcenter Q and the circumradius are given by

$$\frac{a\bar{a}(c - b) + b\bar{b}(a - c) + c\bar{c}(b - a)}{\bar{a}(c - b) + \bar{b}(a - c) + \bar{c}(b - a)} \quad \text{and} \quad \left|\frac{(a - c)(b - a)(c - b)}{\bar{a}(c - b) + \bar{b}(a - c) + \bar{c}(b - a)}\right|.$$

38.19 $z = (c\bar{a} - \bar{c}b)/(a\bar{a} - b\bar{b})$.

39.1 By Definition 33.6.

39.3 By Theorem 39.1 and the discussion in 34.12.

39.5 Because each direct similarity can be factored into a product of one or more of the three forms listed in the proof of Theorem 39.4.

39.7 We have $z' = (z - a)e^{-i\theta} + a$, $z'' = \bar{z}'$, and $z''' = (z'' - a)e^{i\theta} + a$. Hence, since a is real, $z''' = ((\bar{z} - a)e^{i\theta} + a - a)e^{i\theta} + a = (\bar{z} - a)e^{2i\theta} + a$.

39.9 If $z' = a(az + b) + b$, then $ab + b = 0$ and $a^2 = 1$. Since $z' \neq z$, then $a = -1$, whence b is arbitrary. Thus we have $z' = -z + b$, an equation for a halfturn.

39.11 This equation represents a reflection in the real axis followed by a direct similarity, a form into which any opposite similarity can be factored.

39.13 $z' = ((a - b)/(\bar{a} - \bar{b}))\bar{z} + (\bar{a}b - a\bar{b})/(\bar{a} - \bar{b})$.

39.15 If $z' = z + b$ and $z'' = z' + c$, then $z'' = z + (b + c)$, a translation.

39.17 $z' = -\bar{z}$.

39.19 $z' = -z$.

40.1 a) Draw circle $B(A)$ (with center B and passing through point A). Draw circle $A(B)$ to cut circle $B(A)$ at P. Draw circle $P(B)$ to cut circle $B(A)$ at Q, and draw circle $Q(B)$ to cut circle $B(A)$ at C. See the accompanying figure.

c) Draw C', the reflection of C in line AB (by part (b)). Then circles $C(D)$ and $C'(CD)$ meet at the desired points.

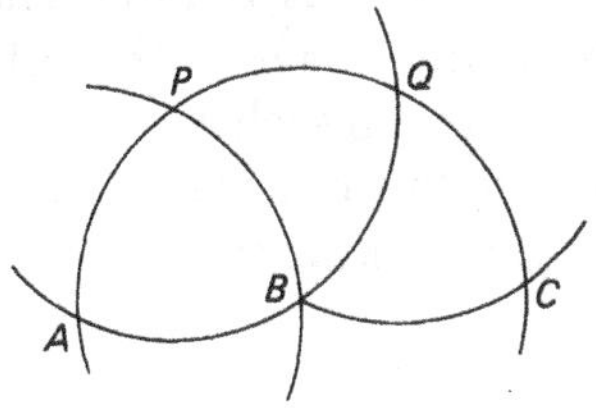

Answer 40.1a

40.3 Choose point C so that sides AC and BC of triangle ABC are cut by the other parallel in points M and L, and let BM and AL meet at O as in the accompanying figure. Then CO bisects AB by Ceva's theorem.

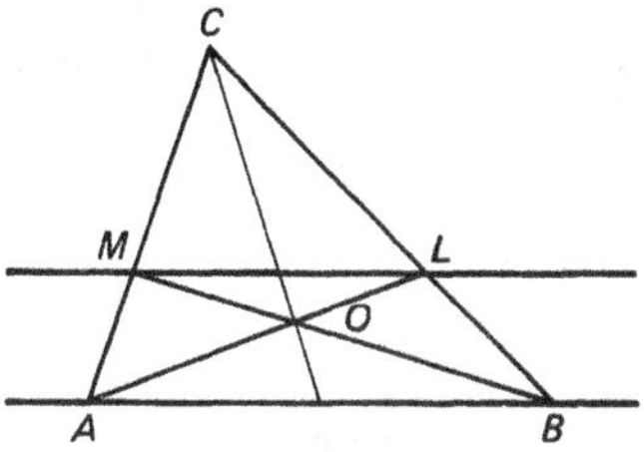

Answer 40.3

40.5 Starting from any one of the six points, there are 5 choices for the second vertex, 4 for the third, 3 for the fourth, 2 for the fifth, and 1 choice left for the last vertex. These $120 = 5!$ possibilities appear in duplicate pairs since $ABCDEF$ and $AFEDCB$, for example, give rise to the same hexagon. The number of hexagons is thus $5!/2 = 60$.

40.6 a) Construct an equilateral triangle.

40.9 Properties of circles and lines, angles between curves (and orthogonality), and cross ratios are some topics studied in inversive geometry.

41.1 a) (6, 8) c) (10, 0)
e) (100, 0) g) (5, 5)
i) (200/53, −700/53) k) (0, 0)

41.2 a) (70/13, 40/13) c) (10, 0)
e) $(-\frac{5}{4}, 0)$ g) (6, 2)
i) (215/58, −175/58) k) (5, 0)

41.3 Let α denote the inversion and let $\alpha(P) = P'$. By Definition 41.4, if $P \neq C$ and $P \neq \infty$, then $CP \cdot CP' = r^2$. Then $CP' \cdot CP = r^2$ and P lies on ray CP'. Hence $\alpha(P') = P$. Also $\alpha(C) = \infty$ and $\alpha(\infty) = C$. Thus α is not the identity, and α will be involutoric when we have proved it to be a transformation. To that end, let P be any point in the inversive plane, and let $\alpha(P) = P'$ using Definition

41.4. Since there is just one point P' on ray CP such that $CP \cdot CP' = r^2$, then P' is unique and α is one-to-one. To show that α is onto, note that $\alpha(P') = P$, so P' is the pre-image of P. Thus α is a transformation of the inversive plane.

41.5 See Theorems and Corollaries 13.17, 13.18, 15.1, and 26.5,

41.7 By Definition 41.4, each point on such a line maps into another point on that line, and by Theorem 41.5, this line maps onto itself.

41.9 See the accompanying figure. The minus signs will change to plus signs.

41.11 See the figure.

41.13 See the figure.

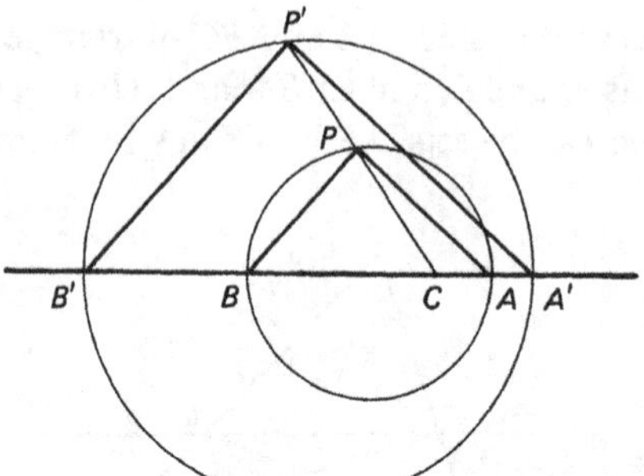

Answer 41.9

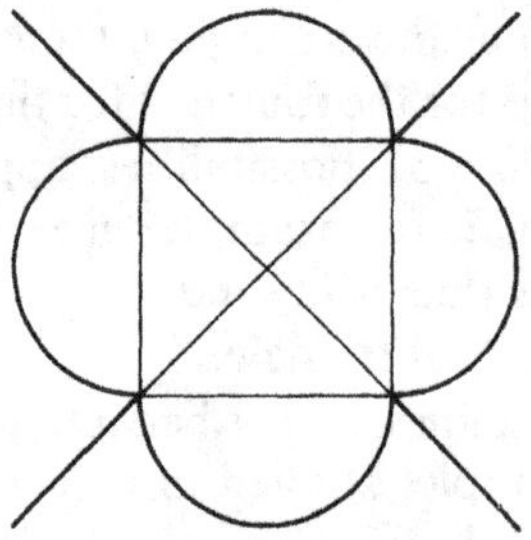

Answer 41.11

41.15 Let circle s cut the circle c of inversion orthogonally at points T and U, and let C be the center of c. Let P be any point other than T and U on circle s and let CP cut s again at P' (Fig. 42.3). Now $CP \cdot CP' = CT^2$ by Exercise 6.18. Hence P and P' are inverse points.

41.17 By Definition 41.4, $CQ' = r^2/CQ$. By Theorem 41.9, triangles CPQ and $CQ'P'$ are similar, so $PQ/Q'P' = CP/CQ'$. Hence we have $Q'P' = PQ \cdot CQ'/CP = (PQ/CP)(r^2/CQ)$. (See Fig. 41.9.)

41.18 a) Let S be the center of the circle s of Fig. 41.14, and let S'' be its image. By Exercise 41.17, $A'S''=(r^2\cdot AS)/(CA\cdot CS)$ and $S''B'=(r^2\cdot SB)/(CS\cdot CB)$. Then $A'S''/S''B'=CB/CA$, since $AS\cong SB$. Since $CB\not\cong CA$, then $A'S''\not\cong S''B'$, so S'' is not the center of circle s'.

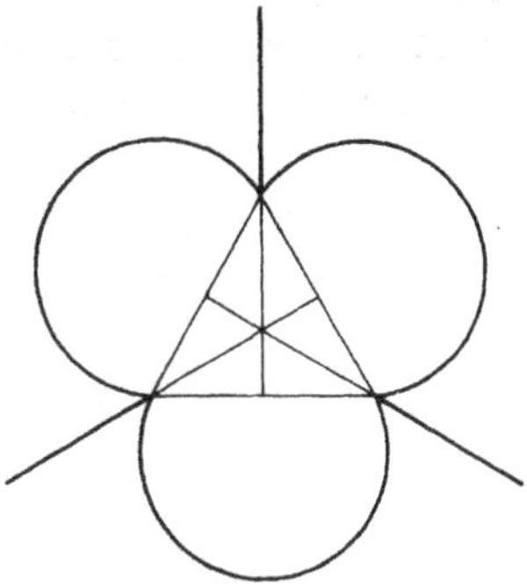

Answer 41.13

42.1 Let circles c and s have centers C and S and let them meet at points P and Q. The circles are orthogonal iff their tangents at P are perpendicular (by definition), iff their radii CP and SP are perpendicular (since each radius is perpendicular to the tangent to that circle), iff either radius coincides with the other tangent.

42.3 Let the two circles meet at P and Q. They are orthogonal iff radii CP and DP are perpendicular. But this is true iff CDP is a right triangle with right angle at P; that is, iff $r^2+s^2=CD^2$.

42.5 For $1/\mathbf{AB}-1/\mathbf{AP}=1/\mathbf{AP}'-1/\mathbf{AB}$ iff $2/\mathbf{AB}=1/\mathbf{AP}+1/\mathbf{AP}'$ iff $2/\mathbf{AB}=(\mathbf{AP}+\mathbf{AP}')/(\mathbf{AP}\cdot\mathbf{AP}')$ iff $\mathbf{AB}=2\cdot\mathbf{AP}\cdot\mathbf{AP}'/(\mathbf{AP}+\mathbf{AP}')$. Also, $CP/CB=CB/CP'$ iff $CP\cdot CP'=CB^2$ iff $CB=(CP\cdot CP')^{1/2}$.

42.7 a) Take one additional link QP such that $m(CA)-m(AP)<2m(PQ)$, and fix point Q so that $m(CQ)=m(QP)$ (see Fig. 42.9). Now P traces a circle (circle $Q(P)$) that passes through C, so P', its inverse, traces a straight line.

42.9 If A and B separate C and D, then one of C and D (say C) is between A and B, and the other (in this case, D) is outside segment AB. Then one ratio ($\mathbf{AC}/\mathbf{CB}$) is positive and the other ratio ($\mathbf{AD}/\mathbf{DB}$) is negative. Now their quotient, which is (AB, CD), is negative. This argument is reversible.

42.11 Use Theorem 42.8 and consider two cases: P inside the circle, and P outside the circle.

42.13 If the points A and B are collinear with the center S of the circle s, then diameter SAB is such a desired line. In this case, no other line passes through A and B, and any circle orthogonal to s and passing through A must also pass through the inverse A' of point A in circle s by Theorem 42.4. But no circle can pass through the three distinct collinear points A, B, and A'. Hence line OAB is the unique solution in this case.

If S, A, and B are not collinear as in the accompanying figure, construct the inverse A' of point A in circle s by Exercise 42.11. Then the circle c through A, A', and B is orthogonal to circle s by Theorem 42.3. This circle may be constructed by taking its center C as the intersection of the perpendicular bisectors of AA' and AB. In this second case, line AB is not a diameter of circle s, and any circle through A and orthogonal to s must also pass through A' by Theorem 42.4. Hence circle c is the unique solution in this case.

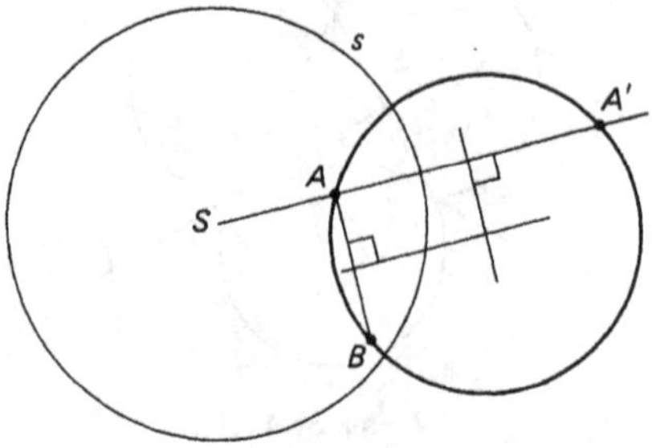

Answer 42.13

42.15 Let the two circles be c_1 and c_2 and let the center of circle s be S. By Theorem 42.4, the inverse P' of P in circle s lies on both c_1 and c_2. Thus $P' = P$ or $P' = Q$. Since the circles cannot be tangent at P (Why?), then P does not lie on circle s. Hence $P' \neq P$, so $P' = Q$.

43.1 Let the inversions be $z' = c + r^2/(\bar{z} - \bar{c})$ and $z'' = c + s^2/(\bar{z}' - \bar{c})$, where r and s are real and nonzero. Then we have

$$z'' = c + s^2/(\bar{c} + r^2/(z - c) - \bar{c}) = c + (s^2/r^2)(z - c),$$

a homothety with center C and ratio s^2/r^2.

43.3 They commute when their circles are orthogonal.

43.5 See the discussion in 38.6. Then use Answer 38.10, replacing z by z' and t by z.

43.7 By Theorem 43.7, the bilinear transformation and its inverse will be equal if $a = -d$.

43.9 The identity map is the bilinear transformation $z' = (1z + 0)/(0z + 1)$. The inverse of a bilinear transformation is the bilinear transformation given by Theorem 43.7. Also, the product of the two bilinear transformations $z' = (az + b)/(cz + d)$ and $z'' = (ez' + f)/(gz' + h)$ is given by

$$z'' = \left(e\,\frac{az+b}{cz+d} + f\right)\Bigg/\left(g\,\frac{az+b}{cz+d} + h\right) = \frac{(ae + cf)z + be + df}{(ag + ch)z + gb + dh},$$

a bilinear transformation. The theorem follows by Exercise 12.12.

43.10 a) $z' = (1 - i)z/(z - i)$. For the other parts of this answer, follow this map of part (a) by each of the maps of parts (c) and (e) in Answer 43.5 to obtain the corresponding answers here.

43.11 a) -1, $(1 + 3i)/2$, ∞, 2
c) ∞, 1, $-i$, 0
43.13 a) No

44.1 The circle through the centers is not orthogonal to any of the three circles, and the center of a circle does not invert into the center of the inverse circle by Exercise 41.18.

44.2 a) Let the two circles c and d meet at P and Q, let the center of the circle s orthogonal to c and d be O, let circle s cut circle c at S and S' and d at T and T', and let OP cut c again at Q' and d again at Q''. Now, see the accompanying figure, $m(OS) = m(OT)$ and $OP \cdot OQ' = OS^2 = OT^2 = OP \cdot OQ''$ by Exercise 6.18. Hence $OQ' = OQ''$, so $Q' = Q''$, and this common point is Q. Hence O, P, and Q are collinear.

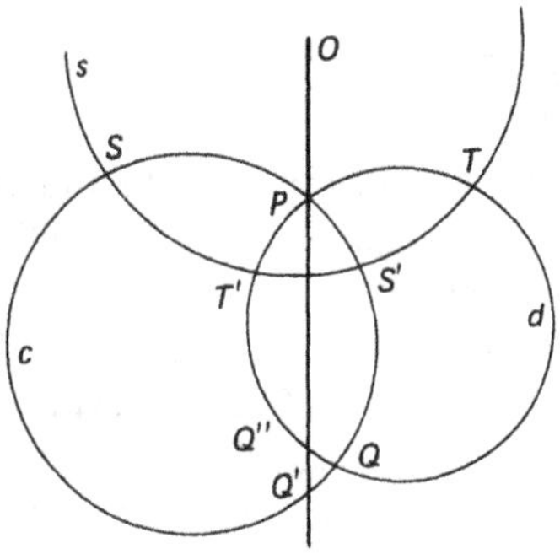

Answer 44.2a

c) The proof of part (a) readily accepts this condition.
g) Since each such point P lies on the radical axis of circles s and u, and circles t and u, then the tangents from P to s and u are congruent, and those to t and u are congruent. Hence those to s and t are congruent. Thus P lies on the radical axis of s and t.
i) See the answer to part (a).
k) See the answer to part (a).

44.3 Find the circle s orthogonal to the three given circles (by Exercise 44.2) and invert in circle s.

44.5 Draw any two circles c and d through both A and B. By Theorem 42.3, they each are orthogonal to s. Now s, c, and d invert into three circles (or lines) s', c', and d', with each of c' and d' orthogonal to circle s'. Hence their intersections A' and B' are inverse in circle s' by Exercise 42.15.

44.7 Let the circle of inversion have center C and radius r, and let the given circle have center T and radius t. Let A and B be the ends of that diameter of circle $T(t)$ that is collinear with C, and let A' and B' be their images. See the accompanying figure. By Exercise 41.17, $A'B' = (r^2 \cdot AB)/(CA \cdot CB) = (r^2 \cdot 2t)/((CT - t)(CT + t) = (r^2 \cdot 2t)/(CT^2 - t^2)$. Since $A'B'$ is a diameter of the image circle, then the radius of the image circle is $r^2 t/(CT^2 - t^2)$.

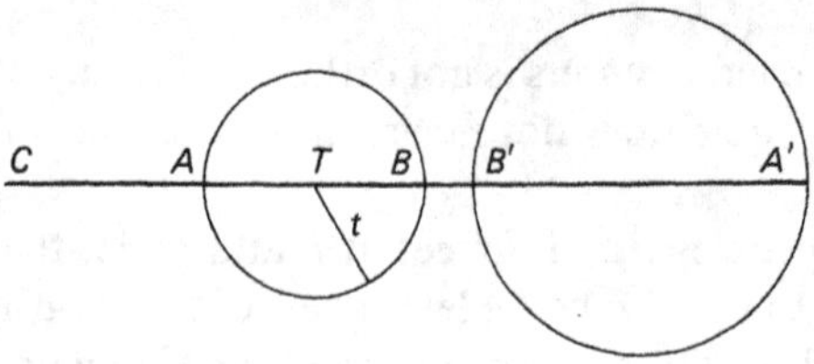

Answer 44.7

44.9 Since $z' = k/\bar{z} = kz/z\bar{z} = kz/(x^2 + y^2)$, then $x' = \text{Re}(z') = k \cdot \text{Re}(z)/(x^2 + y^2) = kx/(x^2 + y^2)$, and $y' = \text{Im}(z') = ky/(x^2 + y^2)$.

44.10 a) $x' = x'^2 + y'^2$
c) $2x' + 1 = 0$
e) $b^2 x'^2 + a^2 y'^2 = a^2 b^2 (x'^2 + y'^2)^2$; no.
g) $x'^2 y'^2 = (x'^2 + y'^2)^2$

44.11 By Exercises 44.5 and 44.6, the inverses of t and u are inverse in s', the inverse of s. But s' is a straight line, so t' and u' are reflections of one another in s', and hence they are congruent.

44.13 a) Let P be any point on a semicircle whose diameter is AOB. Then the circles on A, O, P and on B, O, P are orthogonal.
c) An angle inscribed in a semicircle is a right angle.

44.15 Invert in the circle centered at O and orthogonal to circle s_n. Then line OAB and circle s_n are self-inverse. Circles OA and OB map into lines perpendicular to line OAB, hence parallel to each other, and also tangent to circle s_n. Circles $s_0, s_1, \ldots, s_{n-1}$ map into circles nested between t' and u' as shown in the accompanying figure, hence congruent to one another and to circle s_n. The theorem follows readily from this figure.

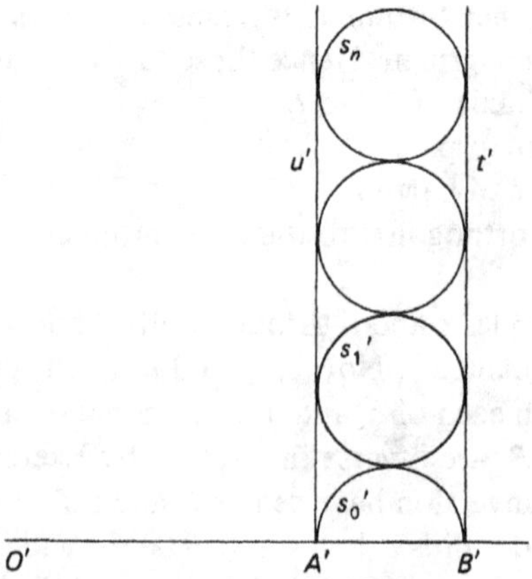

Answer 44.15

44.17 Draw the circle on AB as diameter. Under inversion in center A, this circle maps into a diameter of the inverse s' of circle s. Also diametral line AB of circle s is invariant and maps to a diameter of s'. Hence the inverse of B, lying on these two diameters of s', is the center of s'.

45.1 No

45.3 $a^2 + b^2 = 4p^2q^2 + (p^4 - 2p^2q^2 + q^4) = p^4 + 2p^2q^2 + q^4 = c^2$.

46.1 See Theorem 14.4 and Exercise 14.2.

46.3 See Corollary 15.9.

46.5 a) A reflection in a plane.
c) A central inversion or a glide-reflection or a rotatory reflection.

46.6 a) 1 c) 2 e) 2
g) 2 i) 3

46.7 Isometries (a) and (i) are opposite; (c), (e), and (g) are direct.

46.9 See Theorem 49.3.

46.10 See Theorems 49.10 and 49.13.
a) The plane of the reflection and all lines and points in that plane; all lines and planes perpendicular to the mirror plane.
c) No points, but all lines and planes parallel to the vector of the translation.
e) The axis of rotation and all points on the axis; all planes perpendicular to the axis.
g) In addition to those of part (e), all planes containing the axis; and all lines perpendicular to the axis.
i) The plane of reflection, and in it any line that is parallel to the vector of translation; and any plane that is both parallel to the vector of translation and perpendicular to the mirror.

46.11 a) See Theorem 13.3.
c) This is a corollary to part (a).
e) See Theorem 15.1.

46.13 See Theorem 15.12.

47.3 See Theorem 13.12.

47.5 See Theorem 13.16.

47.7 By Theorem 47.8 and the fact that a reflection is not the identity map.

47.9 Infinitely many. (Consider how many isometries of the plane map a given point A to another point A'.)

47.10 a) Consider, for example, changing only D to $D'(0, 0, -1)$.
c) Apply parts (a) and (b) (perhaps).

47.11 Let the plane through A perpendicular to DD' cut DD' at point P. Then triangles ADP and $AD'P$ are congruent by HL. Since now $m(DP) = m(D'P)$, then this plane is the desired perpendicular bisector of DD'. Similarly this plane passes through B and C. But A, B, C determine a plane. Hence they determine the perpendicular bisector plane.

48.1 For any real constant k, $x + y + z = k$ and $x + y + z = k + \frac{3}{2}$.
48.3 See Theorem 14.5.
48.5 See Theorem 14.12.
48.7 See Theorem 14.15.
48.9 See Theorems 14.9 and 14.10.
48.11 See Theorem 14.14.
48.13 See Theorem 15.8.
48.15 By Theorem 48.15 and Corollary 48.16.
48.17 See Theorem 15.8.

49.1 See Theorem 16.1.
49.3 a) See Theorem 14.14.
c) Let Π be the plane containing the two axes. Factor the rotations into $\sigma_\Pi \sigma_\Delta$ and $\sigma_\Gamma \sigma_\Delta$ for appropriate planes Δ and Γ. Their product is $\sigma_\Gamma \sigma_\Delta$.
49.5 By Theorem 47.9, each direct isometry is a product of two or four reflections in planes. In the former case it is a rotation or a translation (or the identity map). The latter case yields a screw displacement by an argument similar to that for part (d) of Theorem 49.3.
49.7 A rotation about the same axis through twice the angle.
49.9 By Theorem 47.9 it is either a reflection or a product of three reflections. But a product of three reflections in planes forming a pencil is a reflection, and a reflection has fixed points. Also, if the three planes concur in a point, then that point is fixed.
49.11 See Theorem 16.4.
49.12 a) If the mirrors form a pencil, then they represent a single reflection. So suppose just two mirrors meet along a line m. Without loss of generality we may assume that planes Δ and Γ meet in m where the isometry is $\alpha = \sigma_\Pi \sigma_\Gamma \sigma_\Delta$. Now a point P maps into a point P' by the map $\sigma_\Gamma \sigma_\Delta$ iff the perpendicular bisector of PP' passes through line m. But then P is a fixed point in α iff Π is that perpendicular bisector, and the three planes form a pencil. On the other hand, if P is fixed under $\sigma_\Gamma \sigma_\Delta$, then P lies on m, so Π must contain P, and the three planes meet at least in point P.
49.13 A rotation has its axis as a set of fixed points, so this isometry must be either of the two remaining direct isometries: a translation or a screw displacement.
49.15 Let the rotatory reflection be $\alpha = \sigma_\Pi \sigma_\Gamma \sigma_\Delta$ with Δ perpendicular to Γ and to Π. Then $\alpha = \sigma_\Pi \sigma_\Delta \sigma_\Gamma = \sigma_\Delta \sigma_\Pi \sigma_\Gamma$, and the theorem follows. (The theorem is obvious geometrically.)
49.17 For a reflection σ_Γ, let m be any line perpendicular to plane Γ and let Δ be any plane containing line m. Now $\sigma_\Gamma = \sigma_\Gamma \sigma_\Delta \sigma_\Delta$, a rotation of zero degrees about line m followed by a reflection in plane Γ, hence a rotatory reflection. Similarly a central inversion is a rotation (a halfturn) about the line of intersection of its first two mirrors, followed by a reflection in its third mirror, hence it is a rotatory reflection.

50.1 See Theorem 18.2.

50.3 Apply a translation carrying one face to the other.

50.5 In Fig. 50.4, $ABCD$ is a rectangle. Let E be its center. Then a reflection in the plane through E and parallel to the base plane (through A and B) of the parallelepiped carries the rectangle into itself, and specifically, diagonal AC to diagonal BD.

50.7 By Theorem 50.6.

50.9 Let triangle ABC be the isosceles triangle with apex A. The triangle and the plane containing its base are invariant in a reflection in the perpendicular bisector plane of the base BC.

50.11 In the triangle mentioned in Hint 50.11, the traces of the bisector planes are the angle bisectors, which concur at a point I. Now the line through I and the vertex of the trihedral angle is the line of concurrence of the bisector planes.

50.13 See Theorem 19.6.

50.15 This is a corollary to Corollary 50.14.

50.17 Let the lines intersect at half the angle of the rotation in a plane perpendicular to the axis of rotation. See Theorem 14.9.

50.19 Just as a circircular disc is generated by rotating a segment about one of its endpoints.

50.21 a) A reflection (in the xy-plane).
c) A translation (of 1 unit in the z-direction).
e) A glide-reflection (of 1 unit in the z-direction and with mirror the yz-plane).

51.1 See Theorem 21.3.

51.3 See Theorem 22.2.

51.5 Factor the central inversion into a product of reflections in coordinate planes.

51.7 We assume that the z-axis lies in each of the two distinct planes Δ and Π, whose equations are $Ax + By = 0$ and $Ex + Fy = 0$, with $A^2 + B^2 = 1$ and $E^2 + F^2 = 1$. Now equations for σ_Δ are $x' = x - 2A(Ax + By)$, $y' = y - 2B(Ax + By)$, and $z' = z$, and those for σ_Π are $x' = x - 2E(Ex + Fy)$, $y' = y - 2F(Ex + Fy)$, and $z' = z$. Hence their product $\sigma_\Pi \sigma_\Delta$ is given by $x' = Gx - Hy$, $y' = Hx + Gy$, and $z' = z$, where $G = (1 - 2A^2)(1 - 2E^2) + 4ABEF$ and $H = 2AB(1 - 2E^2) + 2EF(1 - 2B^2)$. Verification that $G^2 + H^2 = 1$ is straightforward but tedious. Hence G and H are $\cos\theta$ and $\sin\theta$ for some angle θ.

51.9 Set $A = B = D = 0$ and $C = 1$.

INDEX

Unless stated otherwise, numbers refer to section numbers in the text.

AAS, 18.6
Abelian group, 12.16
Absolute geometry, 40.16
Absolute value, 34.2, 35.17
Addition, vector, 33.6
Affix, 35.12
Ahmes, 1.3
Alexandria, 9.9, 9.13, 23.1, 23.2
Altitudes, concurrence, 7.11
Amasis, King of Egypt, 9.4
Amplitude, 35.17
Angle, 35.17
  dihedral, 50.1
  rotation, 10.3, 14.8
  trisection, 8.2, Exercises 8.13–8.16
Apex, Exercise 18.5
Apollonius, 9.11, 23.9
  problem, 40.11
Arabs, 9.13, 23.2
Archimedes, 9.10–9.11, 23.1
Area, 19.6–19.8, 37.10, Exercise 37.10
  triangle, 6.15–6.21
Argand, J., 31.11
Argument, 35.17
Aristotle, 9.7
Arithmetic mean, 42.6
Arithmetic progression, 42.6
Arithmetic Teacher, 45.12
ASA, 18.5
Associative, 12.6, 33.8, 34.10, 34.14
Axiomatics
  formal, 40.21
  material, 9.1, 40.21
Axis
  imaginary, 32.7
  radical, Exercise 44.2
  real, 32.6, 32.7
  rotation, 48.8

Babylonians, 1.2, 1.7–1.9
Base, 2.6
Basis, 35.1
Bell, E. T., 1.4
Between, 2.4
Bible, 1.7
Bilinear transformation, 43.6
Black Death, 23.4
Boldface, 2.15
Bolyai, J., 40.17
Bombelli, R., 31.5
Brianchon, C. J., Exercise 23.5
  theorem, Exercise 23.5
Brocard, H., 40.14

Cavalieri, 23.9
Cayley, A., 40.14
  line, 40.9
Center
  central inversion, 48.18
  homothety, 24.1, 25.1

Center—*cont.*
inversion, 41.4
ninepoint, 7.16
radical, Exercise 44.2
rotation, 10.3, 14.8
similitude, 24.2, 25.6
Central inversion, 46.2, 48.18
center, 48.18
equation, 51.8
Centroid, Exercise 5.8, 6.2, Exercise 37.18, 50.21
Ceva, G., 40.2
theorem, 5.2, 5.3, 5.10, 5.11, Exercise 5.4, 34.24, Exercise 34.11, 40.2–40.3
Cevian, 5.1
Charles II, King of England, 23.5
Chasles, M., 40.14
China, 1.6
Christina, Queen of Sweden, 23.10
Circle, 19.1–19.5, 38.1–38.5
cross ratio, 42.11
inscribed, 6.2
inversion, *see* Inversion
ninepoint, 6.1, 6.5, 7.13–7.16
notation, 8.4
reflection, 41.4, Exercise 44.6
squaring, 8.2, Exercise 8.16
Circular, 42.1
Circumcircle, 6.2
Cis, 35.17
Clairaut, A., 40.4
Clockwise, 13.15
Closure, Exercise 32.1
Coaxial, 4.6
Coin game, Exercise 19.14
Collinear, 2.6
Commutative, 14.6, 33.8, 34.10, 34.14
Compass, 8.7
equivalence, 8.8
rusty, 8.16–8.19
Complex numbers, 31.1–39.14
conjugate, 35.16
order, Exercise 32.1
polar form, 35.17
polynomial form, 32.5
pure imaginary, 35.18
vectors, 35.9–35.11
Component, vector, 21.4
Composition, 12.3
Concurrent, 2.6
Conformal, 41.7
Congruent, 2.8
Conjugate
complex, 35.16
harmonic, 3.11
isogonal, Exercise 4.11, 20.12, 20.13
isotomic, Exercise 4.11, 5.6
vector, 34.17, 34.19
Constructible, 8.2, Exercise 8.16
Construction, 8.1–8.19, Appendix B
classical, 8.2, Exercise 8.16
impossible, 8.2, Exercise 8.16
mean proportional, 8.10
square root, 8.11
Coplanar, 3.1
Copolar, 4.6
Cotes, R., 31.7
Counterclockwise, 13.15
Cross ratio, 3.6, 3.7, 42.11
Cube, duplication, 8.2, Exercise 8.16
Cyclic order, 10.4
Cyclic quadrilateral, 7.6

Dark Ages, 23.1, 23.3
Dedekind, R., 9.7
De lineis rectis se invicem secantibus, 40.2
De maximus et minimis, 23.7
De Moivre, A., 31.7–31.8, Exercise 31.3
theorem, 31.7–31.9, Exercise 31.2, Exercise 35.7
De Morgan, A., 23.5, 23.8
Desargues, G., 23.11
two-triangle theorem, 4.7, 4.9, 4.10, Exercise 23.1
Descartes, R., 23.9–23.11, 31.6
Determinant, 36.9–36.17
Dihedral angle, 50.1
Direct, 10.4, 13.14, 13.15, 25.12, Exercise 25.18, 41.7, Exercise 46.8, 47.10
Directed distance, 33.3
Directed length, 2.9, 33.3
Directed measure, 2.13
Directed segment, 33.3
Discovery method, 45.10, 45.11

Displacement, screw, 46.3, 49.1
Distance
  directed, 33.3
  inverses of points, Exercise 41.17
Distributive law, Exercise 36.27
Divide, 3.2
  externally, 3.2
  harmonically, 3.11
  improperly, 3.2
  internally, 3.2
Dodge, C., Exercise 45.4
Double point, 12.7
Dual, 4.9, 4.10, 5.10, 5.11, Exercise 17.1
  Pappus' theorem, 4.10–4.12
Duplication of cube, 8.2, Exercise 8.16

e, 35.17
Egyptians, 1.2–1.6
Elements, 2.1, 8.6, 9.2, 9.8, Appendix A
Epsilon-ruler, Exercise 4.16
Equal vectors, 33.2
Equation
  central inversion, 51.8
  glide-reflection, 22.9
  halfturn, 21.11, Exercise 39.9, Exercise 51.7
  homothety, 30.1–30.5
  inversion, 43.1–43.3
  reflection, 22.2–22.6, 39.8–39.10, 39.12, 51.5, 51.10
  rotation, 21.6, 21.9, 39.2, 51.3, 51.4
  similarity, 30.6, 39.5–39.13
  translation, 21.3, 39.1, 51.2
Equicircle, 6.2
Equivalent compasses, 8.8
Eratosthenes, 9.11
Erlanger Programm, 12.22
Euclid, 9.2, 9.8–9.9, 12.23, 40.9
  Elements, 2.1, 8.6, 9.2, 9.8, Appendix A
  fifth or parallel postulate, 40.15–40.16, Exercise 40.8, Appendix A
Euclidean compass, 8.7
Euclidean construction, 8.1–8.19
Euclidean geometry, 12.19, 12.21
Euler, L., 31.9
  line, 29.2, Exercise 29.3
  theorem, 2.16, Exercise 2.16
Eudoxus, 9.7
Eves, H., 8.17, 40.4, Exercise 40.1, Exercise 40.4, Exercise 40.7, 43.10
Excircle, 6.2
Exponential form, 35.17
Extended plane, 3.2
Extended space, 3.2
External center of similitude, 25.6
External division, 3.2

Fagnano's problem, 20.17
False theorem, 7.23, Exercise 7.12
Fermat, P., 23.7, 23.9, 23.10
Feuerbach, K., 7.21, 40.12
  theorem, 7.21, 44.6
Fifth postulate, 40.15–40.16, Exercise 40.8, Appendix A
Fixed point, 10.3, 12.7
Formal axiomatics, 40.21
Fréchet, M., 40.19
Frustum, 1.4, Exercise 1.4
Fundamental theorem of algebra, 31.13

Galileo, 23.6
Game, Exercises 19.14–19.15
Gauss, C. F., 31.13–31.15, Exercise 31.4, 40.17
  plane, 35.12
Geometric mean, 42.6
Geometric progression, 42.6
Geometry
  absolute, 40.16
  Euclidean, 12.19, 12.21
  Klein's definition, 40.20
  non-Euclidean, 40.16–40.17
  plane equiform, 12.20–12.21
  projective, 4.10, 12.21
Gergonne, J., 40.11
  point, 5.4
Glide-reflection, 11.9, 14.16, 46.3, 46.9
  equation, 22.9
Greatest Egyptian pyramid, 1.4
Greeks, 23.2
Group, 12.15, Exercise 12.12, Exercise 12.13, 16.6–16.10, 25.16, 25.17, 26.11, 26.12
  abelian, 12.16

Grundlagen der Geometrie, 40.14

Halfturn, 11.6, 15.5, 46.1, 48.14
  equation, 21.11, Exercise 39.9, Exercise 51.7
Halmos symbol, 2.3 footnote
Hamilton, W., 31.15
Harmonic conjugate, 3.11
Harmonic division, 3.11
  theorem, 42.5
Harmonic mean, Exercise 6.15, 42.6
Harmonic progression, 42.6
Heron, 9.11
  formula, Exercise 2.15, 6.17
Hexagram theorem, 23.12, Exercises 23.2–23.3, 40.9
Hilbert, D., 40.14
Hipparchus, 9.11
Homography, 43.6
Homothety, 24.1, 25.1
  center, 24.1, 25.1
  equation, 30.1–30.5, 39.3
  ratio, 24.1, 25.1
Horblit and Nielsen, 45.12
Horner's method, 1.6
Hundred Years' War, 23.4
Hypatia, 9.11–9.12

i, imaginary unit, 32.5, 35.10–35.18
i, vector, 34.23
Ideal line, 3.2
Ideal plane, 3.2
Ideal point, 3.2, 38.4, 41.3
Idempotent, Exercise 12.7
Identity, 11.3, 12.8
Image, 10.1, 35.12
Imaginary axis, 32.7
Imaginary coefficient, 35.17
Imaginary unit, 35.17
Impossible construction, 8.2, Exercise 8.16
Improper division, 3.2
Incircle, 6.2
Indivisibles, method, 23.14
Inequality, 32.3, Exercise 32.1
  triangle, Answer 34.1
Infinity, 23.13, 38.4, 41.3
  point, 3.1, 3.2
Inscribed circle, 6.2
Internal center of similitude, 25.6
Internal division, 3.2
Invariant point, 10.3, 12.7
Inverse
  see also Inversion
  reflection, 13.12
  rotation, 14.13
  similarity, 24.5
  transformation, 11.4, 12.10
  translation, 14.5
  vector, 34.14
Inversion, 41.1–44.10, 41.4
  center, 41.4
  circle or line, 41.11–41.14
  equation, 43.1–43.3
  line, Exercise 44.6
  plane, 41.3
  power, 41.4
  radius, 41.4
  rotatory, 46.3, 49.9
Involutoric, 11.4, 15.3
Iota, 11.3, 12.8
Irrationality of pi, 40.13
Isogonal conjugate, Exercise 4.11, 20.12, 20.13
Isometry, 10.1–22.10, 46.1–51.11, 13.2, 47.2
  order, Exercise 11.7
Isomorphic, 35.6
Isotomic conjugate, Exercise 4.10, 5.4

Jones, W., 40.5

Kepler, J., 23.6
Klein, F., 12.22, 40.20, Exercise 40.9
  geometry, 12.22, 40.20
Kirkman point, 40.9

Lambert, J., 40.16
Larousse Dictionary, 31.16
Law of sines, Exercise 6.22
Legendre, A., 40.16
Leibniz, G., 23.14
Lemoine, E., 40.14
Length, 2.7
  directed, 2.9
Limits, theory of, 23.15

Line, 2.4, 37.2–37.8
ideal, 3.2
ordinary, 3.2
Linear combination, 35.1
Lobachevsky, N., 40.17
geometry, 40.17, 40.19

Magnitude
constructible, 8.3, Exercise 8.16
vector, 34.1, 34.2
Mahavíra, 31.3
Malfatti's problem, 40.10
Map
see also Transformation
conformal, 41.7
identity, 11.3, 12.8
involutoric, 11.4, 15.3
Mascheroni, L., 40.7–40.8
Material axiomatics, 9.1, 40.21
Mathematics Magazine, 45.12
Mathematics Teacher, 45.12
Matrix, 36.8
Mean, 42.6
Mean proportional, 8.10
construction, 8.10
Measure, 2.7
directed, 2.13
Mechanix Illustrated, Exercise 8.14
Medial property, Exercise 36.27
Medial triangle, 6.2
Median, 6.2, 50.20
theorem, 6.6, 6.7, 50.21
Menelaus, 9.11, 40.2
point, 4.1
theorem, 4.2, 4.3, 5.8, 5.10, 5.11
Method of indivisibles, 23.14
Mirror, 10.4, 13.9
Modern compass, 8.7
Monge, G., 40.6
Multiplication
scalar, 33.10
vector, 34.8–34.14
Mystic hexagram theorem, 23.12, Exercises 23.2–23.3, 40.9

Newton, I., 23.14, 31.7
Ninepoint center, 6.2, 7.16
Ninepoint circle, 6.1, 6.2, 6.5, 7.13–7.16, 29.4
Ninepoint radius theorem, 6.5
Noncommutativity of rotations, 14.15
Non-Euclidean geometry, 40.16–40.17

One, **1**, 34.11
Opposite, 10.4, 13.14, 13.15, 25.12, Exercise 25.18, Exercise 46.8, 47.10.
Order
complex numbers, Exercise 32.1
cyclic, 10.4
isometry, Exercise 11.7
Ordered pair, 31.14–31.16, Exercises 31.4–31.5
Ordinary line, 3.2
Ordinary plane, 3.2
Ordinary point, 3.2
Oresme, 23.9
Orthic triangle, 6.2
Orthocenter, 6.2
Orthocentric quadrangle, 7.18
Orthogonal, 34.17, Exercise 41.15, 42.3
Oughtred, W., 23.5
Overbar, 2.15 footnote

Pacioli, L., 31.4
Pappus, 9.11
ancient theorem, Exercise 44.15
theorem, 4.8–4.12
Papyrus, Rhind, 1.3
Parallel, Exercise 3.2
Parallel postulate, 40.15–40.16, Exercise 40.8, Appendix A
Parallelepiped, 50.3–50.13
Parallelogram, 18.9–18.11
law, 33.7
Parent, A., 40.4
Pascal, B., 23.12
line, 40.9
theorem, 23.12, Exercises 23.2–23.3, 40.9
Peacock, G., 31.2
Peaucellier cell, 42.9
Peaucellier image, 42.9
Pencil, 2.6

Pi, 1.3, 1.7, 1.8, Exercise 1.2, Exercise 1.6, 23.5, Exercise 23.4, 23.13, 40.5
irrationality, 40.13
Pi Mu Epsilon Journal, Exercise 6.15
Plane
extended, 3.2
Gauss, 35.12
ideal, 3.2
inversive, 41.3
ordinary, 3.2
Plane equiform geometry, 12.20–12.21
Plato, 9.7
Plimpton 322, 1.8
Plücker, J., 40.14
line, 40.9
Poincaré, H., 40.19
Point
double, 12.7
fixed, 10.3, 12.7
Gergonne, 5.4
ideal, 3.2, 38.4, 41.3
infinity, 3.1, 3.2, 41.3
invariant, 10.3, 12.7
Menelaus, 4.1
ordinary, 3.2
Polar form, 35.17
Polygon game, Exercise 19.15
Poncelet, J., 40.8, 40.11
Positive, 2.9
Postulate, fifth or parallel, 40.15–40.16, Exercise 40.8, Appendix A
Power of inversion, 41.4
Preimage, 10.1
Principle of permanence of forms, 31.2
Problem, epsilon-ruler, Exercise 4.16
Product
isometry, 10.5
scalar, 33.10
transformation, 12.3
vector, 34.8–34.14
Progression, 42.6
Projective geometry, 4.10, 12.21
Proportional, mean, 8.10
Ptolemy, Claudius, 9.11
theorem, 44.3
Ptolemy, King of Egypt, 9.9, 9.13
Pure imaginary, 35.18
Pyramid, greatest Egyptian, 1.4
Pythagoras, 9.5–9.6
Society, 9.5–9.6
theorem, 1.5, 9.6, Exercise 9.2, 34.12, Appendix A (47–48)
triple, Exercise 45.3

Quadrangle, orthocentric, 7.18
Quadrilateral, cyclic, 7.6

Radical axis, Exercise 44.2
Radical center, Exercise 44.2
Radius of inversion, 41.4
Range, 2.6
Ratio, 3.3
cross, 3.6, 3.7
homothety, 24.1, 25.1
similarity, 26.1
Real, 35.18
Real axis, 32.6, 32.7
Real part, 35.17
Rectangular form, 35.17
Reflection
circle, 41.4, Exercise 44.6
line, 10.4, 13.9, 22.2–22.6, 39.8–39.10, 39.12
plane, 46.1, 47.4, 51.5, 51.10
point, 11.6, 15.5
rotatory, 46.3, 49.7
Rhind papyrus, 1.3
Rich, B., 45.12
Riemann, B., 40.18
Ring isomorphic, 35.6
Roberval, 23.9
Rotation
angle, 10.3, 14.8
center, 10.3, 14.8
equation, 21.6, 21.9, 39.2, 51.3–51.4
inverse, 14.13
noncommutativity, 14.15
plane, 10.3, 14.8
space, 46.1, 48.8
Rotatory inversion, 46.3, 49.9
Rotatory reflection, 46.3, 49.7
Rusty compass, 8.16–8.19

Saccheri, G., 40.16
Salmon point, 40.9

SAS postulate, 13.1, 18.1
Scalar, 33.11
Scalar multiplication, 33.10
Screw displacement, 46.3, 49.1
Segment, 2.4
  directed, 33.3
Sense, 13.15, 47.10
  direct, 10.4
  left, 47.10
  opposite, 10.4
  right, 47.10
Side, triangle, 6.2
Sigma, 15.10, 47.4
Similarity, 24.1, 26.1
  equation, 30.6, 39.5–39.13
  inverse, 24.5
  ratio, 26.1
Similitude, center of, 24.2, 25.6
Sines, law of, Exercise 6.22
Space, extended, 3.2
Square root, construction, 8.11
Squaring a circle, 8.2, Exercise 8.16
SSS, 18.4
Staudt, K. von, 40.14
Steiner, J., 40.9, 40.10
  point, 40.9
Stewart, M., 40.3
  theorem, Exercise 2.10
Subtraction, vector, 33.13
Superposition, 12.23, 45.6

Thales, 9.3–9.4, Exercise 9.1
Theon of Alexandria, 9.11
Theorem
  altitudes, 5.5, 7.11, 28.7
  angle bisectors, 6.10, 20.4
  area, 19.6–19.8, 37.10, Exercise 37.10
  area of triangle, 6.15–6.21
  associativity, 12.6, 33.8, 34.10, 34.14
  Brianchon's, Exercise 23.5
  Ceva's, 5.2, 5.3, Exercise 5.4, 34.24, Exercise 34.11, 40.2, 40.3
  commutativity, 14.6, 33.8, 34.10, 34.14
  De Moivre's, 31.7–31.9, Exercise 31.2, Exercise 35.7
Theorem—*cont.*
  Desargues', 4.7, Exercise 23.1
  distance between inverses, Exercise 41.17
  dual of Pappus', 4.11, 4.12
  duality, 4.10
  equivalent compasses, 8.8
  Euler's, 2.16, Exercise 2.16
  false, 7.23, Exercise 7.12
  Feuerbach's, 44.6
  harmonic division, 42.5
  Heron's formula, Exercise 2.15, 6.17
  Horner's method, 1.6
  isometry, 13.13, 16.14, 17.2–17.17
  medians, 6.6, 6.7, 20.11, 29.1, 50.21
  Menelaus', 4.2, 4.3, 5.8
  mystic hexagram, 23.12, Exercises 23.2–23.3, 40.9
  ninepoint center, 7.16
  ninepoint circle, 7.13–7.14, 29.4
  ninepoint radius, 6.5
  Pappus', 4.8
  Pappus' ancient, Exercise 44.15
  parallelepiped, 50.3–50.13
  parallelogram, 18.9–18.11
  perpendicular bisectors, 7.1, 20.1–20.3
  Ptolemy's, 44.3
  Pythagorean, 1.5, 9.6, Exercise 9.2, 34.12, Appendix A (47–48)
  Stewart's, Exercise 2.10
  Thomsen's relation, 17.17, Exercise 17.17
  two-triangle, 4.7
Theory of limits, 23.15
Thomsen's relation, 17.17, Exercise 17.17
Transform-solve-transform, 18.1
Transformation, 12.2
  see also Map
  bilinear, 43.6
  circular, 42.1
  identity, 11.3, 12.8
  inverse, 11.4, 12.20
Translation, 10.2, 14.1, 46.1, 48.1
  equation, 21.3, 39.1, 51.2
  inverse, 14.5
  plane, 10.2, 14.1
  space, 46.1, 48.1
  vector, 14.1
Trapezoid, 18.12–18.14

Triangle
  area, 6.15–6.21
  inequality, Answer 34.1
  medial, 6.2
  orthic, 6.2
Trichotomy, Exercise 32.1
Trigonometric form
  Ceva's theorem, 5.3
  complex number, 35.17
  Menelaus' theorem, 4.3
Trilateral, 5.10
Trisecting an angle, 8.2, Exercise 8.13–8.16
True, 2.18
"True metaphysics of $\sqrt{-1}$," 31.13–31.14
Two-triangle theorem, 4.7, 4.9, 4.10, Exercise 23.1
Two-Year College Mathematics Journal, 45.12

Unit vector, 34.5
University of Alexandria, 9.9, 9.13

Vector, 10.2, 33.1
  addition, 33.6
  complex numbers, 35.9–35.11
  component, 21.4
  conjugate, 34.17, 34.19
  i, 34.23
  inverse, 34.14
  magnitude, 34.1, 34.2
  multiplication, 34.8–34.14
  one, **1**, 34.11
  orthogonal, 34.17
  product, 34.8–34.14
  subtraction, 33.13
  translation, 14.1
  unit, 34.5
Vertex, 2.6
Vieta, 23.9
Volume, 50.12

Wallis, J., 23.13, Exercise 23.4
Well-defined, Exercise 33.7
Wessel, C., 31.10–31.11

SUPPLEMENT TO PAGES 112–113

In Figures 26.9 and 26.10, point O is the fixed point for the similarity mapping. In Figure 26.10, if you take the homothety with a negative ratio, then the center is still O but the mirror is p.

Additional Results

26.13 Definition A similarity that is not an isometry will be called a *true similarity.*

26.14 Theorem Each true similarity has exactly one fixed point.

26.15 Theorem If two direct similarities have the same fixed point, then they commute.

26.16 Theorem If two true similarities commute, then they have the same fixed point.

26.17 Theorem Two opposite true similarities commute iff that have the same fixed point and their mirrors of reflection (through that fixed point) either coincide or are perpendicular.

26.18 Theorem Two true similarities, one direct and one opposite, commute iff they have the same fixed point and the direct similarity reduces to a homothety (its angle of rotation is either 0° or 180°).